Consciousness Transitions

Phylogenetic, Ontogenetic, and Physiological Aspects

CONSCIOUSNESS TRANSITIONS

PHYLOGENETIC, ONTOGENETIC, AND PHYSIOLOGICAL ASPECTS

Editors

Hans Liljenström

Swedish University of Agricultural Sciences, Uppsala
and Agora for Biosystems, Sigtuna, Sweden

Peter Århem

Karolinska Institutet, Stockholm
and Agora for Biosystems, Sigtuna, Sweden

ELSEVIER

Amsterdam – Boston – Heidelberg – London – New York – Oxford
Paris – San Diego – San Francisco – Singapore – Sydney – Tokyo

Elsevier
Radarweg 29, PO Box 211, 1000 AE Amsterdam, The Netherlands
Linacre House, Jordan Hill, Oxford OX2 8DP, UK

First edition 2008

Library of Congress Cataloging-in-Publication Data
A catalog record for this book is available from the Library of Congress

British Library Cataloguing in Publication Data
A catalogue record for this book is available from the British Library

ISBN: 978-0-444-52977-0

Printed and bound in the United Kingdom
Transferred to Digital Print 2010

Contents

Preface

What is this mysterious phenomenon of consciousness? How is it related to the structure and processes of complex systems, such as our brain? In the last couple of decades, questions related to these have been surging the scientific literature, including biophysics, neuroscience, zoology, evolutionary biology, computational biology and artificial intelligence. We believe a fruitful approach to the scientific problem of consciousness is through the study of transitions between unconscious and conscious states, and between different levels of consciousness. Such studies could include evolutionary, developmental, and physiological aspects, and this is also the focus of the current book. Some specific problems addressed here include: Does consciousness have any evolutionary relevance and if so, which? What are the evolutionary precursors to human consciousness and where do we find them in the phylogenetic tree? How do infants develop and obtain what might be the origin of an explicit human type of consciousness? What are the mechanisms behind anaesthesia and the transition from a conscious to an unconscious state, and which neural structures or processes are involved?

Since the beginning of the 1990s, the Agora for Biosystems has organized courses and meetings on issues concerning fundamental problems in biology, notably on the mind-brain problem. The common theme at the Agora meetings has been various aspects of theoretical biology, with a broad interdisciplinary approach, where we have wanted to illuminate the chosen issue from many angles. The gathering of a set of renowned international scholars mixed with younger researchers at the beautiful and inspiring premises of the Sigtuna Foundation has provided the possibility to probe further into issues of common concern, often with an interesting interdisciplinary blend of personalities. Several of the courses and meetings have resulted in internationally recognized books and special issues of journals, but none has focused on the issue of consciousness transitions *per se*. That such transitions seem profound, both in evolution, and in the development of a child, as well as between conscious and unconscious states in an individual, has become apparent during our series of smaller workshops on the evolution of consciousness, as well as in discussions on ontological and physiological aspects of consciousness in humans and other species.

In the present book, we have collected a number of papers that relate, in one way or the other, to the major research theme of the Agora for Biosystems, the relation between neural and mental processes. Many of the authors have attended one or several of our meetings and courses, while others are included because their contribution to the field seems indispensable for covering phylogenetic, ontogenetic, as well as physiological aspects of the consciousness problem.

The present volume is both an extension of previous investigations and a more focused attempt to analyse the transition issue, using a cross-level approach. In Chapter 1 we introduce and outline some main questions within the field, stressing the difference between consciousness and cognition, not always understood in modern texts about the subject. The next four chapters all have a strongly phylogenetic theme. Most discussions of consciousness have so far focussed, for natural reasons, on human consciousness. van Swinderen in Chapter 2 deviates from this tradition, and analyses, although not explicitly consciousness *per se*, the roots of consciousness in phylogenetically remote species such as insects. Rial and colleagues in Chapter 3 draw the line of phylogenetic origin, based on their analysis, at the origin of reptiles. Århem and colleagues in Chapter 4 take a more cautious position and mainly describe different possible scenarios, discussing modern pro and contra arguments. In Chapter 5 Macphail discusses similar problems with a special focus on the function of consciousness, and comes to the today radical conclusion that only humans possess the capability for conscious experience.

While the cross level theme is presented to varying degree in all contributions, the most ambitious attempt to combine molecular, cellular and systems level approaches is presented in Chapter 6 by Changeaux, one of the pioneers who brought the mind-brain issue to the scientific forefront, but who has also reached a broader readership by books, such as *L'Homme Neuronal* (translated as *Neuronal Man*) and *L'Homme de Vérité* (translated as *The Physiology of Truth*). An ontogenetic theme is presented in Chapters 7, 8, and 9. In Chapter 7, Lagercrantz reviews modern ideas about consciousness in the fetus and the newborn, taking the position that at least neonates can experience consciousness, and thus deviates from Macphail in Chapter 5. In Chapter 8, Cotterill emphasises the role of muscular movements and an exploratory interaction with the environment, for the development and evolution of consciousness, also relating to intelligence and autism. Merker stresses in Chapter 9 the important role of brain stem structures in causing conscious states, basing his arguments on the study of anencephalic children, which he conclude have some form of consciousness. This is a controversial position, which has bearings on the phylogenetic issue forming the main theme in Chapters 2 to 4.

The subsequent three chapters, 10, 11, and 12, discuss various cognitive themes in relation to consciousness. In Chapter 10, Freeman relates the complex neurodynamics of the brain, with examples from the olfactory system, to learning, attention and intention and to state transitions underlying perception. In Chapter 11, Baars' seminal paper of 1983 is reproduced to give a background to the idea of consciousness as "globally distributed information", and where he points at some boundary conditions of conscious content and transitions between unconscious and conscious processes. In Chapter 12 Gärdenfors, like Cotterill, emphasises the interaction with the environment for the development and evolution of consciousness. Gärdenfors focuses on the social interaction and intersubjectivity, forming the basis for a transition to self-consciousness. In the final Chapter 13 Munz takes the point of departure in the philosophy of Karl Popper, one of the important contributors to the philosophical analysis of the brain-mind issue, yet remarkably unknown in this field. Munz is one of the very few philosophers, who have studied under both Popper and his great adversary Ludwig Wittgenstein, and in the chapter he tries to merge their philosophical views, a daring project indeed.

Regrettably, both Peter Munz and Rodney Cotterill died before the publication of this book. Peter Munz in October 2006, just after the submission of his chapter, and Rodney Cotterill on 24 June 2007. Rodney Cotterill was a frequent and popular visitor to the Agora for Biosystems, always cheerful and supportive, sharing his ideas and theories with great enthusiasm. He will be missed by many.

We are profoundly indebted to the Agora Scientific Advisory Board, in particular to Anders Ehrenberg, Hermann Haken, Frank Moss, and Sakire Pögün for their enthusiasm, inspiring discussions, and wise advice. We are very grateful to the former Swedish Council for Planning and Coordination of Research (FRN), The Bank of Sweden Tercentenary Fund (RJ), and the previous national funding agencies for financial support of the Agora for Biosystems and its meetings. We are also indebted to the Royal Swedish Academy of Sciences (KVA) for administrative help and moral support, and to the Sigtuna Foundation (Sigtunastiftelsen) for providing adequate premises in a beautiful and peaceful environment. Among the many people engaged in making the workshops and conferences successful, and the resulting proceedings publishable, we wish especially to thank our Agora co-organizers, Clas Blomberg and Ingemar Lindahl. We are also grateful to Per Aronsson, Soumalee Basu, and Johanna Nilsson for their friendliness and help in taking care of invited guests and practical issues.

Sigtuna, June 2007

Hans Liljenström *Peter Århem*

Contributors

M. Akaârir, Laboratori de Fisiologia, Universitat de les Illes Balears, Spain.

Peter Århem, Dept of Neuroscience, Karolinska Institutet, Stockholm and Agora for Biosystems, Sigtuna, Sweden

Bernard J. Baars, Department of Psychology, State University of New York, Stony Brook, New York, USA.

Ann B. Butler, The Krasnow Institute for Advanced Study and Department of Psychology, George Mason University, Fairfax VA, USA.

Jean-Pierre Changeux, Institut Pasteur and Collège de France, Paris, France.

Rodney M. J. Cotterill, Biophysics Group, Danish Technical University, Lyngby, Denmark. Died on 24 June 2007.

S. Esteban, Laboratori de Fisiologia, Universitat de les Illes Balears, Spain.

Walter J. Freeman, Department of Molecular & Cell Biology, University of California Berkeley, USA.

A. Gamundi, Laboratori de Fisiologia, Universitat de les Illes Balears, Spain.

Peter Gärdenfors, Cognitive Science, Lund University, Lund, Sweden.

Hugo Lagercrantz, Department of Woman and Child Health (KBH) Karolinska University Hospital, Stockholm, Sweden.

Hans Liljenström, Dept of Biometry and Engineering, SLU, Uppsala and Agora for Biosystems, Sigtuna, Sweden

B. I. B. Lindahl, Department of Philosophy, Stockholm University, Stockholm, Sweden.

Euan Macphail, Department of Psychology, University of York, Heslington, York, UK.

Paul R. Manger, School of Anatomical Sciences, Faculty of Health Sciences, University of the Witwatersrand, Johannesburg, South Africa.

Björn Merker, Stockholm, Sweden.

Peter Munz, Victoria University, Wellington, New Zealand. Died in October 2006.

M. C. Nicolau, Laboratori de Fisiologia, Universitat de les Illes Balears, Spain.

R. V. Rial, Laboratori de Fisiologia, Universitat de les Illes Balears, Spain.

Bruno van Swinderen, The Neurosciences Institute, San Diego, USA.

Chapter 1

Beyond Cognition – On Consciousness Transitions

Peter Århem and Hans Liljenström

1. What are the problems?

What is the functional role, if any, of consciousness and how is it related to cognition? Is consciousness an epiphenomenon, or does it have any survival value? In order to survive, living organisms need to react and adapt to a changing environment. They must have the capability to learn and to solve problems impressed upon them by the environment. Thus, in a wide sense, all organisms must have knowledge about the environment. Seen in this evolutionary perspective, a traditional definition of what is meant by knowledge - understanding gained by actual experience; the state of being aware of something or of having information; something learned and kept in the mind – seems too narrow to be useful. Learning, knowledge and problem solving, in this wider sense, entered upon earth with life, about 3.8 billion years ago. Obviously, if we take this perspective, learning and knowledge are not necessarily mediated via a nervous system. However, a nervous system increases the speed and range of learning, as well as the flexibility in the interaction with the environment. It gives the organism an increased survival probability. The kind of knowledge that is associated with processes in the nervous system is what is traditionally called cognition. Its origin could be traced back to the first nervous systems, i.e. when the first coelenterates appeared about 700 million years ago (Anderson, 1989).

There are no compelling reasons to believe that the logic of the knowledge (information) processing associated with the early nervous system, in principle, would deviate from non-neural knowledge processing. Both types are presumably based on a combination of stochastic and selection processes, and normally not so much on instruction processes (Maturana and Varela, 1992). In higher animals, knowledge acquisition depends to a higher degree on instructions from parents and other individuals, but this does not imply any major difference.

Consciousness Transitions: Phylogenetic, Ontogenetic, and Physiological Aspects
Edited by Hans Liljenström and Peter Århem

However, we find it reasonable to believe that conscious cognition, in principle, differs from unconscious cognition, that the emergence of conscious cognition was a major transition in the evolution of life. The degree of consciousness, we believe, is associated with the degree of complexity, and conscious cognition would require a rather complex nervous system, that could not be found in any primitive animal. We assume there has been an evolution of consciousness in smaller or larger steps, but in parallel and in interaction with the evolution of the nervous system. At present, we cannot know at what stage in evolution, with what organism, the first signs of conscious cognitive processes appeared, but it can be argued that mammals, and possibly birds, possess this quality. These animals are believed to have Gestalt perception of objects and are able to think in abstract symbols, to a lesser or higher degree (for example, a chimpanzee can do this to a much higher degree than a mouse). They can be assumed to have subjective experiences, although not necessarily be aware of themselves as individuals.

However, the fundamental question of the function of consciousness remains unanswered. So far, there is no strong argument for believing that consciousness would have any additional survival value for an organism. It is conceivable that all cognitive actions could be possible without the contribution of a conscious or qualia dimension. This notion may seem to weaken the arguments just mentioned about consciousness in non-human species, which has been developed in more detail by Macphail (1998, Chapter 5 this volume). Nevertheless, the fact that consciousness is an outstanding feature associated with our human physiology, and that outstanding physiological features often seem to be adaptive, and that we can trace a general phylogenetic continuity of physiological features, speaks against such a radical view.

We are thus left with a number of tentative scenarios, most of which place the phylogenetic origin of consciousness earlier than the emergence of humans; at the reptilian-mammalian transition, at the reptilian-mammalian and reptilian-avian transitions, or at the amphibian-reptilian transitions, or even earlier (see van Swinderen, Chapter 2 this volume; Rial *et al*, Chapter 3 of this volume Århem *et al*, Chapter 4 of this volume). The majority view is that consciousness emerged continuously at some pre-human stage. Macphail, on the other hand, argues that self-consciousness and the human language emerged together, and that other, simpler forms of consciousness (feeling-consciousness or awareness) originated as a consequence of an existing self-consciousness. The majority view also assumes a major transition when self-consciousness and a specific human language emerged, but that it has been preceded by another major transition, the first consciousness transition. Whatever position taken, the self-consciousness transition and the emergence of a specific human language implied a critical, goal-directed and scientific thinking which has changed the world dramatically. Hence, as earlier stated (Århem and Liljenström, 1997), we believe that consciousness is a central feature of higher cognitive processes. This means that studies of cognition without taking consciousness into account will be rather sterile, and even misleading. In taking this position, we differ from many authors in the field. For example, we think that discussing cognitive processes purely in terms of computations (Hopfield, 1994; Dennett, 1991) will not be fertile in the attempt to understand human thinking. Our approach, in this respect, is more in line with ideas stressing that the subjective aspect of our mind cannot be fully understood in terms of computations (Penrose, 1989, 1994; Edelman, 1992; Searle, 1992; Chalmers, 1996).

In order to avoid the problem of origin, some have postulated that consciousness exists as an independent feature of the universe (Rensch, 1968; Eccles, 1989; Chalmers, 1996). Others claim it is meaningless to talk about consciousness at all, at least in scientific terms (Quine, 1975). Further, many of those who recognize that consciousness is a quality different from anything else in nature, still do not think it has any evolutionary value, or any causative effects on matter (Huxley, 1898; for an overview of these positions see Popper and Eccles, 1977, and Churchland, 1988).

We believe that an evolutionary perspective suggests that consciousness indeed has causative effects, that conscious cognitive functions actually are more advantageous to an animal than purely unconscious cognition would be. For example, if the brain has evolved to become efficient with respect to energy, information processing rate, and/or information accuracy, as has been suggested (see e.g. Levy and Baxter, 1996; Liljenström, 1997; Laughlin and Sejnowski, 2003), it is conceivable that consciousness may also serve to make the neural information processing more efficient. Consciousness could perhaps be guiding in the selection of the neural processes most relevant for the task at hand, or even determine which strategy that is most efficient, depending on circumstances. In any case, we believe that consciousness is a biological problem, and that it would be fruitless to analyse the problem without any relation to biology. This point of view forms the background to the current chapter.

2. Knowledge in an evolutionary perspective

As already mentioned, we think a traditional definition of knowledge is too narrow and not encompassing all aspects. Recent studies of unconscious learning (see Macphail, 1998) points to the insufficiency and even irrelevance of such traditional definition of knowledge. We think it is more fertile to see learning as an adaptation to the environment, at several time scales, and knowledge being acquired through experience of an individual, or in a species over several generations. Knowledge would imply any information gained through an interaction with the environment and that potentially could be used for making such an interaction more advantageous for the organism.

This leads us to the view that knowledge is (at least) as old as life itself, and that knowledge acquisition was essential already for the first organisms, to increase their survival probability. It leads us to see knowledge acquisition and learning as biological processes that have existed and evolved gradually with the life forms. A major step was taken with the introduction of a nervous system, which enabled a tremendous increase in the way an organism could interact with the environment. The electro-chemical processes that take place in the intricate networks of nerve cells can lead to a great variety of internal states and "mental" processes of an animal. In lower animals, the nervous system allows only for primitive cognitive processes, simple learning and recalling.

Presumably, these processes would be more or less unconscious. As the nervous system evolves and attains a higher degree of complexity and organization, its processes become more advanced and sophisticated, such as imaging, reasoning, believing, and willing. Eventually, possibly only with the humans, symbolic language becomes an important and integrated part of the cognitive functions. (Even many of these more advanced activities do not have to be conscious; experiments with so-called split-brain patients show that such activities also can go on unconsciously, at least for the dominating brain half (Sperry, 1977)). The cognitive functions are sometimes grouped into pre-attentive and higher cognitive functions, respectively. The latter type of functions are usually considered individual, or "private", in character, which means that it is often seen as a separate category and in principle not open for scientific study. This is not our position. We do not believe in sharp borders excluding rational investigations of difficult matters. This applies also to what is included in the concept of mind, often seen as the totality of all cognitive functions, presumably associated with the collective spatio-temporal pattern of neural activity in the brain. Although, at this pre-Copernican stage of cognitive science, the classification of the elements within the mind is extremely difficult and open for widely deviating opinions, mind can be regarded as involving a set of processes, including sensation, perception, imagination, emotion, memory, thinking, cognition, and reasoning (see Gärdenfors, Chapter 12 this volume). Some of these may be conscious, while others may not.

2.1 Different forms of knowledge

It may be fruitful to distinguish between knowledge about long-term and short-term events and processes, to recognize the difference between knowledge about slow, seemingly deterministic, law-like changes in nature such as the day and night, seasons etc. and knowledge about fast, seemingly indeterministic features of the world (Popper). The long-term knowledge that is gathered under generations is mediated by changes in DNA structure, whereas short-term knowledge, gathered during a life time primarily is mediated by changes in the nervous system connectivity (but also through the immune system). The first form of learning is what has been called phylogenetic learning, the second belong (together with immunologic learning etc.) to what has been called ontogenetic learning. The latter form of learning corresponds to what we traditionally mean with learning and is the subject for treatises in epistemology. However, by not taking phylogenetic learning into account we gravely misunderstand the nature of knowledge and learning. This has been misleading in many textbooks treating the subject.

This is more clearly seen when using Kant's well-known terminology. In his terminology, long-term knowledge corresponds to *a priori* knowledge, knowledge that exists independent of our senses. The ontogenetic learning mediated by the nervous system, i.e. traditional learning, corresponds to *a posteriori* knowledge. It was a revolutionary view when Kant postulated that a *priori* knowledge exists, and it has not really been taken seriously even in modern learning theories. The mechanism of this form of learning seemed rather mysterious until Konrad Lorenz proposed a tentative solution in

1941, using an evolutionary approach (see Sjölander, 1997). In this perspective it is not only a logical consequence that *a priori* knowledge exists, it is also likely that *a priori* knowledge dominates over *a posteriori* knowledge. The whole organism is sculptured by phylogenetic learning. Popper (1994) provocatively suggests that 99% of our knowledge is *a priori*, and very little passes through our senses, and even when it does it is mostly *a priori*, inherent in the construction of the sense organs. We may also say that all *a priori* knowledge is unconscious, and so is most of the *a posteriori* knowledge. It is only a minor part of the *a posteriori*, or ontologically learnt knowledge that actually is conscious. In this perspective, the presently dominant empiricist theory of knowledge, assuming that all knowledge comes from our own senses, is not correct. This mistake seems related to a mistake concerning the role of information in cognition. Often, when discussing the brain's capacity for information processing, one treats the environment as if there were many signals (implying information) with noise superimposed. Then, one discusses the capability for increasing the signal to noise ratio. However, the information of the "signal" only becomes apparent, in some sense is created, in the brain. Much of the sensory input is "nonsense", until it reaches higher (cortical) areas where it is attaining meaning. What the brain does is to make meaning out of sensory input (Freeman, 1991). This process involves a lot of filtering, clustering, separation, association, etc. Information alone can never produce mind, no matter how much of it is gathered. Mind is anything but static. Rather, it seems to be a process that depends on the (organized) amount of information/knowledge processed per unit time. A high flow rate of information may not be sufficient for mind, but it is very likely a prerequisite for it. Also the dynamical state of the brain is of great importance for how the information is being processed, and what the result of this processing will be.

2.2 The optimization problem in learning

All forms of knowledge are used by the organism in its interaction with the environment. There is usually some knowledge processing before an event in the environment (a stimulus) will result in an adequate response behaviour. Different strategies have been used during evolution to optimize the response patterns. Different requirements are imposed concerning the speed of the response, the energy used, or concerning the balance between flexibility and stability, not all easily compatible (Liljenström, 1997).

One evolutionary line has maximized speed and stability at the cost of the flexibility in the response pattern. This line is amply exemplified by species among the unicellular organisms, monerans and protists, and among multicellular animals by insects, notably social *Hymenoptera* species. Another evolutionary line has opted for more flexible response patterns at the cost of speed or stability. It is this strategy that requires centralized nervous systems, and it is within this evolutionary line that we find the most developed central nervous systems, epitomized in humans. A key role in this evolutionary line has the interneurons, the class of neurons mediating the impulses between sensory and motor neurons. The larger number of interneurons, the larger the theoretical possibility for flexible responses to environmental impressions. The highest number of interneurons of any nervous system is, not unexpectedly, demonstrated by the human nervous system. It has a ratio, sensory neurons – interneurons – motorneurons of 10:100,000:1 (Maturana and Varela, 1992).

Also for the central nervous system there are certain optimization problems to be solved. Presumably, the nervous system has evolved to process and store information in an efficient way, i.e. it should be optimized for (i) maximal processing rate and (ii) maximal information storage capacity. It seems likely that the nervous system has evolved primarily for a fast information transfer, from sensation (of the environment) to action. A high information storage capacity would supposedly be of a secondary and phylogenetically younger origin, evolved for more advanced behaviour in a more complex interaction with the environment. In conclusion, biological learning or adaptation processes show different evolutionary strategies depending on the organism and its interaction with its natural environment. In relative terms, phylogenetic learning is of a very high accuracy, although slow, whereas ontogenetic/traditional learning is of medium accuracy, and relatively fast. The immediate reactions to the environment demonstrated by the neurodynamics *per se* have low accuracy, but are very fast. Here, we will focus on the evolutionary strategy leading to the neural processes of the brain, and in particular to conscious cognition.

3. Evolution of cognition

3.1 The neural correlate of cognition

The evolution of cognition can be assumed to parallel the evolution of centralized nervous systems. The first primitive nervous systems, emerging with the first coelenterates (phyla *Cnidaria* and *Ctenophora*) evolved about 700 million years ago (Anderson, 1989), and seem correlated with relatively simple cognitive behaviours. Two organizational features of the evolution of early central nervous systems are essential for understanding the evolution of cognition: (i) the tendency of neurons to aggregate in groups (ganglia) and (ii) the tendency of the anterior ganglia to increase in relative volume (encephalisation) (Maturana and Varela, 1992).

This will be very clear when we consider the origin of the vertebrate central nervous system, which will take the cognition to new complexity levels. Within the paraphyletic reptilian group a three layered cortex emerges, with the dorsal portion becoming isocortex in mammals, the medial portion becoming hippocampus and the lateral portion the olfactory cortex. At the same time the reptile brain also evolve along another line, into the avian brain where the dorsal ventral ridge plays a crucial role. Thus brain evolution at the reptilian stage reaches a bifurcation point where different strategies, different *Bauplane,* come into play (see Butler *et al*, 2005; Århem *et al*, Chapter 4 this volume; Macphail, Chapter 5 this volume). It is tempting to think that this bifurcation is reflected in different cognitive strategies, in turn reflected in different forms of consciousness (see Butler *et al*, 2005). The mammalian brain adds three novel cortical layers to the three reptilian ones, forming a characteristic six-layered structure and allowing multimodal and higher sensory areas to develop. Within the mammalian radiation we find new cortical inventions; with the emergence of primates we find a granular prefrontal cortex, with the emergence of the great apes (subfamily *Anthropidea*), we, surprisingly, find a new type of neurons, spindle cells, with unknown function, and with the human brain we find the unique language areas, associated with the unique human language. This is the neural background on which we must project the evolution of cognition, and the evolution of consciousness.

Another feature of the nervous system, presumably of functional importance for cognition, and possibly for consciousness (?), is that the nervous system depends on electrical events rather than on other physical-chemical reactions. It is likely that this is, at least partly, due to the superlative efficiency in speed an electrical information processing system is capable of. This is an issue little studied.

The central basic event in the nervous system is the transmission of electrical signals, waves of electricity that are conducted along nerve fibres and transported over synapses. The underlying mechanisms of impulse conduction were unveiled by Hodgkin and Huxley in their classical voltage clamp studies on the giant squid axon in the early fifties, and followed up by Frankenheauser (Dodge and Frankenhaeuser, 1958) for the more complex vertebrate nerve fibre. The cause of the conducted impulses was found to be time and potential dependent Na^+ and K^+ currents through the membrane. During the eighties, due to the development of the new patch clamp technique (Neher and Sakmann, 1976), it became evident that the ion currents passed through water filled pores in the membrane (see Hille, 2001). Further studies revealed that the pores were constituted by specific membrane proteins, ion channels. Several ion channel types have now been classified and cloned with molecular biological methods. There are in principle two types of channels, channels controlled by the electrical potential over the membrane and channels controlled by ligands, neurotransmitters. The potential controlled channels are often classified according to the dominant ion passing through its pore (examples are Na, Ca and K channels), while the ligand controlled channels are classified according to its activating ligand (examples are acetylcholine, glutamate and GABA channels). The number of channels described increases continuously – the human genome contains 143 genes giving rise to voltage-gated or voltage-gated-like channels, making this super family of signal transduction proteins the third largest (Yu and Catterall, 2004). The fundamental building plan for this channelom depicts the standard channel as a protein (or an aggregate of proteins), consisting of four membrane-spanning domains with a selectivity filter close to the extracellular side and a gating mechanism at the intracellular side.

An essential observation from an evolutionary view point is that most channel types are much older than the nervous system. Molecular biology studies suggest that Ca and K channels originated perhaps earlier than 1400 million years ago (Hille, 2001). Na channels and the first ligand activated channels originated with the first primitive nervous systems, that is with the first coelenterates (phyla *Cnidaria* and *Ctenophora*) about 700 million years ago (Anderson, 1989). This means that with the first simple nervous systems all components required for an advanced centralized nervous system was at place, suggesting that organization and complexity is the key to advanced cognition.

3.2 Stages in cognition

The nervous system informs the organism about the environment in which it lives and moves, processes the sensory data, relates it to previous experience, and transforms it into appropriate actions or memories. Already at early stages in nervous system evolution the processing of sensory information can involve discrimination and categorization, possibly with the aid of some rudimentary learning and memory capacity. Even relatively simple organisms, such as worms, slugs, and insects are able to learn and store information in their nervous systems. Much of what is known about simple learning and memory at a molecular level is based on the studies of a sea slug, *Aplysia californica.*

At this stage of nervous system development there is no compelling reason to assume that the cognitive processes are associated with conscious experience. However, at later stages there are clearly reasons to assume such an association, although, as discussed above, they are not conclusive until humans arrive (see Macphail, 1998, and Chapter 5 this volume). Here we will briefly discuss the evolution of cognition independent of its relation to consciousness. At some point in the evolution the nervous system is complex enough to allow for "Gestalt" perception, i.e. the ability to "see" an object as a whole. This ability should involve the more or less simultaneous (spatial) binding of neural activity in different parts of a cortical area (Crick and Koch, 1990, Crick, 1994). Although such binding has been discussed primarily with regard to vision, all sensory modalities of higher animals can be said to function with perceptions based on patterns of nerve activities. The ability to form "objects" or percepts of neural activity patterns has apparently increased tremendously during evolution from reptile to mammal. Through such perceptual binding, internal representations of the external world are formed, resulting in an overall "model" of the world. It is a precondition to forming a category of objects, which is one of the early concepts. An increased complexity of the evolved nervous system also enables the organism to make better predictions about the future environment, based on previous and present experience. The success of these predictions depends largely on the learning and memory capacity of the animal.

At still higher evolutionary levels, reasoning, planning, and abstract thinking come into play. The behaviour of the brain (and rest of body) becomes less dependent on direct peripheral stimuli, although that will always be of great survival value.

Finally, in evolutionary terms, the ability to "internalize" the world has come to include the formation of temporal patterns. That is, binding together sequences of neural events that mostly would correspond to some external sequences of events. That temporal binding allows us to understand the relation between cause and effect and to experience "the arrow of time", something characteristic for human reasoning. Interestingly, non-human primates seem much less capable of causal reasoning, suggesting that this capacity adds to the unique human mind features (Gärdenfors, Chapter 12 this volume).

For humans, the internal model of the world also includes other individuals and their minds. This is what sometimes, somewhat vaguely, is summarized in the concept intersubjectivity, more analyzable in terms of four components; the capacity to represent emotions of others, the capacity to represent the attention of others, the capacity to represent intentions of others and the capacity to represent belief or knowledge of others

(Gärdenfors, Chapter 12 this volume). Models of other individuals are important for determining the significance of their behaviour, and for predicting the "next step" in that behaviour. A sense of "I-ness" can develop when experienced events are put into a sequential context in the inner model of the world. It is obvious that most cognitive functions depend upon memory, but the "I" needs memories well structured in time, in order to be perceived. The perception of an "I" also requires a detachment of an internal representation of self from that of non-self (Stoerig, 1996). It should here be stressed again, that although all these higher cognitive functions clearly are associated with conscious experience, the question about their logical relationship remains open.

3.3 Cognition and temporal efficiency

Clearly, there is a close relation between the time and space scales of the nervous system and the cognitive capacities. If the relevant information transfer concerns events and processes at a subcellular level, the appropriate time scale could be very small, in the order of nanoseconds or less. This is the time scale of conformational changes in proteins, ion channel openings etc. On the other hand, if the relevant information transfer is not within but between cells, the upper limit for the rate of information transfer is determined by such parameters as interspike interval (the inverse of firing rate), membrane time constant, and synaptic and axonal conduction delays. The values of these parameters vary for different types of cells and fibres and signal travelling distance, but are all in the range of one to tens of milliseconds. The shortest time for information transfer, from sensation to action, would be found in simple nervous systems, such as in insects. It is known that e.g. flies can react to a single cell response and change its direction of flight within a few milliseconds, as a reaction to a single or a few action potentials (Downer, 1988). Certain ants are supposed to catch their preys within a single millisecond upon detection. In larger networks, such as in the mammalian brain, where normally millions of cells are involved in any type of activity and where there are several synaptic steps between sensory system and motor control, the shortest time is much larger, typically in the order of a hundred milliseconds or more. Some specific pathways, like those involved in simple instinctive sensation-reaction schemes, may be very fast. In those cases, however, the process would probably not be considered as a cognitive process.

For a relatively simple nervous system, such as that of an insect, the capacity for storing new information is low. Instead, most of its wiring is determined genetically, which is possible for such simple but life sustaining actions as feeding, mating etc. The action control program can be specified by a small number of instructions. For example, a single specific molecule, such as a sex pheromone, can trigger the flight instructions and the mating behaviour of an insect. In the more complex nervous system of, for example, a mammal, there is no way to genetically encode all the possible behaviours in the natural environment of such an animal. Instead, there must be a great capacity for learning and adaptation during the lifetime of an individual. An increase in the number of cells and synapses, and hence in the number of processing steps, will presumably make the system slower, but will also allow for a richer and more robust behaviour.

(The complexity of the nervous system and the possible behavioural programs of an animal are also most likely related to the maximal lifespan of that species).

3.4 The computational correlate of cognition

We cannot easily study evolution, let alone different options that evolution might have had during the history of life. However, with computer models we can simulate various evolutionary steps and study the effects of altered complexity and organisation of the neural structures and processes. Computer models can help us understand the relationship between structure, dynamics, and function, and allow us to test different possibilities. However, it is important to bear in mind that all variables and parameters introduced in a model should have a counterpart in the real system. With appropriate computer models we can make simulation "experiments" that would be inconceivable, or would take an unrealistic long time to carry out with the real system. For example, if a model parameter can be related to the concentration of a certain transmitter substance or neuromodulator, its effects on the dynamics and functions of the system can be investigated in detail using computer simulations.

When making a computer model of a biological system like the brain, or some part of it, one has to find an appropriate balance between realism and abstraction. The amount of available data of the structure and function of the system is far too large to be included in any one model or simulation. The challenge is to make the proper simplifications, extracting those features of the system that are essential for the particular aspects one intends to model. If a model is simple enough and only has a few parameters, analytical solutions may be found. However, in most cases where some realism is sought, the models become far too complex, incorporating thousands of equations and many more parameters. Then, only numerical solutions are possible. Simulations of large and detailed mathematical models are time-consuming, but with the computational power of today's computers, this is no longer a serious bottle-neck in computational neuroscience. It is in this perspective the extensive use of cortical network models for clarifying cognitive functions should be seen. They are ways to understand mechanisms of certain oscillatory activity, of determining oscillatory states and attractors, of understanding associative memory states in associative networks. However, it must be realized that the question of what conscious states per se are or what the physical correlate to consciousness is not likely to be addressed by such studies although many assertions to the contrary.

Most connectionist models so far are based on the convergence to steady state "point attractors" (as in the Hopfield net), which may not be appropriate for any real neural system with a rich dynamics. The different (functional) "states" of the brain would then make up a multidimensional "energy landscape" (Hopfield 1982, 1984). In the simplest case, this landscape is supposed to be more or less fixed with "valleys" and "ridges" that are statically determined by the network connections, and where the valleys correspond to point attractor memory states. However, a more realistic picture would be that of a roaring sea, which constantly is changing, and where the memory states rather would correspond to the rolling waves. A more complex dynamics, with oscillations and chaos-like behaviour is found for models with a more realistic architecture, for example in models of the olfactory system (Wilson and Bower, 1989; Liljenström, 1991; Freeman, 2000). Compared to, for example, the visual system, the olfactory system (primarily the bulb and the olfactory, or piriform, cortex) is more primitive and much simpler, but also has a well-characterized neurodynamics (see Freeman, 1991; Chapter 10 this volume). By studying such simpler parts of the brain, one hopes also to get an understanding of the functioning of other parts of the brain. Even if olfaction *per se* is not what one associates with conscious cognition, the way the olfactory system processes sensory (odour) information, including its associative memory properties, may hint to higher cognitive functions (see Freeman, 2000; Chapter 10 this volume).

In order to investigate whether the neurodynamics of cortical structures has any significance for perception and associative memory we have developed a computer model of the olfactory cortex (Liljenström, 1991). With this model we study to what extent its complex dynamics can facilitate an efficient neural information processing. We have used a network architecture analogous to the three-layered structure of the real cortex, and simple network units corresponding to populations of neurons. The model includes, implicitly or explicitly, adaptation at all relevant time scales: phylogenetical, ontogenetical and neurodynamical.

The network units are modelled with a continuous input-output function, corresponding to the average firing frequency of neural populations. Three different types of network units (cell populations) are organised in three layers. The top layer consists of inhibitory "feedforward interneurons", which receive inputs from an external source (the "olfactory bulb") and from the excitatory "pyramidal cells" in the middle layer. The bottom layer consists of inhibitory "feedback interneurons", receiving inputs only from the excitatory units and projecting back to those. The two sets of inhibitory units are characterized by their different time constants. In addition to the feedback from inhibitory units, the excitatory units receive extensive inputs from each other and from the "olfactory bulb". All connections are modelled with time delays for signal propagation, corresponding to the geometry and fibre characteristics of the real cortex.

Neuronal adaptation is implemented as an exponential decay of the output, proportional to the time average of previous output (Liljenström and Hasselmo, 1995). To allow for learning, i.e. adaptation at an intermediate time scale, the weights are incrementally changed according to a learning rule of Hebbian type. Noise effects on learning and associative memory tasks are studied by continuously changing the level of the additive or multiplicative noise.

With computer simulations of this model of the olfactory cortex, we are able to reproduce much of the rich dynamics of the real cortex (Liljenström, 1991; Liljenström and Hasselmo, 1995). In particular, input patterns corresponding to different odours result in distributed patterns of primarily oscillatory activity in network. Such a response is presumed to be related to perception of an odour object. Short input pulses can give rise to waves of activity moving across the simulated cortical surface, similar to what is seen in experimental studies.

The complex dynamics of this model is primarily a result of time constants and network connectivity: feedforward and feedback inhibitory loops, as well as short and long range excitatory-excitatory connections, all modelled with distance-dependent delays. Details at the neuronal level (such as action potentials or cellular compartments) do not seem necessary for the dynamic phenomena studied.

The model further exhibits associative memory properties similar to those thought to exist in the real cortex, making use of the dynamic behaviour, described in (Liljenström, 1991). The oscillating patterns of activity that appear in the network in response to external input patterns are stored as (near) "limit cycle" memory states, rather than as "fixed point" memory states, typical for most neural network models (e.g. Hopfield, 1982, 1984). Simulated neuromodulatory effects (of acetylcholine) on the oscillatory behaviour and on the memory properties of this system agree well with experimental results and current hypotheses (Liljenström and Hasselmo, 1995). In particular, we have shown that the cholinergic suppression of neuronal adaptability and of intrinsic synaptic transmission can facilitate learning by sustaining the oscillatory activity of an input pattern and reducing interference with previously stored patterns. By changing the gain parameter, corresponding to a change in neuronal excitability, the system can display all types of dynamic behaviour: point attractor, limit cycle or strange attractor behaviour (Wu and Liljenström, 1994). Simulation results showed that the oscillatory and aperiodic ("transient chaos") dynamic behaviour can improve system performance by reducing learning and recall (convergence) time in associative memory tasks. This is in accordance with Freeman's view on "stochastic chaos" or "constrained noise", as described for the olfactory bulb (see Chapter 10 this volume).

In addition to the neuromodulatory effects, neuronal and synaptic noise adds to the complex behaviour of the system and may increase the rate of information processing further. In fact, optimal noise levels can be found for which the recall time reaches a minimum. Noise can also induce transitions from one (attractor) state to another (Liljenström and Wu, 1995).

With the type of computer models we have discussed here, attractor neural networks with biologically inspired connectivity, it is possible to capture features of real neural systems and get simulation results that in many respects closely resemble experimental results. In addition to the type of simulations presented here, these models can also be used for prediction of temporal sequences, which is currently under study. The models can be useful for finding out which neural properties and network architecture are responsible for the characteristic dynamics of the real system. By exploring the parameter space for e.g. connection strengths, delays, gains, etc. we can point at plausible solutions to neurobiological problems, and put constraints on the possibilities of the natural system. Although we cannot make any conclusive statements based on the models alone, theories for certain cognitive functions, such as perception, learning and memory can be tested, and critical experiments can be suggested. Here, we have pointed at the relationship between the network structure and the complex dynamics it displays, using a neural network model of the olfactory cortex. Furthermore, we have linked the dynamics of the system to its function, and shown that the complex behaviour can enhance the performance, by increasing fidelity and rate of information processing.

4. Evolving consciousness

We have discussed the evolution of increasingly complex cognitive functions, spanning from highly conscious to unconscious cognitive activities. All cognitive processes are clearly not conscious. On the contrary, most cognitive processes, are unconscious, but at some degree of complexity some cognitive processes become conscious. Furthermore, it is apparent that consciousness appears at different levels of complexity. There are reasons to believe that full human consciousness shows more levels than that of any other species. For instance, it has not been shown that any other species shows more than rudimentary forms of self-consciousness, i.e. consciousness/awareness of a self and the capability to reflect about it (see Gärdenfors, Chapter 12 this volume).

Whether lower and higher forms emerged simultaneously or separately and gradually is still unknown. Macphail (1998, Chapter 5 this volume) argues for a simultaneous emergence; both self consciousness and awareness emerged with the first humans. In contrast, we tentatively suggest, as others have done (Popper and Eccles, 1977; Edelman and Tononi; 2000), that lower forms of consciousness emerged earlier in the phylogenetic process than higher forms. Furthermore, we think consciousness is central for higher cognitive functions. In spite of the more or less gradual evolution of conscious processes, we think the emergence of conscious cognitive processes implied a major transition in the evolution of life.

We conjecture that conscious cognition enables a more efficient goal-directed behaviour and allow for more efficient prediction, expectation, wills, plans, goals, hopes etc beyond the immediate perception. This would have consequences for choosing mating partners, for securing good living conditions for offspring etc. We conjecture it would be far more efficient than unconscious "planning". Conscious cognitive functions may also permit a more efficient complex interaction with other individuals and with the environment.

Predictions and expectations that bring the cognitive processes into a (distant) future also open up for a more advanced planning and setting up of goals. Will is a driving force in nature. Thus, these conscious functions can indeed interact with the complexity of the nervous system, leading to further complexity by speeding up evolution towards higher complexity.

It is possible, or perhaps even likely, that unconscious cognition can reasonably well be described in terms of neurophysiology and neural computations. The question is, however, more problematic when it concerns conscious cognition. As will be discussed below there are arguments for an interactionistic relation between conscious cognitive processes and neurophysiological processes, described in classical physical terms. This means that consciousness is not possible to fully describe in physical terms, but that it interacts with physical processes. The problem of this relation has been termed the hard mind-matter problem (Chalmers, 1996). The problem of clarifying the neural processes associated with cognitive processes, but not the nature of the relation, has consequently been termed the easy problem. This difference is not always recognized in the discussion about consciousness, and may cause confusion. As discussed above, we assume that conscious cognition evolved because it gave the organism survival advantages. This perspective will have important implications, especially concerning the hard problem, as will be discussed below.

We do not think that a strict definition of consciousness is necessary, or even desired, at this stage. As Crick and Koch (1990) points out, the lack of an exact explicit definition of consciousness needs not be an obstacle in the analysis. It is better to avoid a precise definition because of the dangers of premature definition. We have to accept subjectivity as part of human consciousness, that a first person view is implied.

Although we think there are different levels of consciousness, ontogenetically and as well as phylogenetically, we think there are strong arguments for a specific human consciousness. The main reasons are based on the high degree of complexity and organization of the human brain, as well as on the complexity of human language (Hocket and Altman, 1968; Thorpe, 1978; Popper, 1994; Århem *et al*, Chapter 4 this volume; Macphail, Chapter 5 this volume; Munz, Chapter 13 this volume)).

An evolutionary perspective leads to the question when the first signs of consciousness appear in phylogeny, with what organisms. If that could be known it could provide a link to the nature and role of consciousness in life. Regardless of when it first appeared in the history of life, it seems reasonable that its appearance and evolution depends on an increasing complexity of the nervous system. At present, we also have no method to determine a point in evolution when certain cognitive functions became conscious, i.e. when conscious cognition first appeared. In order to do so we need anatomical, physiological and physical correlates to such processes. Such correlates have been proposed.

In the following we will briefly discuss this. Somewhat simplified, we can be said to discuss the anatomical, the physiological and the physical correlates of consciousness. What brain structures are associated with consciousness? What physiological processes? What physical states?

4.1 The anatomical correlate of consciousness

What parts of the nervous system are directly associated with conscious cognition? What level of neural organization is related to conscious cognition? The brain consists of several subsystems and cortical areas, and there are several levels of organization and processing, ranging from molecules to the whole central nervous system. Structure and function, as well as type of processing, can vary greatly for different parts of the brain. In "higher" cortical areas, such as the visual cortex or the prefrontal cortex, processing appears to take place in webs of strongly interacting networks. This type of distributed organization may be typical for higher cognitive functions in general and thus for consciousness (Churchland and Sejnowski, 1988). Conscious activity, as measured with EEG techniques, seems characterized by low amplitude, unsynchronized, high-frequency oscillations, while high amplitude, synchronized, low-frequency oscillations characterize loss of consciousness, as observed in connection with general anaesthesia and non-dream sleep. It is generally assumed that these strongly interacting networks, forming the neuronal correlate to consciousness, are identical with thalamo-cortical circuits (Seth *et al*, 2005). But there is less consensus as to what parts of cortex form the neuronal correlate to consciousness. Or which thalamic nuclei. Before entering into this complex of issues we should note that there is a crucial difference between neurons and circuits that are necessary for creating consciousness and neurons and circuits that actually are sufficient for generating consciousness, sometimes called generator neurons (Butler *et al*, 2005). This is an important distinction, seldom made in current literature.

Ambitious attempts to specifically identify the brain areas and brain processes directly associated with conscious events have been made (Eccles, 1990; Crick, 1994; Edelman and Tononi, 2000; Cotterill, 1998 and Chapter 8 this volume). We can recognize two lines of approaches, roughly classified as a behavioural-cognitive and an anatomical-physiological approach (Århem *et al*, Chapter 4 this volume). It can be argued that the anatomical-physiological approach at present is a safer and more fertile road to success than the cognitive-behavioural. Nevertheless, the road is still sprinkled with problems. The only – from a philosophical point of view – safe starting point for consciousness studies is the consciousness of the human brain and the studies are consequently mainly limited to imaging techniques and to clinical case studies; either of accidental lesions – for instance those causing the blindsight phenomenon (see Stoerig and Brandt, 1993) – or of surgical operations – like that causing the split brain phenomenon (Sperry, 1980), or of pharmacologically induced consciousness effects – such as general anaesthesia (Århem *et al*, Chapter 4 this volume; Changeux, Chapter 6 this volume).

By combining results and conclusions from such studies we are presented with a number of hypotheses, such as those developed by Crick and Koch (Crick, 1994; Crick and Koch, 1995; Rees *et al*, 2002; Koch, 2004) and by Edelman and Tononi (Edelman, 2003; Edelman and Tononi, 2000; Tononi and Edelman, 1998).

The hypotheses of Crick and Koch concern the visual system and limit the generator neurons to temporal, parietal and prefrontal cortex. Furthermore, they seem to limit the generator structure to a subpopulation of neurons in the fifth cortical layer. A prominent feature is that the primary visual cortex is not a generator structure (Crick and Koch, 1995; Rees *et al*, 2002). Further, they have recently added the idea that claustrum plays a crucial role in producing visual consciousness (Crick and Koch, 2005).

Edelman and Tononi stress the general features of consciousness — such as complexity and unity (Tononi and Edelman, 1998) — and assert that consciousness is associated with activity in the temporal and frontal associative and motor regions of the cortex: a "dynamic core", characterized by "re-entrant" interactions within limited portions of the CNS.

In summary, the anatomical ideas about neural generators of conscious experiences show great variation with respect to their extent and dynamics. The cortico-thalamic system plays a central role in all the major hypotheses (for a discussion see Århem *et al*, 2003). Emphasis has been placed on components of the temporal and parietal association areas and the anterior cingulate, premotor, and prefrontal cortical areas (Crick and Koch, 1995; Tononi and Edelman, 1998; Cotterill, 1998; Edelman and Tononi, 2000; John, 2001; Rees *et al*, 2002). The only exception from the idea of a key role for the cortico-thalamic system is the interesting suggestion by Merker (2007, Chapter 9 this volume).

4.1.1 Testing the anatomical hypotheses: The case of general anaesthesia

A way to test and to illustrate these principal ideas on the anatomical correlate of consciousness is to analyse the effects of general anaesthetics on the human brain. General anaesthetics form a large group of chemically unrelated compounds that reversibly induce unconsciousness. "General anaesthesia" is basically an umbrella term for a complex state comprising the endpoints amnesia, immobility, analgesia and hypnosis, all depending on different mechanisms and different molecular and anatomical targets (Århem *et al,* Chapter 4 this volume; Changeux, Chapter 6 this volume). General anaesthetics, thus allows us to explore brain activity at different stages in the transition between consciousness and unconsciousness.

What do such studies reveal about which structures of the brain are critically affected by general anaesthetics and which structures are critically associated with the loss of consciousness? Analysing the rather extensive studies performed, a rather complex picture comes to the fore: three families of hypotheses about the target structures for general anaesthesia have emerged. Either general anaesthetics primarily affect cortical neurons, as suggested by conclusions from neocortical slice studies as well as from simulation studies (Steyn-Ross *et al*, 2004). Or they primarily affect thalamic intralaminar relay neurons, as suggested by the fact that neocortical slices including thalamic neurons show EEG activity associated with unconscious EEG activity (delta waves) (Antkowiak and Miller, 2005). Or general anaesthetics primarily affect reticular neurons, suggested by the fact that there are many similarities between the processes associated with general anaesthesia and sleep. All these suggestions are by and large compatible with the general thalamo-cortical consciousness hypothesis but also with the more precise hypotheses discussed above (Crick and Koch, 1991; Edelman and Tononi, 2000). Activity in thalamus and cortex seem critical for consciousness, perhaps more precisely activity in the parietal cortex, and possibly in the underlying claustrum (Alkire and Miller, 2005). However, we are still far from identifying generator circuits and generator neurons.

In analysing the different hypotheses of the target structures for general anaesthetics we note that the hypotheses besides being structure specific they also are *process specific* or *neuron specific.* In cortical hypotheses the general anaesthetics are often assumed to simultaneously attack a large dynamic network of cells, while in the reticular system hypotheses they are assumed to attack a smaller and more specified neuronal target. Within the frames of all these theories we in addition find two competing ideas: that of direct *suppression* and that of *disrupting coherence.* General anaesthetics are assumed to directly suppress the activity, or the decreased activity is due to disruption, of coherent neuronal activity (Cariani, 2000).

4.1.2 Testing the anatomical hypotheses: Comparative studies

Above we tested the anatomical hypotheses by applying them to the human brain. Another approach is to apply them to non-human brains, assumed to be associated with conscious activity (which, however, according to Macphail (1998, Chapter 5 this volume), do not exist). Do these brains comprise the critical structures postulated by the discussed hypotheses? A special case is the avian brain, traditionally assumed to differ structurally from the mammalian brain blueprint, nevertheless producing a complex cognitive behaviour well in par with mammalian cognition (Butler *et al*, 2005; Macphail, Chapter 5 this volume). Such comparative studies show, by and large, that the discussed hypotheses well explains a tentative avian consciousness. Most of the hypotheses outlined appear to be valid for the avian brain. The critical structures assumed to be necessary for consciousness in mammalian brains have their homologous counterparts in avian brains, perhaps even the claustrum (see Århem *et al*, Chapter 4 this volume). Such comparative studies also compel us to investigate where in the phylogenetic process consciousness emerges. As shown by Århem *et al* (Chapter 4 this volume) and by Rial *et al* (Chapter 3 this volume) we find that we are presented with a number of possible scenarios, but with no conclusive answers.

All scenarios are built on chains of fragmentary indices and which scenario will prevail depends on when and how its weakest links are eliminated. (I) Either are only humans conscious. The merit of this scenario is that it takes the uniqueness of human language into account (Macphail, 1998; Chapter 5 this volume). (II) Alternatively, all mammals, but only mammals, are conscious. This scenario offers structural markers of consciousness, the six-layered cortex, and perhaps the claustrum. (III) Alternatively, both mammals and birds are conscious. This means that consciousness must have emerged twice. Nevertheless, it has many supporters (Eccles, 1994; Sjölander, 1997; Edelman and Tononi, 2000). (IV) Alternatively, mammals, birds, and reptiles are conscious, thus assigning consciousness exclusively to amniotes. This would introduce the problem to find the crucial brain structure lacking in the amphibian brain and not in the reptilian one. Interestingly, some of the structural features, distinguishing the amniotic brain from the amphibian brain, are also present in a number of taxa of fishes, including many teleosts and some sharks as well as skates and rays.

Other scenarios are of course possible and have also been suggested; ranging from various scenarios assuming consciousness in vertebrates in general, with a focus on cognitive-level behavioural studies on fish, to more global ones, assuming a gradual evolution of consciousness from a diminutive start in phylogenetically very primitive organisms, as posited by e.g. Delbrück (1986) and Griffin (2001).

4.2 The neurophysiological correlate of consciousness

The most detailed attempts discuss the type of neuronal coding and type of neurons involved in conscious visual perception (Crick and Koch, 1990; Koch and Crick, 1994; Koch, 2004; see also Hodgson, 1991). Experimental investigations, including recent studies of the blindsight phenomenon (see Stoerig and Brandt, 1993), point to the crucial importance of primary visual projection areas of the cortex (V1; but for another view, see Crick and Koch, 1995), and to certain firing patterns (bursts or 40 Hz oscillations) as a crucial activity binding the neurons associated with visual perception together (see Singer, 1990; Engel *et al*, 2001). Koch and Crick (1994) further speculate that certain bursting neurons in the lower layers (V and VI) of visual cortex play special roles in visual perception. By such nerve system markers of visual awareness we have, in principle, the possibility to determine at what point in evolution such awareness arose.

So far the discussion has focused on simpler forms of conscious experience. It can be argued that higher forms, like self-consciousness, are restricted to humans. Arguments for this view are easily obtained from for instance linguistic studies. Of all biological communication systems the human language is unique in its complexity. No other language has, for example, a developed descriptive function, and even less an argumentative function. As language seems to be essential for the experience of a self, the cortical language areas must therefore be of interest when discussing areas associated with self-consciousness. At least one comparative study of language areas among primates has been made. This showed that the orangutan cortex lacks language areas (Mauss, 1911), lending support to the view that language areas are essential for higher forms of consciousness. This would mean that self-consciousness, the experience of a conscious I, indeed is restricted to humans, and is associated with activity in the cortical language areas.

Assume now that our anatomical studies have been successful, and that area V1 is the cortical correlate to visual experience and the language areas (or prefrontal cortex) for the conscious experience of a self. Have we thereby solved the problem of consciousness? Do we understand how conscious experience arises? Of course not. We still do not know what processes in these areas are involved. In order to do so we have to identify critical processes involved. At present we have no detailed ideas of what they can be. However, there are some attempts to determine these processes. As mentioned above Crick and Koch (1990) suggest that the 40 Hz oscillatory activity is the critical activity. They also suggest that certain cells in the lower cortical layers are instrumental in causing the synchronous activity. The evolutionary implication of this hypothesis is that brains with 40 Hz are conscious and brains that do not show such oscillatory activity are not. It should be mentioned that at present the hypothesis has not been supported by further experimental evidence.

Another line of investigation focuses on brain structures that are possible to amplify microevents to global synchronized activity. This line leans heavily on the idea that quantum mechanical processes are required to explain mind-brain interaction with out breaking physical conservation laws. In this line Eccles (1992) proposes that certain synapses on cortical pyramidal cells are the critical structures. He further suggests that aggregates of pyramidal cells, dendrons, are the functional units in this mind-brain interaction. Using this criterion he concludes that consciousness is restricted to mammals and possibly to birds (Eccles, 1994).

In this context other amplification mechanisms should also be mentioned. Such a mechanism has been pointed out by Johansson and Århem (1994; see also Lindahl and Århem, 1995; Århem 1996; Århem and Lindahl, 1996) and studied computationally by Liljenström and Århem (1996). It is based on the experimental findings that a subset of hippocampal neurons may function as cellular random generators. Such activity may control transitions between chaos-like states and global oscillatory states in hippocampal network models.

4.3 The physical correlate to consciousness

By identifying structures and processes underlying conscious cognition and experiences we have, in principle, tools to follow the evolution of cognition and consciousness. But, as indicated above, we do not have the solution to the question how conscious cognition arises. We have not solved the classical mind-matter problem. A prerequisite for approaching an answer is of course first to understand what physical states are directly associated with conscious states. If a certain spiking pattern in certain neurons is critically associated with conscious events, we must ask what physical feature of the spike pattern is relevant: Is it the dynamic configuration of charged particles? Or is it the associated electromagnetic field? Or is it still something else?

Some mean that the neural correlate should be found in the quantum mechanical events at a subcellular level (Donald, 1990; Penrose, 1989, 1994; Conrad, 1996), whereas perhaps the most common view is that the spike trains emitted from active neurons are best related to mental events. Still others believe that it is the electromagnetic fields resulting from the electrical activity of the neurons that are the true correlates to mental processes (Popper *et al*, 1993). That should imply that the electrical activity measured with e.g. EEG or with electrodes measuring the local field potentials in the extracellular fluid, actually would reflect some of the information-processing going on in the brain. If so, the 40 Hz oscillations found to be correlated with visual awareness (Singer, 1990; Crick and Koch, 1990; Engel *et al*, 2001) could indeed be a direct correlate to mental activity, and not merely a sign of synchronized cellular activity, as many researchers seem to suggest.

Since, as we have discussed, most experimental approaches are not primarily aimed at such basic questions, i.e. about the nature of the association between conscious and neuronal events, they are therefore somewhat vague on this point. But neither do the dominating philosophical theories specify the issue. What specified physical state or event in the brain is critically correlated with a specified mental state or event? Is it a specific spatio-temporal pattern of cortical action potentials? Or is it the dynamic configuration of charges in a crucial brain area? Or the electromagnetic field caused by the dynamic flow of charged particles? In order to be more fruitfully discussed and to allow constructive criticism, any proposed solution of the consciousness problem has to be specific.

5. Philosophical implications

What is the nature of the relation between neural activity and conscious events? This question, which extends beyond the question about the neural correlate of consciousness, is the basic philosophical issue. We here have to recognize that there are different forms of consciousness: primitive, primary, minimal, feeling-consciousness and higher order consciousness (see e.g. several chapters in this volume). In this chapter we label all these forms as consciousness. Some philosophers reserve the word consciousness for higher order consciousness (Munz, Chapter 13 this volume), assuming that the major transition occurs between low and high order consciousness, and that this is the major problem to solve. This is not our position. We think, in line with many others (Eccles, 1992; Edelman, 1992; Crick, 1994) that the major transition and the major problem are to understand how electro-chemical events become experience, how matter becomes imagination (Edelman and Tononi, 2000). We think, given that understanding, the solution of the problem of higher level consciousness should be easier to resolve.

As mentioned repeatedly above, the basis for the present discussion is the assumption that the problem of consciousness is a central issue for the discussion of cognition (see Århem and Lindahl, 1993). The modern consciousness discussion started with Descartes' interactionistic hypothesis; mental events and brain events were seen as separate and interacting substances, the separation being based on the notion of extension (see Popper and Eccles, 1977). A well-known problem with this solution is to understand how extended substance can interact with unextended substance within the framework of Descartes' view of causality, i.e., action by contact. Discussions of this problem (the mechanistic argument; Lindahl and Århem, 1994) led to a number of alternative solutions, classified as forms of psycho-physical parallelism. Versions of this solution came to dominate the philosophical landscape, and do so even today in the shape of the identity theory (see Churchland, 1988). Most neurobiology oriented discussions of the problem also seem to be based on some form of identity theory (Changeux, 1983; Crick, 1994). In this theory, mental states or events are assumed to be, in some respect, identical to certain physical states or events, i.e., certain brain states or events.

It is clear that the identity theory says rather little, if relevant physical states are not determined. This is usually not done. The identity theory, as well as parallelistic theories, has a major weakness. It is not easily compatible with the theory of evolution. The reason is the following: According to the theory of evolution, outstanding features of organisms have evolved because they have a survival value for the organism. They causally affect the physical states of the organism. Consciousness is an outstanding feature of man, and presumably of other species as well and must thus be causally effective and interact with physical events of the organism. This argument was probably first used by William James (1879; see Richards, 1987), and has been developed further by Popper (Popper and Eccles, 1977) and Hodgson (1991). We think this is a strong, although of course not conclusive, argument for an interactionistic view upon the mind-brain problem. Indeed, this interaction may be a driving force in the evolution, giving a bias to the probabilistic laws at work. In the form of conscious cognition, it may provide a basis for the execution of some kind of a will, where an appropriate choice/selection is made out of a set of possible actions.

Other recently proposed interesting arguments against parallelist and identity mind-brain theories, concern the nature of human thinking. They all criticize a basic thesis of the identity theory, namely that conscious processes are assumed to be fully described by algorithms. The rather specific case of mathematical thinking (understanding) has been discussed by Penrose (1989, 1994), who concludes that mathematical understanding involves non-computable, non-algorithmic, components. A more general approach has been taken by Hodgson (1991), analyzing formal and plausible reasoning. His conclusion is stated in terms of a criticism of mechanism, the view that the world is completely describable in terms of physical quantitative laws (i.e., algorithms). The common conclusion is thus that conscious cognition cannot in principle be fully described by algorithms, as argued by some proponents for the identity theory and parallelism (Dennett, 1991)

These considerations suggest an interactionistic solution to the consciousness problem. As mentioned, however, parallelist and identity solutions still dominate. We have already mentioned the mechanistic argument against Descartes' version of interactionism (Lindahl and Århem, 1994) – that something immaterial cannot influence something material. This argument is based on a Cartesian concept of matter, as something extended and impenetrable, and a Cartesian notion of causality, confined to action by contact. However, the mechanistic argument should have lost its power already with the development of the Newtonian concept of action at a distance, and with the modern concept of force, introduced by Faraday and Maxwell (see Jammer, 1957; Popper and Eccles, 1977; Popper *et al*, 1993; Lindahl and Århem, 1994). Nevertheless, it is still used; even modern arguments are to some extent based on the Cartesian notions of matter and causality (see Lindahl and Århem, 1994). Perhaps the time is now ripe to replace this old dichotomy of mind/matter with the interaction between non-computable/computable processes (Århem and Lindahl, 1996).

6. Conclusions

In summary, an evolutionary perspective leads us to the view that adaptation and (phylogenetic and ontogenetic) learning is a widespread and old property of living organisms. It was an integral and essential part of early life forms that appeared about 3.8 billions years ago. We suggest that knowledge processing mediated by a centralized nervous system, i.e. cognition, shows the same principal features as non-neural adaptive processes.

Similarly, we hypothesize that consciousness appears, to different degrees, at different stages in evolution. Both cognition and consciousness, and in particular conscious cognition, depends heavily on the organisation and complexity of the organism, primarily with regard to the nervous system.

The evolutionary perspective also leads us to suggest an interactionistic solution to the hard mind-brain problem, i.e. the question of subjective experience, where we suggest a shift in the discussion towards an interaction between computational and non-computational (algorithmic/non-algorithmic) processes. This view is supported by a logical analysis of mathematical understanding and plausible reasoning (see Penrose, 1989). Conscious processes seem not identical with any known physical process. It may simplest be regarded as an emergent phenomenon, or even as a unique property of the universe (cf. Chalmers, 1996).

In this perspective, consciousness, as a biological phenomenon, becomes understandable. It may be regarded as a driving force in evolution, amplifying and improving the adaptive and cognitive processes of an organism. When we know more about the neural processes correlated with conscious cognition, we may be able to start tracing the evolution of consciousness in more detail.

References

Alkire MT, Miller J (2005) General anesthesia and the neural correlates of consciousness. *Prog Brain Res* **150**:229-244.

Antkowiak B (2002) In vitro networks: cortical mechanisms of anaesthetic action. *Br J Anaesth* **89**:102-111.

Anderson PAV (Ed) (1989) *Evolution of the First Nervous Systems.* New York: Plenum Press.

Århem P (1996) Vertical information flow in the brain: On neuronal micro events and consciousness. *BioSystems* **38**:191-198.

Århem P, Lindahl BIB (Eds) (1993) Neuroscience and the problem of consciousness: Theoretical and empirical approaches. *Theor Med* **14**:77-194.

Århem P, Lindahl BIB (1997) On consciousness and spontaneous brain activity. In: P Århem, H Liljenström and U Svedin (Eds). *Matter Matters?* Berlin: Springer, pp 235-253.

Århem P, Johansson S (1996) Spontaneous signalling in small central neurons: mechanisms and roles of spike-amplitude and spike-interval fluctuations. *Int J Neural Systems* **7**:369-376.

Århem P, Liljenström H (1997) On the coevolution of cognition and consciuousness. *J Theor Biol* **187**:601- 612.

Butler AB, Manger PR, Lindahl BIB, Århem P (2005) Evolution of the neural basis of consciousness: a bird-mammal comparison. *BioEssays* **27**:923-36.

Cariani P (2000) Anesthesia, neural information processing, and conscious awareness. *Consc Cogn.* **9**:387-95.

Chalmers D (1996) *The Conscious Mind: In Search of a Fundamental Theory.* Oxford: Oxford University Press.

Changeux J-P (1983) *L'Homme Neuronal.* Paris: Fayard.

Churchland PS (1986) *Neurophilosophy: Toward a Unified Science of the Mind-Brain.* Cambridge, MA: The MIT Press.

Churchland PM (1988) *Matter and Consciousness: A Contemporary Introduction to the Philosophy of Mind.* Cambridge, MA: The MIT Press.

Churchland PS, Sejnowski TJ (1988) Perspectives on cognitive neuroscience. *Science* **242**:741-745.

Conrad M (1996) Cross-scale information processing in evolution, development and intelligence. *Biosystems* **38**:97-109.

Cotterill R (1998) *Enchanted Looms – Conscious Networks in Brains and Computers.* Cambridge: Cambridge University Press.

Crick F (1994) *The Astonishing Hypothesis.* New York: Charles Scribner's Sons.

Crick F, Koch C (1990) Towards a neurobiological theory of consciousness. *Sem Neurosci* **2**:263-275.

Crick F, Koch C (1995) Are we aware of neural activity in primary visual cortex? *Nature* **375**:121-123.

Crick FC, Koch, C (2005) What is the function of the claustrum? *Phil Trans R Soc B.* **360**:1271-1279.

Delbrück M (1986) *Mind from Matter?* Palo Alto, CA: Blackwell Scientific Publ.

Dennett D (1991) *Consciousness Explained.* Boston: Little, Brown and Company.

Dodge F, Frankenhaeuser B (1958) Membrane currents in isolated frog nerve fibre under voltage clamp conditions. *J Physiol* **143**:76-90.
Donald MJ (1990) Quantum theory and the brain. *Proc R Soc Lond A* **427**:43-93.
Downer J (1988) *Supersense – Perception in the Animal world.* London: BBC books.
Eccles JC (1989) *Evolution of the Brain: Creation of the Self.* London: Routledge.
Eccles JC (1990) A unitary hypothesis of mind-brain interactions in the cerebral cortex. *Proc R Soc Lond B* **240**:433-451.
Eccles JC (1992) Evolution of consciousness. *Proc Natl Acad Sci USA* **89**:7320-7324.
Edelman, G., 1992, *Bright Air, Brilliant Fire.* Allen Lane, The Penguin Press, London.
Edelman GM, Tononi G (2000) *A Universe of Consciousness – How Matter Becomes Imagination.* New York: Basic Books.
Engel AK, Fries P Singer W (2001) Dynamic predictions: oscillations and synchrony in top-down processing. *Nat Rev Neurosci* **2**:704-16.
Freeman WJ (1975) *Mass Action in the Nervous System.* New York: Academic Press.
Freeman WJ (1979) Nonlinear gain mediating cortical stimulus-response relations. *Biol Cybern* **33**:237-247.
Freeman WJ (1991) The physiology of perception. *Sci Am* **264**:78-85.
Freeman WJ (1996) Random activity at the microscopic neural level in cortex sustains and is regulated by low-dimensional dynamics of macroscopic cortical activity. *Int J Neural Systems* **7**:473-480.
Freeman WJ (2000) *Neurodynamics. An Exploration of Mesoscopic Brain Dynamics.* London: Springer.
Freeman WJ, Skarda CA (1985) Spatial EEG patterns, non-linear dynamics and perception: the Neo-Sherringtonian view. *Brain Res Rev* **10**:47–175.
Griffin DR (2001) *Animal Minds: Beyond Cognition to Consciousness.* Chicago: The University of Chicago Press.
Hille B (2001) *Ionic Channels of Excitable Membranes.* Sunderland, MA: Sinauer.
Hockett CF, Altmann SA (1968) A note on design features *In:* TA Sebeok (Ed) *Animal Communication.* Broomington: Indiana University Press.
Hodgson D (1991) *The Mind Matters.* Oxford: Clarendon Press.
Hopfield J (1982) Neural networks and physical systems with emergent collective computational abilities. *Proc Natl Acad Sci USA*, **79**:2554-2558.
Hopfield J (1984) Neurons with graded response have collective computational properties like those of two-state neurons. *Proc. Natl. Acad. Sci. USA,* **81**:3088-3092.
Hopfield J (1994) Physics, computation, and why biology looks so different. *J Theor Biol* **171**:53-60.
Huxley TH (1898) *Method and Results: Collected Essays.* Vol I. London: Macmillan.
James W (1879) Are we automata? *Mind* **4**:3-4.
Jammer M (1957) *Concepts of Force: A Study in the Foundations of Dynamics.* Cambridge, MA: Harvard University Press.
Johansson S, Århem P (1994) Single channel currents trigger action potentials in small cultured hippocampal neurons. *Proc Natl Acad Sci USA* **91**:1761-1765.
John ER (2001) A field theory of consciousness. *Consc Cogn* **10**:184-213.
Koch C (2004) *The quest for consciousness: A neurobiological approach.* Eaglewood: Roberts and Company Publ.

Koch C, Crick F (1994) Some further ideas regarding the neuronal basis of awareness. In: C Koch and JL Davis (Eds) *Large- Scale Neuronal Theories of the Brain.* Cambridge, MA: MIT Press,

Laughlin SB, Sejnowski TJ (2003) Communication in neuronal networks. *Science* 301:1870-1874.

Levy WB, Baxter RA (1996) Energy-efficient neural codes. *Neural Comput* **8**:531–543.

Liljenström H (1991) Modeling the dynamics of olfactory cortex using simplified network units and realistic architecture. *Int J Neural Syst* **2**:1-15.

Liljenström H (1997) Cognition and the efficiency of neural processes. In: P Århem, H Liljenström, U Svedin (Eds) *Matter Matters? On the Material Basis of the Cognitive Activity of Mind.* Berlin: Springer.

Liljenström H, Hasselmo M (1995) Cholinergic modulation of cortical oscillatory dynamics. *J Neurophysiol* **74**:288–297.

Liljenström H, Wu X (1995) Noise-enhanced performance in a cortical associative memory model. *Int J Neural Syst* **6**:19–29.

Lindahl BIB, Århem P (1994) Mind as a force field: Comment on a new hypothesis. *J Theor Biol* **171**:111-122.

Macphail, EM (1998) *The Evolution of Consciousness.* Oxford: Oxford University Press.

Maturana HR, Varela FJ (1992) *The Tree of Knowledge - The Biological Roots of Human Understanding.* Boston: Shambhala Publications.

Mauss T (1911) Die faserarchitektonische gliederung der Grosshirnrinde. *Journal für Psychologie und Neurology* **8**:410-467.

Merker B (2007) Consciousness without a cerebral cortex: A challenge for neuroscience and medicine. *Behav Brain Sci* **30**:63-81.

Neher E, Sakmann B (1976) Single-channel currents recorded from membrane of denervated frog muscle fibres. *Nature* **260**:799–802.

Penrose R (1989) *The Emperor's New Mind.* Oxford: Oxford University Press.

Penrose R (1994) *Shadows of the Mind.* Oxford: Oxford University Press.

Popper KR (1953) Language and the body-mind problem: A restatement of interactionism. *Proc XIth Int Cong Philos* **7**:101-107.

Popper KR (1978) Natural selection and the emergence of mind. *Dialectica* **32**:339-355.

Popper KR (1982) *Quantum Theory and the Schism in Physics.* London: Unwin Hyman.

Popper KR (1990) *A World of Propensities.* Bristol: Thoemmes.

Popper KR (1994) *Knowledge and the Body-Mind Problem.* London: Routledge.

Popper KR, Eccles J (1977) *The Self and Its Brain.* Berlin: Springer-Verlag.

Popper KR, Lindahl BIB, Århem P (1993) A discussion on the mind-brain problem. *Theor Med* **14**:167-180.

Quine WVO (1975) Mind and verbal dispositions. In: S Guttenplan (Ed) *Mind and Language.* Oxford: Clarendon Press.

Rees G, Kreiman G, Koch C (2002) Neural correlates of consciousness in human. *Nature Reviews* **3**:261-270.

Rensch B (1968) *Biophilosophie auf Erkenntnistheoretische Grundlage.* Stuttgart: Gustav Fischer.

Richards RJ (1987) *Darwin and the Emergence of Evolutionary Theories of Mind and Behavior*. Chicago: The University of Chicago Press.

Searle J (1992) *The Rediscovery of Mind.* Cambridge, MA: MIT Press.

Singer W (1990) Search for coherence: a basic principle of cortical self-organization. *Concepts in Neurosci* **1**:1-26

Sjölander S (1997) On the evolution of reality – some biological prerequisites and evolutionary stages. *J Theor Biol* **187**:595-600.

Sperry R (1980) Mind-Brain interaction: mentalism, yes; dualism, no. *Neuroscience* **5**:195-206.

Steyn-Ross ML, Steyn-Ross DA, Sleigh JW (2004) Modelling general anaesthesia as a first-order phase transition in the cortex. *Prog Biophys Mol Biol* **85**:369-85.

Stoerig P (1996) Varieties of vision: from blind responses to conscious recognition. *Trends Neurosci* **19**:401-406.

Stoerig P, Brandt S (1993) The visual system and levels of perception: Properties of neuromental organization. *Theor Med* **14**:117-136.

Thorpe WH (1978) *Purpose in a World of Chance. A Biologist's View.* Oxford: Oxford University Press.

Tononi G, Edelman GM (1998) Consciousness and complexity. *Science* **282**:1846-1851.

Wilson MA, Bower JM (1989) The simulation of large scale neural networks. In: C Koch and I Segev (Eds) *Methods of Neural Modeling: From Synapses to Networks*. Cambridge MA: MIT Press.

Wu X, Liljenström H (1994) Regulating the nonlinear dynamics of olfactory cortex. *Network: Computation in Neural Systems* **5**:47–60.

Yu FH, Catterall WA (2004) The VGL-chanome: A protein superfamily specialized for electrical signaling and ionic homeostasis. *Sci. STKE* 2004 re15.

Chapter 2

The Remote Roots of Consciousness in Fruit-fly Selective Attention?[†]

Bruno van Swinderen

1. Introduction

Fly cognition, encased in only about 200,000 neurons, will one day be well understood and predictable. Yet, will our knowledge of the workings of a fly's brain transfer to that stew of perception and experience that we term ''consciousness'' in humans? If consciousness is self-awareness, then flies are very unlikely to be conscious in the sense that we humans appreciate the concept. It might therefore seem impossible to study consciousness in an organism where the phenomenon does not exist. For want of adequate behavioural measures of self-awareness, we often substitute checklists of ''intelligent'' behaviour. Many ''simple'' animals display what appears to be intelligent behaviour, but that in itself does not qualify as a marker of consciousness. For example, bees display impressive capabilities of navigation (Giurfa and Menzel, 1997) contextual learning (Fauria *et al*, 2002.) visual binding (of illusory contours) (Horridge *et al*, 1992) and even forms of abstraction (Fry and Wehner, 2002; Zhang *et al*, 2004). Certain jumping spiders have been shown, by their eye movements, to be capable of planning ahead their complicated approach to a prey (Tarsitano and Andrew, 1999). Even fruit flies display evidence of contextual learning (Liu *et al*, 1999) second-order conditioning (Brembs and Heisenberg, 2001) and object invariance (the ability to recognise an object as being the same or different regardless of where it impacts the eyes) (Tang *et al*, 2004). Yet, it is not by documenting such examples of intelligence that we will ever convince psychologists (for example) that invertebrates such as flies can be used to study mechanisms relevant to consciousness. Providing a checklist of a dozen or more anthropocentric criteria may be the wrong approach here. We could, hypothetically,

† "The remote roots of consciousness in fruit-fly selective attention" by Bruno van Swinderen appeared in *BioEssays* 27:321-330 (2005). Reprinted with permission of Wiley-Liss, Inc., a subsidiary of John Wiley & Sons, Inc.

Consciousness Transitions: Phylogenetic, Ontogenetic, and Physiological Aspects
Edited by Hans Liljenström and Peter Århem

demonstrate flies burying their dead and we would still be faced with the fact that simple robots could be programmed to do the same without ever touching upon mechanisms of consciousness.

An alternative approach is to be agnostic about an insect's potential subjective experience and instead tackle quantitative variables associated with changes in consciousness. Rather than asking whether a fly has consciousness, which can only be entertained philosophically and is not the point of this essay, we might use "the fly" *Drosophila melanogaster* to study key changes in behavioural responsiveness, which can be approached mechanistically. Just as *Drosophila* has proved useful toward understanding phenomena such as nervous system development (Levine *et al*, 1995) or circadian rhythms (Hall, 2003) *Drosophila* may also prove useful toward understanding general anaesthesia, sleep and selective attention—three consciousness-related variables accessible to fly research. Unlike some nebulous checklists for consciousness (language, play, abstraction and the like), these variables benefit from very well-defined criteria for their evaluation (quantitative behavioural responsiveness levels) and they consequently seem entirely knowable at a mechanistic level. The reversible loss of arousal can be defined and measured by criteria independent of the conscious reports associated to them by humans, and these measures have been shown to be remarkably similar across animal phyla (Campbell and Tobler, 1984; Nash, 1999). Animals as divergent as insects and humans require daily sleep (Shaw, 2003) and can be reversibly rendered "inert" with virtually identical concentrations of anaesthetic compounds (Campbell and Nash, 1994). Furthermore, studies of fruit flies and other invertebrates show that sleep and general anaesthesia do not require the brain structures traditionally associated with consciousness, such as the cortex and thalamus (van Swinderen and Andretic, 2003). This suggests that some fundamental mechanisms supporting arousal emerged prior to the evolution of a cortex. By focusing on the phenomenally complex human brain in the study of consciousness, we may overlook the possibility that all brains, from invertebrates to humans, may have a common way of governing arousal.

Sleep and general anaesthesia in *Drosophila melanogaster* are already well-established, albeit by a still relatively small group of researchers. Sleep research in flies was recently reviewed in this journal (Cirelli, 2003) and general anaesthesia in flies was also recently reviewed (van Swinderen and Andretic, 2003). In the current essay, I will focus on the flip-side of arousal in the fly, selective attention—a trickier problem. A discussion on selective attention is tricky because it tends to conjure up the "whacky" problem of consciousness (which can be safely ignored in sleep and anaesthesia studies). Rather than ignore this problem, I will reconsider the relationship between attention and consciousness in the light of recent work on selective attention in the fruit fly. I hope to make the case that selective attention can be studied on several levels (behavioural, electrophysiological and genetic) in *Drosophila*, at least on par with the other more sophisticated non-human animals commonly used in cognitive studies (e.g. rats or monkeys). I will outline certain similarities with human attention which are, as for sleep and general anaesthesia, striking enough to suggest, again, the possibility of a common mechanism. Since *Drosophila* lends itself very well to investigating biological mechanisms, a genetic approach to relating different states of arousal in the fly is proposed to complement the traditional studies of such states in humans and other mammals.

2. Behavioural selection

Attention has often been considered a component of the larger phenomenon of consciousness (James, 1890; Crick and Koch, 1990). As a result, any discussion of attention is often burdened with the accumulated paraphernalia of consciousness studies, such as free-will, thalamocortical loops, global workspaces, language and souls. Subjectively, it appears obvious that some attentional processes are under our conscious control: we selectively direct our attention to various targets in a fully conscious and motivated manner, as if attention were a "spotlight" of sorts. In contrast, our attention is often drawn, unconsciously at first, to salient stimuli such as loud noises, bright lights or movements.

Attention may thus appear to be evoked sometimes and motivational other times. The traditional view of attention-like processes in animals such as insects is that they are driven by immediate external events, whereby salient stimuli evoke a fixed motor pattern that could be interpreted as a form of selective discrimination. For example, a male fruit fly will court and mate with a female fly once pheromonal, visual and mechanosensory cues coalesce in the right pattern, above a certain threshold (Greenspan and Ferveur, 2000).

Yet, insects do not always respond to the same cues in the same way, partially because of the additional dimension of learning and memory. Insects (and most animals) actively probe their environment in an experience-dependent manner, and they react selectively to stimuli according to the behavioural relevance or "salience" of the stimulus, which is modulated by memory (Waddell and Quinn, 2001). There are several convincing demonstrations of experience-dependent selection (or learning) in flies, which include associative conditioning (Tully and Quinn, 1985) incidental learning (Dill and Heisenberg, 1995) and even second-order conditioning (Brembs and Heisenberg, 2001). But, whether these demonstrations of learning can be interpreted as involving selective attention-like mechanisms requires more than the measure of behavioural choice. One requires ways of monitoring simultaneously the suppression or uncoupling of other stimuli as the choice is being made; there may never have been any selection at all.

3. Behavioural suppression

Several studies on selective "attention" in insects have focused on the response to visual stimuli in flies. Initial experiments highlighted the optomotor responses found throughout the animal kingdom: flies (and humans) orient towards the direction of apparent motion, thereby minimising the sequential activity of photoreceptor groups (retinal slip) and stabilising their worldview (Poggio and Reichardt, 1976; Reichardt and Poggio, 1976). Optomotor studies in flies were given a boost with the development of biofeedback devices, where the angular position of an image rotating around a fly in a flight arena could be controlled by the animal's behaviour (Fig. 2.1). During tethered "closed-loop" flight, flies demonstrate their "perception" of objects by using body torques in order to hold a dynamic object still in part of their visual field (Goetz, 1975; Heisenberg and Wolf, 1984). Flies are able to control optic flow with precision, but

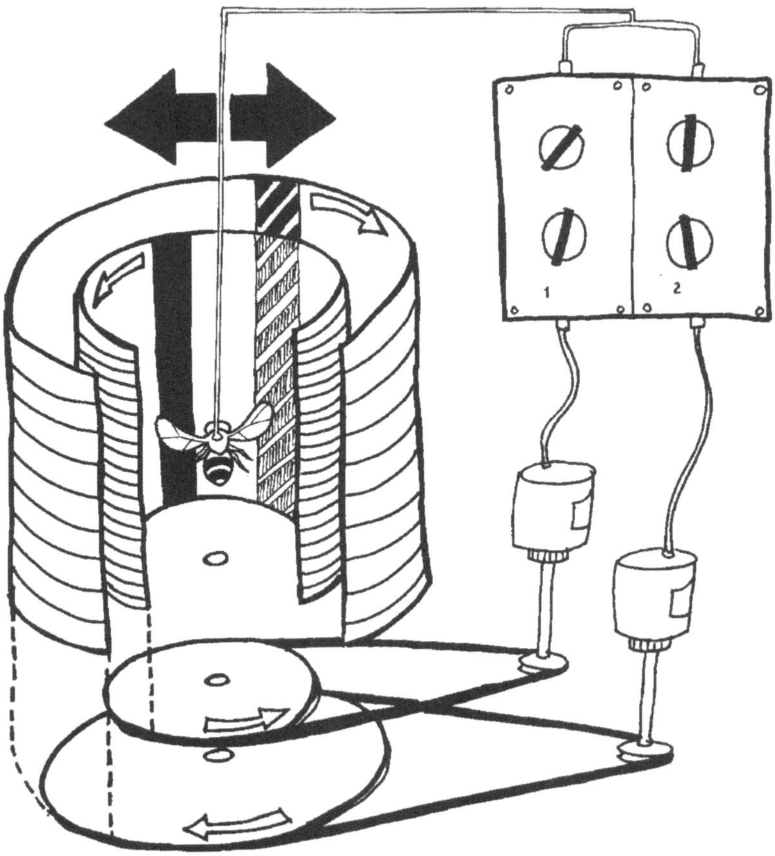

Fig. 2.1 Flight arena concept. A tethered fly is positioned inside an arena of distinct, layered images rotating around the fixed animal. Flies can control the angular position of the rotating image layers under behavioural feedback (''closed loop'') conditions during flight. The different images are tuned to require different control parameters for fixation (stabilisation), thereby evoking object-specific behavioural patterns. Flies lock their behavioural responses (by body torques or wing beats) to parts of the dynamic visual field and ignore others, such as other moving bars or background patterns (Wolf and Heisenberg, 1980). Experimental control of behavioural feedback parameters (box with dials) tests the animal's ability to adapt specifically to changing fixation constraints for one of the objects, while ''ignoring'' the other. Similar closed loop flight arena experiments can also be performed with feedback from wing beat differentials in a virtual-reality arena (Lehmann and Dickinson, 1997).

interestingly, do not always ''choose'' to do so, sometimes letting images slip by for seconds before ''grabbing'' them again behaviourally. Although such capricious control resembles changes in ''attention'', these behaviours were originally viewed as afferent processes based on multistable networks residing primarily in the optic lobes and thorax, bypassing the central brain — in other words, a ''reflex''(Kalmus, 1964).

Most researchers studying behaviour in *Drosophila melanogaster* might counter this unsatisfying explanation with rich anecdotes suggesting that some form of selective ''attention'' over-rides optomotor responses. For example, my colleague Herman Dierick who works on fly aggression describes how, following a fight bout between two males in a cage full of flies, the victorious male will sometimes keep track of (orient itself to) the defeated male, which dive-bombs around amongst a constellation of other flies. Clearly, the retinal slip caused by all the other flies, some which are looming closer, is ignored and an optomotor response is selected for only one memorable individual. Relevant to such casual observations is recent work showing that visual pattern recognition in tethered flies is invariant for retinal position (Tang *et al.*, 2004).

The need to over-ride certain optomotor responses is also evident during self-motion. This was studied in detail by Collett and colleagues in the hoverfly *Syritta pipiens*, where the optomotor response was shown to be controlled by shifting portions (''poles'') of the fly retina, depending on flight style (sideways, forward) or on whether the animal was tracking an object or not (Collett, 1980a,b). Collett concludes that an animal must in some very broad sense ''know'' what the consequences of its actions are going to be for it to appropriately select responses across certain retinal patches and ignore others. Subsequent electrophysiological investigations in the larger blowfly identified unique classes of neurons in the optic lobes that respond only to small moving objects on textured backgrounds, and other neurons that respond only to background motion (Borst and Haag, 2002; Egelhaaf *et al*, 2002). Nevertheless, behaviourally relevant selection of optic output must somehow occur, whether that optic source is confined to patches of retina or ensembles of specialised neurons, while other large swathes of retinal action are ''ignored''. In away, the ongoing dynamics between retinal output selection and suppression in relation to experience begin to resemble what we might call attention. A careful study of this phenomenon requires a controlled visual environment presenting multiple image choices, and means by which the fly can ''report'' its visual selections.

4. Fly virtual reality

The closed-loop flight arena (Fig. 2.1) provides an elegant paradigm for studying visual selection in tethered flies, since the interplay between optic flow and behaviour can be manipulated. Distinct, rotating objects are tuned to require different control parameters for fixation, thereby evoking object-specific behavioural patterns. Flies lock their behavioural responses (by body torques or wing beats) to parts of the dynamic visual field and ignore others, such as other moving bars or background patterns (Wolf and Heisenberg, 1980). Fixation behaviour in the arena is of variable latency and vigour, and behavioural responses are rarely stereotypical, except by being generally exploratory in nature (Dill and Heisenberg, 1995; Goetz, 1004). The virtual reality environment of this paradigm allows for the design of some interesting experiments testing the limits of optomotor flexibility in non-natural situations. For example, if the closed-loop electronics are reversed such as to counter optomotor responses (e.g, turning with images makes them flow even faster), flies will quickly adapt to this odd situation and still manage to stabilise their object of fixation, violated optomotor reflexes notwithstanding (Heisenberg and Wolf, 1984). Such experimental control of closed-loop fixation has convincingly demonstrated that visually driven behavioural responses become extremely flexible when feedback allows the fly to control its visual environment. Further, the capacity for visual selection to be steered (via conditioning experiments) by other sensory modalities, such as odour or heat, suggests that such selection in the fly is not confined to the optic lobes (Wolf and Heisenberg, 1991; Guo and Gotz, 1997). As in humans, fly visual selection recruits different parts of the brain in associating disparate stimuli or stimulus modalities.

Attention in humans is partitioned serially in time between simultaneous stimuli (Shapiro, 2001). Flies have been shown in open-loop (no-feedback) flight arena experiments to respond to simultaneous images in a similar way (Heisenberg and Wolf, 1984; Wolf and Heisenberg, 1980). A stationary object in the fly's visual field will reduce the response to a (second) suddenly moving object by 50%, compared to responses to the suddenly moving object alone. This suggests that the behavioural response is subject to a limited resource partitioned among potential percepts. Similarly, two objects oscillating at the same frequency but in anti-phase on laterally opposite sides of the arena will evoke a response pattern that is very similar to that seen for a single-object oscillation, except with spontaneous switches from one phase to the other. In humans, competing images presented to either eye separately cannot be reconciled, and are therefore perceived in alternation one at a time (Srinivasan *et al*, 1999). Although the human effect involves a completely different neuroanatomy than the effect in flies, a common mechanism is not excluded.

5. Alternating selection

The preceding examples of fly selection and suppression demonstrate a critical aspect of fixation common to many animals: it is rarely sustained for one target percept for very long, but rather, it shifts or alternates among targets. Any investigation of behavioural selection must consider this phenomenon, as it confounds the motive in selection. Alternating selection has been examined extensively in behavioural choice experiments in flies (closed loop flight or walking assays) where it has been referred to as "sustained spontaneous alternation" (Goetz, 2004; Horn, 1978, Schuster *et al*, 2002). Flies will persistently walk (or fly) back and forth between two salient visual targets, "attending" to each for a while until the other somehow wrenches it away. It is not clear whether such sudden alternations in behaviour are at all goal-directed in flies. The timing of alternations is not strictly periodic and often seems random, yet it might appear maladaptive for flies if their behaviour were released without a good reason and only as an accident of a random-number generator in their brains. Sudden shifts in behaviour have been studied in flies on shorter time scales as well, where they appear as flight saccades, occurring in the context of object fixation in closed-loop flight (Heisenberg and Wolf,

1984). The saccades do not occur in response to anything detectable (the visual stimuli are not saccading), yet they seem to be employed in a meaningful way. They reveal possibly novel visual stimuli outside the centre of fixation, and they probe parts of the visual field that might be associated experimentally with other sensory modalities such as odours or heat. Martin Heisenberg and colleagues have suggested that behavioural saccades, or sudden spontaneous alternations, may act on behaviour in the same way as mutations act on evolution: chance events that produce changes that will be selected according to their advantage to the organism (Heisenberg and Wolf, 1984). Such alternating behaviour resembles ancient mechanisms such as the tumbling versus swimming behaviour of certain bacteria (Fenchel, 2002). As with genetic mutations, there is no "goal" involved here per se, but just naturally selected outcomes of such spontaneous behaviour within a specific environment. In animals with learning and memory capabilities, spontaneous alternations might dwell on those stimuli that prove to be more salient (relevant) to the animal for its survival. Motivation is, in this sense, history. Sudden spontaneous alternations in flies might thus support the consolidation of salience which, in turn, sustains "attention". The serial nature of human attention seems just as necessary in flies for salience to be efficiently established; connections must be drawn between single causes and effects, and spontaneous saccades may force all animals into making those connections, or perish.

6. Salience and memory

Salience is intertwined with memory, especially short-term or "working" memory. Salient stimuli draw attention, often via change from a previous state, and working memory provides a mechanism for an animal to detect change. Yet, the mechanisms subserving selective attention and memory must be related or intertwined because each seems causal to the other: memory steers attention, yet attention results in memory formation. Although this mutually causal relationship may seem to create a problem of "infinite regress" in understanding its genesis (the chicken or the egg?), the relationship instead suggests that a form of neural evolution occurred within the lifetime of an animal (Edelman, 1987). Likewise, if human consciousness emerges—even in part—from perceptual processes intertwined with memory, this makes it difficult to study as a final product in the adult without considering its evolution from more basic constituents (e.g. saccades and neuronal group selection; Edelman, 1987) throughout the early life of a person. As in phylogenetic analysis (deducing evolutionary relationships between species), this problem calls for techniques that allow us to study the sequence of changes throughout the development of an attentive brain. One would be in a good position to study mechanisms of attention in an animal offering an array of tools with which to track its neural development and plasticity alongside behaviour.

7. Mushroom bodies

The *Drosophila melanogaster* nervous system can be systematically manipulated throughout its life by various genetic tools, which facilitates our ability to investigate the mechanisms supporting the phenomena displayed by the animal in behavioural assays (Waddell and Quinn, 2001). For instance, mutations affecting a variety of signalling, structural and transcription genes modulate learning behaviour (Dubnau, 2003). Although attention-like mechanisms would appear to be a pre-requisite for learning, these genetic resources have rarely been examined in discrimination paradigms (instead of association). Classical conditioning experiments in *Drosophila* mutants have shown that synaptic output from brain structures called ''mushroom bodies'' is crucial for the retrieval of associations between specific olfactory stimuli and electric shocks (Dubnau *et al*, 2001). These bilateral structures (see Fig. 2A) are involved in memory formation and retrieval in most insects studied for learning assays (de Belle and Heisenberg, 1994; Connolly *et al*, 1996; Heisenberg, 1998; Straussfeld, 1998). The mushroom bodies have also been associated with visual selection by examining, in visual fixation paradigms, some of the same *Drosophila* learning mutants gleaned from odour conditioning

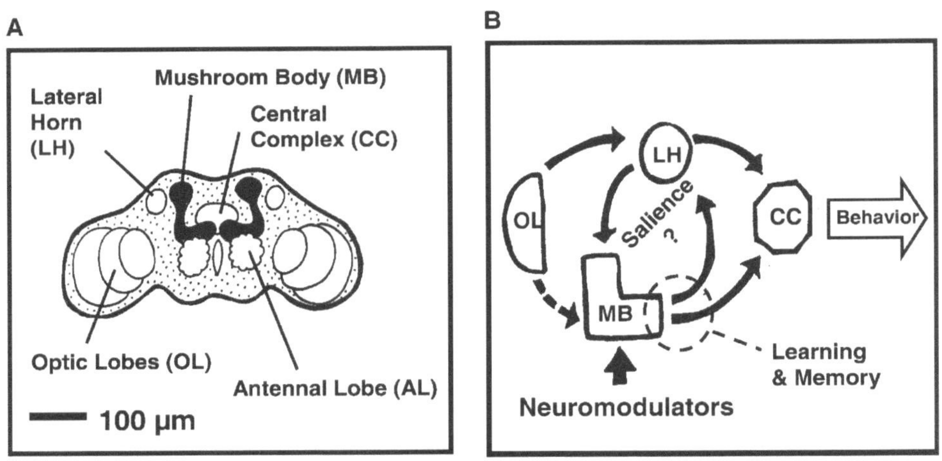

Fig. 2.2 *Drosophila neuroanatomy*. A: Schematic of the brain of *Drosophila melanogaster*, with select structures indicated. The two bilaterally symmetrical mushroom bodies (MB) are shown darkened. The medial protocerebrum is roughly in the area bracketed by both mushroom bodies, including the central complex (CC). Below the brain is a 100 mm scale. B: A speculative model relating memory/learning mechanisms with salience/discrimination in the *Drosophila* brain. Visual input to the optic lobes (OL) reaches the lateral horn (LH) and mushroom bodies (MB) by direct and indirect pathways (Straussfeld, 1998) respectively. Some forms of learning and memory are mediated by neuromodulators and cAMP signalling pathways, which modify synaptic output from the MB (Dubnau, 2003) Output from the MB and LH converge in the central complex (CC) to control motor behaviour (Martin *et al*, 1999). Connections between the LH and MB (Heisenberg, 2003) could gate the level of salience of a visual stimulus via an oscillatory mechanism similar to that found in the neuroanatomy of the locust, albeit for smell (Perez-Orive *et al*, 2002).

screens (Schuster *et al*, 2002; Gong *et al*, 1998; Wu *et al*, 2000). This argues for a certain overlap between association and discrimination, as might be expected. When tested in visual selection paradigms, learning and memory mutants in *Drosophila* are either unable to fixate on visual stimuli or fail to seek novelty, depending on whether they are flying or walking, respectively (Schuster *et al*, 2002; Gong *et al*, 1998; Wu *et al*, 2000). Comparative studies of discrimination and association with such mutants offer insights into the connections between perception and learning in flies, thereby outlining the framework for a speculative attention mechanism (Fig. 2.2B).

Visual selection experiments in the closed-loop flight arena have shown that the mushroom bodies not only guide the dynamics of tracking behaviour, but also provide the ability for the animal to extract salience from ''messy'' compound stimuli. This includes contextual generalisation and selecting (discriminating) among contradictory cues in classical conditioning experiments (Liu *et al*, 1999; Tang and Guo, 2001). A fly's selective behaviour is not a linear function of stimulus parameters but, as in humans, will ''jump to conclusions'' about the salience of a stimulus. This can be demonstrated by conditioning flies to avoid a stimulus with two parameters (shape and colour) that can then be mixed and matched and quantitatively changed in subsequent choice tests. Wild-type (normal) flies assign salience reliably to one parameter (e.g. colour) even as it changes, but beyond a certain level of change, salience (evidenced by the fly's behavioural selection) suddenly flips entirely to the other parameter (e.g. shape). However, with genetically controlled defects in the memory-forming mushroom bodies, flies instead respond to changing choice parameters in a more linear and, presumably, maladaptive fashion (Tang and Guo, 2001). In such mutants, the likelihood of flies switching parameters is a simple function of the parameters' quantitative characteristics. In a normal fly, switching is not a graded response to quantitative aspects of stimuli, but rather ''pops out'' at a dynamic threshold, somewhat like our own sudden awareness of an object or a memory (like the classic hidden Dalmatian picture in psychology textbooks). It is conceivable that such ''pop out'' salience effects in the fly are not unique to conditioning experiments in the laboratory, but rather are serially ongoing with respect to the animal's interaction with its dynamic environment. The challenge is to begin reconstructing the molecular and neural machineries that allow insects to discriminate and assign salience in such an experience-dependent manner. For a number of reasons, the power of genetics has not been well applied to this problem, even though the memory-forming mushroom bodies and the array of learning mutants (Dubnau, 2003) seem to be promising starting points in this endeavour.

8. The missing piece

Although attention-like mechanisms may accomplish similar functions in insects and mammals, the obvious association of attention with consciousness in humans complicates and perhaps even hinders the investigation of attention analogs in animals. Thus, the phenomenon is called ''selective attention'' in mammals, but ''selective discrimination'' in flies. The problem is more than just a semantic one: by relegating fly behavioural selection to the dubious ''reflex'' category of selective discrimination, we presuppose that there must exist a fundamental difference between mechanisms of behavioural selection in insects and higher animals, a difference that must closely parallel an imaginary consciousness divide, and which hence must be tied to mammalian neuroanatomy. Yet, even monkeys are in the same bag as flies with respect to demonstrations of selection: neither can report to us verbally what they were attending to in the past. We rely on their behaviour to infer their ''attentional'' state.

One problem with behavioural studies of attention, especially in invertebrates (which lack the neuroanatomical crutches available in monkey research), is precisely that they often involve behaviour alone. Demonstrations of attentional suppression by behavioural methods alone are never entirely convincing for a number of reasons. First, an animal by definition cannot be in two places at once. By observing a fly's choice behaviour, we have little way of knowing whether the alternative choice was even on its radar (except perhaps in some of the elaborate flight arena experiments discussed earlier). Yet, evidence of suppression is just as important as evidence of selection in order to study ''attention'' in an animal. Another problem with purely behavioural studies of attention is that we know that attention in humans can be dissociated from a motor output; we do not have to be orienting or moving toward a stimulus in order to attend to it. Perhaps we have been reluctant to call selective fly behaviour ''attention'' precisely because it is a motor behaviour, while our (adult) attention is so clearly independent of what our muscles are doing. Furthermore, it is often difficult to distinguish an animal's ability to assign salience to a stimulus from its ability to perform the motor task at hand: a fly must walk or fly adequately, for example. A thorough genetic dissection of the *Drosophila* brain might produce animals fully able to assign salience but unable to show us that they have, thus sabotaging any adequate investigation of the phenomenon. This is probably the main reason why some of the sophisticated behavioural selection phenotypes, discussed earlier, have not been further investigated genetically. Finally, any selective behaviour, regardless of how non-linear and ''intelligent'' it is, can always be described as a very elaborate motor program. The missing piece here is evidence of attention occurring within the brain of an insect, independent of motor behaviour.

9. Selective attention in the fly brain

A recent study of visual selection in *Drosophila* showed that several properties of selective attention are represented by 20–30 Hz local-field-potential (LFP) activity in the fly brain (van Swinderen and Greenspan, 2003) LFP recordings were made in the medial protocerebrum (mpc), a region of the fly brain bracketed by the mushroom bodies (see Fig. 2.2A), during a similar behavioural feedback paradigm as shown in Fig. 2.1, but with brain recordings performed alongside the behavioural selection. Object fixation evoked transiently increased 20–30 Hz activity in closed-loop flight, suggesting that this brain activity was associated with the onset of selection. Critically, transient changes in fly 20–30 Hz brain activity were also found to be associated with selection in animals that were not flying or overtly tracking objects behaviourally: Brain activity was correlated selectively to ''tagged'', distinct images made salient by conditioning (with heat or odour), at the expense of other simultaneous images, without any observable correlated behaviour such as twitches or flight dynamics (a schematic explaining how such selection in fly brain activity can be identified is shown in Fig. 2.3). Such brain poten-

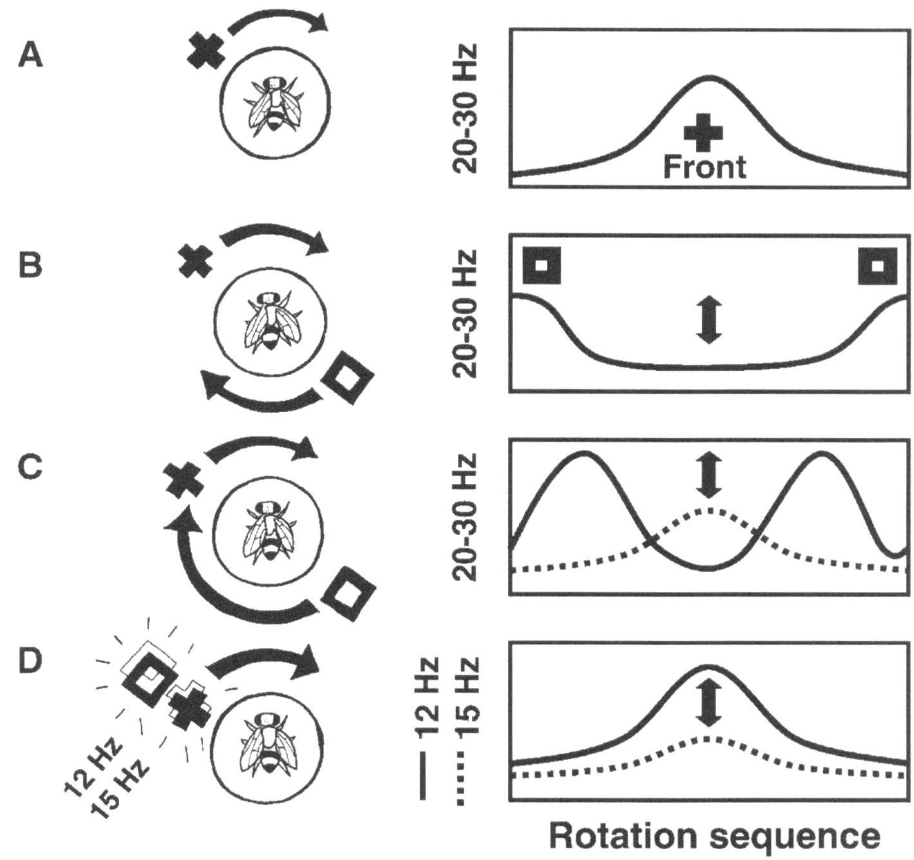

Fig. 2.3 Measuring selective attention in the fly brain by ''tagging'' visual objects. A: Salience. A single object (a cross) rotating around the recording preparation evokes 20–30 Hz brain activity, which increases in amplitude when the object is in front of the fly's visual field. The 20–30 Hz response amplitude can be modulated by salience associated with the rotating object, such as odour, heat or novelty. B: Position tag. Introducing a second visual object (a box, 1808 from the first) evokes a 20–30 Hz response mapping to an opposing position on the rotation sequence, while suppressing the response to the first object (suppression is indicated by the double arrow). C: Period tag. Two distinct objects can also be tagged by making one rotate faster around the fly than the other (shown by the longer curved arrow). 20–30 Hz brain activity can then be mapped and contrasted (double arrow) according to the period of either object. D: Frequency tag. Simultaneous objects can be made to flicker at distinct frequencies (e.g. 12 versus15 Hz) as they rotate around the fly. Salience of either object will modulate the amplitude of either frequency recorded in the fly's brain (Bruno van Swinderen, unpublished data).

tials bring to mind gamma (30–80 Hz) activity in the human EEG, which is also transient and modulated by salience (Zeman, 2001) as well as neural correlates of attention in the primate visual cortex (Treue, 2001). A neural correlate of salience in the fly brain provides an answer to the confounding behavioural issues raised earlier, and encourages a genetic strategy towards understanding how choices are made in the fly. Not too unexpectedly, considering previous work outlined above, the first fly mutants analysed for this phenotype indicate that synaptic output from the mushroom bodies is critical for the normal attentive brain response (van Swinderen and Greenspan, 2003). Recent experiments also show that neuromodulators such as dopamine appear to be crucial as well in governing visual selective attention in the fly. Attenuating synaptic output from dopaminergic neurons impairs behavioural fixation (Ye *et al*, 2004) as well as the 20–30 Hz brain response to visual salience (Andretic, van Swinderen and Greenspan, unpublished data). The parallels with human deficiencies are intriguing: without being able to retrieve certain forms of memory, and without a ''balanced'' neuromodulatory system assigning value (Friston *et al*, 1994.) to stimuli, selective attention is probably impaired in flies as well as humans.

10. Arousal in the fly brain

Fruit flies sleep, as was demonstrated recently by behavioural criteria such as increased arousal thresholds during prolonged immobility at night, or sleep rebound following mechanical sleep deprivation (Hendricks *et al*, 2000; Shaw *et al*, 2000). However, as was the case for selective attention, an independent measure of arousal (sleep versus wake) in the fly brain seems necessary to convince us that such a state in flies is more than just a correlate of motor behaviour. Human studies have such independent measures in the EEG (Zeman, 2001). Immobile, unresponsive humans sitting on a couch watching TV do not display the delta waves characteristic of mammalian sleep; motor behaviour and arousal state can be uncoupled in the short term. Brain activity recordings from the *Drosophila* medial protocerebrum (mpc) showed that this is true for flies as well (Nitz *et al,* 2002.) Although sleep in flies is characterised by decreased brain activity (1–100 Hz), the increased brain activity recorded during wakefulness can be independent of movement. Thus, wakefulness, like visual selection, has a neural correlate in fly brain activity independent of the motor behaviours usually utilised to identify that state.

Arousal state in an animal is nevertheless primarily defined by behavioural responsiveness. Thus, sleep and general anaesthesia are characterised by increased arousal thresholds (decreased behavioural responsiveness to irritating stimuli). However, selective attention is also characterised by increased arousal thresholds, namely, for all the simultaneous stimuli that are being ignored. It appears that during wakefulness in general, the brain is suppressing the perception of stimuli almost as much as during sleep, with the exception of that one dynamic, yet narrow window of selection. Recent work on arousal states in *Drosophila* showed that behavioural responsiveness to mechanical stimuli can be as low in a moving (i.e., awake) fly as during sleep (van Swinderen *et al.*, 2004). One cannot conclude that all such heightened arousal thresholds indicate sleep since, in many cases, flies are moving as much on average as they ordinarily do while awake and responsive. Similarly, our own arousal thresholds to someone calling our name, for example, may be as high while we concentrate on an exam as when we sleep. In humans, sleep and wake states are very different conscious states (Zeman, 2001), so we often assume that these must involve very different brain mechanisms, rather than a continuum of suppression or uncoupling. We assume that the consciousness of our waking (or dreaming) states must therefore stem from something entirely different and unique that is absent during deep sleep. Such neo-dualism stemming from categorising our own sleep/wake dichotomies may be misleading insofar as understanding these states, and perhaps deters us from investigating mechanistic continuity with simpler animals. Yet, in both flies and humans, brains appear to be disabling perception of most stimuli at any given time, just more so during sleep. Understanding what governs the dynamics of suppression in fly brain activity may be more relevant to consciousness studies than searching for signs of (human) intelligence in other animals.

11. Conclusions

General anaesthesia, sleep and selective attention represent different states of arousal that have been associated with changes in consciousness in humans. Nevertheless, these states are well-defined by behavioural responsiveness criteria that can be identified and measured in *Drosophila*, suggesting that via genetic analysis their mechanisms may be understood. On a broader level, genetic analysis allows different behavioural phenomena such as these to be drawn together in correlation. Genetic studies never assume that a single change in a gene "explains" a phenomenon such as attention or sleep. Rather, changes in distinct phenomena such as attention or sleep can be related precisely because all is held constant in the organism except for one "lesion" at a time affecting a specific network of cells. Thus, it is possible that a suite of molecularly related changes will affect anaesthetic sensitivity, sleep, attention, learning and memory "all at once" (i.e., within the same isogenic mutant strains). Or, some changes might increase sleep need without compromising anaesthetic sensitivity. In any case, genetic approaches to behavioural responsiveness in *Drosophila* should organise these phenomena to produce a model of arousal that could be compared to electrophysiological and pharmacological data from the same or other species.

One model might describe different forms of arousal as a continuum defined by suppression or uncoupling dynamics in the brain. As discussed, even states of heightened arousal such as attention may be characterised by the suppression of stimuli evident during low arousal, so an ancient mechanism involving suppression of stimuli may be fundamental here. The differential patterns of suppression resulting in anaesthesia, sleep, wakefulness, selective attention and even learning could occur via related global mechanisms recruiting (or targeting) most of the brain. Data from researchers currently working on every possible behavioural state in *Drosophila* will surely address whether arousal is essentially one mechanism or many different phenomena in the fly. It would appear, however, that the dynamic suppression mechanisms in the brain that may control arousal state would be operant in just about any *Drosophila* behaviour. It may therefore be informative to document the degree of overlap in mutations or microarray data among different behavioural studies where, in effect, an arousal difference may have been evident alongside the particulars of a specific behavioural difference. Alternatively, comparative genetic studies can be designed to identify common pathways modulating responsiveness to stimuli under different behavioural arousal paradigms.

One form of arousal in *Drosophila*—visual selection—displays key properties associated with attention, suggesting that the quotation marks traditionally guarding this ''human'' term may some day be discarded. The implications are of course provocative: attention may be widespread among animals. We might even go further and entertain the possibility that our consciousness stems from insect-like attention mechanisms. Instead of only coordinating the consecutive salience of immediate external stimuli, such attention mechanisms may have evolved to encompass as well the neural substrates of memories peculiar to different animals. Those peculiar to humans will always be different to a fly's or a monkey's, even though similar suppression mechanisms may be operant. Thus, human consciousness may come out of attention, rather than the other way around. With this reversed and admittedly simplified view, our issues with consciousness do not stand in the way of our studying aspects of it in model organisms: we can study selective attention. The latter is at least accessible to science, unlike the so-called ''hard'' questions pertaining to ''qualia'' (e.g. the inevitable redness of red). A student of consciousness recently reminded us that another ''hard'' question, ''Why is there something instead of nothing'', has not prevented us from studying the physical universe (Edelman, 2003).

A strong argument can be made for directing our efforts into fully understanding the *Drosophila melanogaster* brain. Only the size of the period at the end of this sentence, it is the smallest brain shown to display arousal variables similar to our own. Sleep, selective attention and responses to drugs such as general anaesthetics or psychostimulants closely parallel human responses (van Swinderen and Andretic, 2003; Andretic *et al*, 1999). The varied array of genetic tools available to the large *Drosophila* community makes every functional network of neurons in the fly potentially accessible to manipulation, and a brain screen for the neuroanatomy controlling selective attention in the fly is currently underway by, at least, this author. One may ask, then, why not go even simpler, like with *Caenorhabditis elegans* and its 302 neurons? *C. elegans* behavioural responsiveness is also targeted (reversibly) by general anaesthetics at the same concentrations that abolish consciousness in humans (Crowder *et al*, 1996). The worm, however, has not been shown (yet) to require sleep and rigorous demonstrations of selective attention in nematodes are difficult to imagine, l*et al*one neural correlates thereof. Perhaps relevant is the fact that *C. elegans* lacks the complexity of a central, differentiated, brain-like network. In contrast, animals with more complex brains, such as monkeys or rats, are ''closer'' to humans than flies are. Furthermore, brain research on mammalian models is already very advanced and more sophisticated in many respects than *Drosophila* brain research. The argument for closeness to humans, however, may be true neuroanatomically speaking, but does not necessarily hold for the function of sleep, the reversible mechanism of general anaesthesia and suppression phenomena in general. Fly arousal, including perception, now seems just as accessible to study as in any animal below verbal humans, with both behavioural and neural correlates. Although brain electrophysiology is in its infancy in *Drosophila*, much ground can be gained from extensive experimental and computational studies on blowfly vision (Egelhaaf *et al*., 2002), honeybee learning (Menzel, 2001) and locust olfaction (Perez-Orive *et al*,

2002) for example. Combined with recent developments in recording from single neurons in the intact *Drosophila* brain (Wilson *et al*, 2004) we can potentially understand thoroughly how a small brain works to govern arousal. We may even be surprised to discover that selective attention is supported by similar mechanisms in humans and flies. Rather than indicating consciousness in flies, such similarities would suggest that the roots of our consciousness are indeed remote.

Acknowledgements

I thank my colleagues at the Neurosciences Institute for creating an environment where humans and flies can be investigated side by side, especially Ralph Greenspan (a fly person) for the ideas, Doug Nitz (a rat person) for the science, and Gerald Edelman (a human person) for the context.

References

Andretic R, Chaney S, Hirsh J (1999) Requirement of circadian genes for cocaine sensitization in *Drosophila*. *Science* **285**:1066–1068.

Borst A, Haag J (2002) Neural networks in the cockpit of the fly. *J Comp Physiol A Neuroethol Sens Neural Behav Physiol* **188**:419–437.

Brembs B, Heisenberg M (2001) Conditioning with compound stimuli in *Drosophila melanogaster* in the flight simulator. *J Exp Biol* **204**:2849–2859.

Campbell DB, Nash HA (1994) Use of *Drosophila* mutants to distinguish among volatile general anaesthetics. *Proc Natl Acad Sci USA* **91**:2135–2139.

Campbell SS, Tobler I (1984) Animal sleep: a review of sleep duration across phylogeny. *Neurosci Biobehav Rev* **8**:269–300.

Cirelli C (2003) Searching for sleep mutants of *Drosophila melanogaster*. *Bioessays* **25**:940–949

Collett TS (1980a) Some operating rules for the optomotor system of a hoverfly during voluntary flight. *J Comp Physiol* **138**:271–282.

Collett TS (1980b) Angular tracking and the optomotor response: An analysis of visual reflex interaction in a hoverfly. *J Comp Physiol* **140**:145–158.

Connolly JB, Roberts IJ, Armstrong JD, Kaiser K, Forte M, *et al* (1996). Associative learning disrupted by impaired Gs signalling in *Drosophila* mushroom bodies. *Science* **274**:2104–2107.

Crick FHC, Koch C (1990) Towards a neurobiological theory of consciousness. *Sem Neurosciences* **2**:263.

Crowder CM, Shebester LD, Schedl T (1996) Behavioural effects of volatile anaesthetics in Caenorhabditis elegans. *Anesthesiology* **85**:901–912.

de Belle JS, Heisenberg M (1994) Associative odour learning in *Drosophila* abolished by chemical ablation of mushroom bodies. *Science* **263**: 692–695.

Dill M, Heisenberg M (1995) Visual pattern memory without shape recognition. *Philos Trans R Soc Lond B Biol Sci* **349**:143–152.

Dubnau J (2003) Neurogenetic dissection of conditioned behaviour: evolution by analogy or homology? *J Neurogenet* **17**:295–326.
Dubnau J, Grady L, Kitamoto T, Tully T (2001). Disruption of neurotransmission in *Drosophila* mushroom body blocks retrieval but not acquisition of memory. *Nature* **411**:476–480.
Edelman GM (1987) *Neural Darwinsm.* New York: Basic Books.
Edelman GM (1989) *The Remembered Present.* New York: Basic Books.
Edelman GM (2003) Naturalizing consciousness: a theoretical framework. *Proc Natl Acad Sci USA* **100**:5520–5524.
Egelhaaf M, Kern R, Krapp HG, Kretzberg J, Kurtz R, *et al.* (2002) Neural encoding of behaviourally relevant visual-motion information in the fly. *Trends Neurosci* **25**:96–102.
Fauria K, Dale K, Colborn M, Collett TS (2002) Learning speed and contextual isolation in bumblebees. *J Exp Biol* **205**:1009–1018.
Fenchel T (2002) Microbial behaviour in a heterogeneous world. *Science* **296**:1068–1071.
Friston KJ, Tononi G, Reeke GN Jr., Sporns O, Edelman GM (1994) Value-dependent selection in the brain: simulation in a synthetic neural model. *Neuroscience* **59**:229–243.
Fry SN, Wehner R (2002) Honey bees store landmarks in an egocentric frame of reference. *J Comp Physiol A Neuroethol Sens Neural Behav Physiol* **187**:1009–1016.
Giurfa M, Menzel R (1997) Insect visual perception: complex abilities of simple nervous systems. *Curr Opin Neurobiol* **7**:505–513.
Goetz K (1975) Hirnforschung am Navigationsystem der Fliege. *Naturwissenschaften* **62**:468–475.
Goetz K (2004) Exploratory strategies in *Drosophila.* Schildberger K, Elsner N (Ed). Tutzing, Germany: International Symposium, pp 47–59.
Gong Z, Xia S, Liu L, Feng C, Guo A (1998) Operant visual learning and memory in *Drosophila* mutants dunce, amnesiac and radish. *J Insect Physiol* **44**:1149–1158.
Greenspan RJ, Ferveur JF (2000) Courtship in *Drosophila. Annu Rev Genet* **34**:205–232.
Guo A, Gotz KG (1997) Association of visual objects and olfactory cues in *Drosophila. Learn Mem* **4**:192–204.
Hall JC (2003) Genetics and molecular biology of rhythms in *Drosophila* and other insects. *Adv Genet* **48**:1–280.
Heisenberg M (1998) What do the mushroom bodies do for the insect brain? *Learn Mem* **5**:1–10.
Heisenberg M, Wolf R (1984) *Vision in Drosophila.* Berlin: Springer-Verlag.
Heisenberg M (2003) Mushroom body memoir: from maps to models. *Nat Rev Neurosci* **4**:266–275.
Hendricks JC, Finn SM, Panckeri KA, Chavkin J, Williams JA, *et al* (2000) Rest in *Drosophila* is a sleep-like state. *Neuron* **25**:129–138.
Horn E (1978) The mechanism of object fixation and its relation to spontaneous pattern preferences in *Drosophila melanogaster. Biol Cybern* **31**:145–158.
Horridge GA, Zhang S-W, O'Carroll D (1992) Insect perception of illusory contours. *Philos Trans R Soc Lond B Biol Sci* **337**:59–64.

Liu L, Wolf R, Ernst R, Heisenberg M (1999) Context generalisation in *Drosophila* visual learning requires the mushroom bodies. *Nature* **400**:753–756.
Levine RB, Morton DB, Restifo LL (1995) Remodeling of the insect nervous system. *Curr Opin Neurobiol* **5**:28–35.
James W (1890) *The Principles of Psychology.* New York: Henry Holt & Co. 19.
Kalmus H (1964) Animals as mathematicians. *Nature 202*:1156–1160.
Lehmann FO, Dickinson MH (1997) The changes in power requirements and muscle efficiency during elevated force production in the fruit fly *Drosophila melanogaster*. *J Exp Biol* **200**:1133–1143.
Martin JR, Raabe T, Heisenberg M (1999) Central complex substructures are required for the maintenance of locomotor activity in *Drosophila melanogaster*. *J Comp Physiol [A]* **185**:277–288.
Menzel R (2001) Searching for the memory trace in a mini-brain, the honeybee. Learn Mem **8**:53–62.
Nash HA (1999) General anaesthesia. *Curr Biol* **9**:R83–R85.
Nitz DA, van Swinderen B, Tononi G, Greenspan RJ (2002) Electrophysiological Correlates of Rest and Activity in *Drosophila melanogaster*. *Curr Biol* **12**:1934–1940.
Perez-Orive J, Mazor O, Turner GC, Cassenaer S, Wilson RI, Laurent G (2002) Oscillations and sparsening of odour representations in the mushroom body. *Science* **297**:359–365.
Poggio T, Reichardt W (1976) Visual control of orientation behaviour in the fly. Part II. Towards the underlying neural interactions. *Q Rev Biophys* **9**:377–438.
Reichardt W, Poggio T (1976) Visual control of orientation behaviour in the fly. Part I. A quantitative analysis. *Q Rev Biophys* **9**:311–375, 428–438.
Schuster S, Strauss R, Gotz KG (2002) Virtual-reality techniques resolve the visual cues used by fruit flies to evaluate object distances. *Curr Biol* **12**:1591–1594.
Shapiro K (2001) *The Limits of Attention: Temporal Constraints in Human Information Processing.* New York: Oxford University Press.
Shaw P (2003) Awakening to the behavioural analysis of sleep in *Drosophila*. *J Biol Rhythms* **18**:4–11.
Shaw PJ, Cirelli C, Greenspan RJ, Tononi G (2000) Correlates of sleep and waking in *Drosophila melanogaster*. *Science* **287**:1834–1837
Srinivasan R, Russell DP, Edelman GM, Tononi G (1999) Increased synchronization of neuromagnetic responses during conscious perception. *J Neurosci* **19**:5435–5448.
Straussfeld NJ (1998) Evolution, discovery, and interpretations of arthropod mushroom bodies. *Learn Mem* **5**:11–37.
Tang S, Guo A (2001) Choice behaviour of *Drosophila* facing contradictory visual cues. *Science* **294**:1543–1547.
Tang S, Wolf R, Xu S, Heisenberg M (2004) Visual pattern recognition in *Drosophila* is invariant for retinal position. *Science* **305**:1020–1022.
Tarsitano MS, Andrew R (1999) Scanning and route selection in the jumping spider Portia labiata. *Anim Behav* **58**:255–265.
Treue S (2001) Neural correlates of attention in primate visual cortex. *Trends Neurosci* **24**:295–300.
Tully T, Quinn WG (1985) Classical conditioning and retention in normal and mutant *Drosophila melanogaster*. *J Comp Physiol [A]* **157**:263–277.

Waddell S, Quinn WG (2001) What can we teach *Drosophila*? What can they teach us? *Trends Genet* **17**:719–726.
van Swinderen B, Andretic R (2003) Arousal in *Drosophila. Behav Processes* **4**:133–144.
van Swinderen B, Greenspan RJ (2003) Salience modulates 20-30 Hz brain activity in *Drosophila. Nature Neurosci* **6**:579–586.
van Swinderen B, Nitz DA, Greenspan RJ (2004) Uncoupling of brain activity from movement defines arousal States in *Drosophila. Curr Biol* **14**:81–87.
Wilson RI, Turner GC, Laurent G (2004) Transformation of olfactory representations in the *Drosophila* antennal lobe. *Science* **303**:366–670.
Wolf R, Heisenberg M (1980) On the fine structure of yaw torque in visual flight orientation of *Drosophila melanogaster. J Comp Physiol* **140**:69–80.
Wolf R, Heisenberg M (1991) Basic organization of operant behaviour as revealed in *Drosophila* flight orientation. *J Comp Physiol [A]* **169**:699–705.
Wu Z, Gong Z, Feng C, Guo A (2000) An emergent mechanism of selective visual attention in *Drosophila. Biol Cybern* **82**:61–68.
Ye Y, Xi W, Peng Y, Wang Y, Guo A (2004) Long-term but not short-term blockade of dopamine release in *Drosophila* impairs orientation during flight in a visual attention paradigm. *Eur J Neurosci* **20**:1001–1007.
Zeman A (2001) Consciousness. *Brain* **124**:1263–1289.
Zhang S, Srinivasan MV, Zhu H, Wong J (2004) Grouping of visual objects by honeybees. *J Exp Biol* **207**:3289–3298.

Chapter 3

The Evolution of Consciousness in Animals

R.V. Rial, M.C. Nicolau, A. Gamundí, M. Akaârir, C. Garau and S. Esteban

1. Introduction: The definition of animal consciousness

Subjectivity tells me that I am conscious. The interaction with other people and the need to reject solipsism tells me that they are also conscious. However, the problem of an objective definition of consciousness increases several orders of magnitude when trying to recognize the physical basis of consciousness and the existence of conscious entities different from human beings. For a long time, the uniqueness of the human consciousness has even been a religious dogma. Nowadays, science has developed objective and operational definitions of consciousness and even dared to ask whether conscious animals exist.

Considering the definition of consciousness first, two levels are generally acknowledged. Primary consciousness might be defined as being responsive to the environmental changes and being able to report them (Seth *et al.*, 2005). In this sense a mechanical or electronic sensing device showing stimulus-related state changes could be considered as having primary consciousness. However, to avoid considering that a photocell is conscious, the definition generally asks for a multimodal perceptual capability. Multimodality allows creating a meaningful representation of the environment, the "Merkwelt" of Von Uexküll (1909). The perceived world, interacting with the efficient world (the "Wirkwelt"), constitutes a closed whole, which is the environment of the species (the "Umwelt").

1.1 Consciousness and the liabilities of mobility

From a similar, but more elaborated viewpoint, Merker (2005, 2007) proposed that consciousness appeared in response to the need to develop a central control taking behavioural decisions based on motivational, sensory-spatial and motor circumstances. To

Consciousness Transitions: Phylogenetic, Ontogenetic, and Physiological Aspects
Edited by Hans Liljenström and Peter Århem

make such decisions, the organism should distinguish the sensory changes produced in the environment from those produced as a consequence of its own behavioural output. Some animals can produce such distinctions only using peripheral controls, for instance inhibiting the sensory input during motor activity. However, a central unification of the correction mechanisms in animals with multiple spatial senses (vision, audition, olfaction, somesthesis) would end in producing a stable world-space trough a multisensory representation of the environment. This "reality space", stripped from the confounding effects of self-produced movements, is necessary to produce goal directed behaviour. The central motor control would only need to decide what to do, unencumbered by compensatory complexities, which always would work behind the scenes. According to this viewpoint, to see, to hear, to feel or otherwise to experience something, is to be conscious with independence of the changing scenery due to the movements of the sensory organs and irrespective of whether, in addition, one has secondary consciousness, i.e., one is aware that is seeing, hearing, etc. (Edelman, 1989). "*Such additional awareness, in reflective consciousness or self-consciousness is one of many contents of consciousness available to creatures with sophisticated cognitive capacities*" (Merker, 2007).

1.2 Consciousness and sensation

A different viewpoint has been proposed, postulating that consciousness is the evolutionary result of sensation and therefore retains the properties of sensation, i.e., quality, intensity, affectivity and duration. The following paragraphs summarise the relationships between sensation and consciousness according to Cabanac (1996).

The quality of a sensation describes the nature of the stimulus, which in turn depends on its physical qualities, but also on the senses used to detect them, on the structure of the afferent pathways and on the analysing centres. Intensity and duration also have obvious definitions, but the affectivity, the pleasure-displeasure dimension is more difficult to grasp. It is so because pleasure depends (1) on the nature on the stimulus, (2) on the internal state and (3) on the past history of the subject. A given stimulus can arouse pleasure or displeasure according to the combination of these three parameters and alliesthesia is the faculty of a sensation to move up and down in the affective dimension. Sensory pleasure describes the usefulness of the stimulus in terms of the capability to improve the physiological fitness and to correct a physiological deficit. The usefulness of a sensation could be appreciated by its short term survival value, but it can be extended to long term by learning.

The usefulness of a stimulus changes with time. In consequence, the pleasure that it provides is always transient, lasting only as long as the physiological variables have not yet returned to normality. Thereafter, the stimulus turns indifferent or even unpleasant. Comfort is the state in which all sensations have a nil affective dimension. While sensory pleasure is a dynamic state, depending on the actual usefulness of the stimulus, comfort is stable, lasting indefinitely while the conditions of the subject also remain stable.

The environment is continuously providing the various sensory systems with multiple sensations of different quality and the brain, trying to optimise pleasure and comfort, must develop a procedure to rank the levels of pleasure or displeasure produced by each one. Ranking sensations is necessary before they lead to behaviour. This is so because behaviour is the final common path of brain activity; it should be unique for every brain state. In consequence, the brain must develop a common currency to compare the usefulness of every sensory input and maximization of sensory pleasure is the motivation for behaviours adapted to physiological maintenance. Pleasure is thus the common currency of the brain and the capability to evaluate and maximize the pleasure is postulated as the result of having conscious capability for evaluating and comparing the usefulness of different incoming stimulus.

If, as postulated, consciousness evolved from sensation, it must keep the four properties and every conscious state might be defined in terms of quality, intensity, affectivity and duration.

The quality of a conscious state is not difficult to explain. Conscious experiences can be classified as sensations, perceptions, illusions, hallucinations, premonitions, emotions, memory recalls, etc., and each category can be divided in sub-categories: sensations can be visual, auditory, internal external, and so on.

The intensity of consciousness is more difficult to explain. However, conscious states of different intensity are normally identified, for instance, in terms of arousal, from complete unconsciousness during deep sleep, through partial arousal after awakening with sleep inertia, to full arousal during a dangerous experience. But intensity can also explain that several consciousness levels may be working at the same time with different intensities, for instance, one can perform complex mental activities when driving a car.

The affective dimension of consciousness poses no difficulty and it is easy to understand that a conscious experience can be unpleasant, indifferent or pleasant and a discussion on the value of sensory pleasure could be helpful here. Like a sensation, a conscious state can be indifferent, as for instance thinking in a new acquaintance, but it can change to pleasant after developing friendship and to unpleasant if friendship is lost. Also like a sensation, the pleasure of a conscious state is transient and happiness is to joy what comfort is to pleasure. Joy is thus the index of a useful conscious event.

The last property of conscious states is duration, but the temporal nature of consciousness is self-evident.

The postulate of congruence and the evolutionary continuity between sensation and consciousness has a practical advantage in the sense that it allows for the development of relatively simple empirical tests for consciousness. For instance, it seems easy to develop tests to recognize whether an animal has the capability to rank different sensations, i.e., to measure their pleasant-unpleasant dimension.

1.3 Consciousness and emotion

According to the relationships between sensation and consciousness described, the affectivity dimension is essential in consciousness. Accordingly, other researchers have proposed that emotions are central in consciousness. (MacLean, 1990, Pankseep 1998, 2005) Next, a brief account of the evolution of the main systems controlling the emotional behaviour is provided.

The motivational systems of the brain have evolved during millions of years to ensure survival. These systems cater for the perception of environmental events and for the reaction to them, generating specific active states that are powerful drivers of behaviour. In mammals, the elaboration and expression of these systems enables the communication between ancestral motivational and modern cognitive parts of the brain. However, the neurochemical coding of these networks seems to be highly preserved in evolution as shown by the role of dopamine, serotonin and opioid peptides in reward, plasticity, aggression, depression, pain and pleasure. The long history of these neurotransmitters and the constancy of their functions contrast with the position of many neurophysiologists, which reject the assumption of emotions in animals as intolerable anthropomorphisms (Pankseep 2005).

The emotions can be defined as motivational states shaped by natural selection that allows modulation of physiological and behavioural responses ensuring survival, reproduction and fitness. Motivations subserve homeostatic needs and can be activated or not, depending on external and internal factors. In contrast, emotions are the readout of active motivational systems, i.e., manifestations of actual homeostatic needs. In this way, emotions act as amplifiers of the internal or external signals of homeostatic significance.

Motivations and emotions are controlled by specific brain networks which are increasingly understood. A neuroanatomical structure basic in this context is the limbic system, a set of neural structures placed beneath the telencephalic pallium of mammals. The conceptual basis of the limbic system has been questioned and has been subjected to a considerable revision (LeDoux, 2000). The "Triune Brain" hypothesis (MacLean, 1990) considers that the limbic system is in the middle of three components. It was developed in early mammals and layered over a striatal "reptilian brain". In a latter stage, modern mammals developed the third component, the neocortex. Nowadays, the Triune Brain hypothesis cannot be sustained under comparative and evolutionary neuroanatomical grounds and a good deal of evidence shows that the limbic system evolved long before the advent of amniote vertebrates, let alone early mammals (Butler and Hodos 1996, pp 86 and 455). Despite this fact, recent reports still consider that the triune brain is "*a useful structure-function framework for more recent thinking about the evolution of emotional systems*" (Kelley 2005). However, the view that the limbic system is an early development of the vertebrate brain gives support to the antiquity of the emotions and to the idea that they exist not only in reptiles, but also in their predecessors.

Three main neurotransmitter-neuromodulator systems are in control of emotional states. Dopamine (DA) is proposed to play a major role in motor activation, appetitive motivation, reward processing and cellular plasticity. DA, DA receptors and associated synthesizing and metabolising enzymes have been found in all species studied, including invertebrates and vertebrates and evidence has been found showing that most physiological roles of DA are preserved along the evolution (Pani and Gesa, 1997). The major functions of DA lie in the motor control, in appetitive motivation, in cellular plasticity and in the production of reward responses, up to the point of having been called the "pleasure chemical". Although recent evidence has challenged this view, DA continues to be recognized as a major controller in hedonic responses (Cannon and Bseikri, 2004).

Like DA, serotonin also has an extremely long evolutionary history and the first membrane receptor for serotonin has been traced back to 800 million years (Peroutka and Howell, 1994). Nowadays, serotonin and serotonin receptors have been found in worms, molluscs, crustaceans, insects and mammals. Serotonin is involved in the regulation of many behavioural functions, including arousal, motor pattern generation, sleep, learning, food intake, mood and social behaviour. In vertebrates, the brain serotonergic neuronal groups reside in midbrain and pontine regions, but they have extensive ascending and descending projections, from the cortex to the spinal cord, including limbic system, basal ganglia and hypothalamus, having been proposed as showing the most extensive neurochemical network of the vertebrate CNS. It is evident that serotonin has a wide range of functions, but studies performed show that serotonin plays a critical role in the regulation of aggressive behaviour and mood in animals as diverse as lobsters, crayfish and humans (Huber *et al.*, 1997).

The opioid peptides comprise a family with three main constituents: endorphins, enkephalins and dynorphins. In addition, several types of receptor exist for opioids. Like DA and serotonin, opioids and their receptors have been found in all animal groups, from invertebrates to humans. In vertebrates, opioids can be found in several brain networks and are involved in emotional regulation, modulation of pain and stress, endocrine regulation and food intake (Van Ree *et al*, 2000). Increased opioid function is associated with positive affective states, relief of pain, euphoria, well being, relaxation, social attachment, and pleasurable food related states (Panksepp, 1998).

Pain is a sensation with physical and affective components. The first effect of a noxious stimulus is physical pain, but it also causes a negative emotional state. Opiates can act at spinal levels blocking the physical sensation of pain, but can at cortical levels induce a state of disinterest in pain sensation. Undoubtedly, pain is adaptive playing a powerful role to avoid dangerous stimulus, but the well known phenomenon of stress analgesia is also useful in allowing escape from danger (Iversen and Iversen, 1981). Thus, in a first response, the stressed individual has an opiate attenuated pain response. Thereafter, the pain sensation is amplified, allowing the development of powerful avoidance learning mechanisms for future occasions.

In summary, dopaminergic, serotonergic and opioid systems are universal in the animal kingdom exerting an extremely powerful control of emotions. These functions procure the modulation of the affective properties of the sensory systems, that is, they play a fundamental role in alliesthesia and, following the proposal of Cabanac described, in the ranking of different sensations and in the development of a common currency between them.

2. Evolution and adaptation

Actual live beings are the result of evolution, a process based on the production of random variations in the descent of the organisms and selection of the best adapted variants. Using evolutionary methods, it has been possible to draw evolutionary trees, from unicellular organisms to complex plants and animals. Similarly, evolutionary trees have been drawn for the development of particular traits, for instance, the changes suffered by the tetrapod limb, from the initial fin to the human hand or the avian wing.

This review aims at developing an evolutionary tree of consciousness. We know the final step in its development (human beings) but we do not know the first step. Two main questions will be addressed: When did consciousness appear? Is it a unique property of the human mind? To reach an answer, we should first ask which adaptive advantage animals acquired by consciousness, but we should also look backwards in the evolutionary tree. Behaviour of sufficient complexity to consider the existence of consciousness could be found in three groups: arthropods, molluscs and vertebrates. The following paragraphs will analyse first, the question on the adaptiveness of consciousness. Then, a summary on the empirical proofs supporting (or rejecting) the existence of consciousness in the arthropods and molluscs will be provided. Finally, a much larger attention will be given to vertebrates.

2.1 Is consciousness adaptive?

It is almost a postulate that Mother Nature does not produce unnecessary organs and/or functions. This simplistic point of view is sometimes called Panglossianism (Dennet, 1995) and is most probably wrong. Many traits present in live beings could be mere by-products of other truly adaptive traits. For instance, the oxygen loading capability of haemoglobin is a true adaptation, but its red colour has no importance and animals exist in which blue hemocyanin has similar oxygen loading properties. Also, many traits can have a simple explanation and calling for adaptation is unnecessary. For instance, a nonsensical reasoning would explain that the fall of ripe fruits from tree branches to the ground was evolved as an adaptation to facilitate the rooting of the fallen seeds and the production of new trees. A third case in which adaptation should not be called for as explanation is in traits resulting from growth and differentiation. Another disparate example would be explaining the small size of newborns in comparison to their mother's size as an adaptation to allow easy maternal care.

It is nowadays accepted that adaptation is an onerous concept which should be advocated only when a positive proof is provided (Williams, 1966). It is in fact too easy to find *a posteriori* explanations for everything, and it is even easy to imagine adaptive unfalsifiable explanations both for the presence and the absence of a given trait. Thus, according to the null hypothesis, the adaptiveness of consciousness should in principle be negated.

However, attaining a positive proof of adaptiveness is extremely difficult. One should search two variants of the trait under consideration and study their differential biological efficiency in number of viable descent. Indeed, such a proof is practically impossible for consciousness: first, we are unable to unambiguously ascertain whether a given animal has consciousness; second, we probably lack paired control and experimental animals, identical in every aspect, but differing in their conscious capability. Finally, the number of generations which should be studied to recognize the differential reproductive efficiency of the two variants could be virtually impossible for eventual conscious species, which obviously should have a long life cycle.

An indirect proof of adaptiveness could be obtained from the argument of complexity and co-adaptation between the different parts needed to produce a particular function (Dawkins, 1986). The most used example of adaptive complexity and co-adaptation is the eye, in which the structure of the brain and the complex optical mechanisms of vision must have co-evolved to achieve an undoubtedly useful activity. With respect to consciousness, it also seems to be an onerous activity, in anatomical and functional terms.

The nervous system evolved from simple neural nets without brain, such as in coelenterate animals, to complex brains such as those of humans. Undoubtedly, mind is a result of brain activity; therefore, consciousness should be an emergent property resulting from the evolution of complexity in the nervous systems.

However, although complexity is a quantitative property, we do not know whether consciousness varies quantitatively in animals or if it is an all-or-none capability. Accordingly, consciousness could have developed gradually, in direct proportion to the complexity of nervous systems, but it could also be an emergent capability, which suddenly appeared when the brain reached a certain complexity threshold. In fact, secondary consciousness seems, at first sight, to be unique of humans, probably tied to the speech capability, i.e., both seem all or none properties of humans and could have appeared simultaneously as emergent properties of the human brain.

If the uniqueness of the human secondary consciousness and speech capability would be confirmed, then the lack of adaptive value of both could perhaps be supported. It is interesting to note that the selective value of human speech has been questioned by Noam Chomsky (1988), probably the most distinguished linguist in the world. Looking at a group of people passionately debating and frequently interrupting each other, one can ask about the benefits for every speaker, who should spend his energy in a way more productive for himself. The lack of a well identified advantage in speaking could thus similarly support the absence of a function for secondary consciousness. Both could be epigenetic traits developed as by-products of other truly adaptive traits.

However, few doubts can be cast on the existence of intermediate evolutionary steps in the development of language, each one with its own adaptive value. The number of well recognized intermediate stages of speech in animals is growing. First, the non-vocal language ability of the chimpanzee, for instance using the deaf-mute language signs or using sign keyboards (Rumbaugh and Savage-Rumbaugh, 1986) should be recalled. The brain of apes has thus important grammatical abilities not matched by the ability of their pharynx. In addition, a sophisticated animal language has been discovered even in non-anthropoid primates. The vervet is a social cat-sized monkey living in the African savannahs and rainforests. Struhsaker was the first to note different types of predator-triggered alarm calls in a troop of vervets; alarm calls that caused different defensive measures in the rest of the troop. After his observations, Cheney and Seyfarth (1990), using tape and film recordings together with sound frequency analysis, demonstrated the existence of objective differences in the calls advertising for leopards, eagles and snakes; calls that are undistinguishable for the human ear. The discovery of this vocabulary was soon extended to social interactions and at least ten different words were described, the already mentioned "leopard", "eagle" and "snake", but also "baboon", "other predatory mammal", "unfamiliar human", "dominant monkey", "watch other monkey" and "see rival troop". As Diamond (1992) pointed out, it should not come as a surprise if the vocabularies of wild chimpanzees and gorillas eclipsed that of vervets. Thus, these intermediate steps in the evolution of language are highly suggestive of a gradual evolutionary development of language, a capability which otherwise seems obviously adaptive for vervets. Thus, several steps could be recognized in the evolution of language. Returning to consciousness, it seems really difficult to reject its existence in vervets which show elaborated and purposeful linguistic abilities.

Another sign of adaptation is the presence of costs in the development and the production of the trait under consideration. Every trait has always costs and benefits, and a given trait remains only if the benefits exceed the costs. Continuing with the relationships between language and consciousness, the human pharynx experienced a considerable enlargement to allow the production of articulate language, which considerably increased the risk of choking and suffocation. Thus, if language would have no adaptive function, the individuals with enlarged pharynx would have disappeared in front of other having spared of such costs. Hence, the evolution seems to have accepted a considerable cost to be able to speech (and to be conscious), a cost which should have been balanced by a superior advantage.

3. Consciousness in animals

When trying to achieve an empirical proof for the behavioural ability of some animal, important concerns appear from two opposite ends. The first one is anthropomorphism. According to the principle of parsimony, human-like abilities should not be advocated in animals when simpler mechanisms could explain the observed behaviour. On the contrary, an excessive confidence in the design of tests to discriminate animal abilities could lead to false negative results. In the wild, the animal confronted to a vital need could be quite able to use some particular ability, but in the artificial conditions of the laboratory, the animal responses could show only modest achievements and even no capability at all. Thus, it should be kept in mind that the experimental results represent only a lower limit for the animal performance. Regarding consciousness, no positive proof of existence could have been obtained after some experiments, but the search could have been made using the wrong experimental arrangement and the animal could not have shown the ceiling of its true ability.

3.1 Consciousness in arthropods

The capability to be responsive to the environment, i.e., to produce accurate reports is quite usual in simple animals and even in non-conscious entities. Photocells, for instance, readily report the amount of environmental light, a report which could be equalled to simple and unconscious reflexes. Probably at the same level of photocells, dozens of moths can be seen turning around a street light every summer night, a performance which accurately reports on the light the moths see. However, moths remain trapped in a tropotactic movement, endlessly turning to the direction of the more illuminated eye. If by chance they begin turning leftwards, their left eye will continue receiving more light forever and the animal will turn until exhaustion. They are trapped in an artificial light source which produces divergent light beams, opposite to the parallel beams of moonlight which can never cause rotational behaviour.

The concept of accurate report is ambiguous in the sense that accuracy is a quantitative function. A moth circling a light is accurately telling that it perceives light, but such report has, perhaps, only two bits of information and a simple "exclusive or" logic could be derived from the truth table representing the behaviour of the moth (Table 3.1).

Obviously, the moth circling behaviour is a paradigmatic example of the absence of compensation for the movements produced by the same animal and according to the proposal of Merker (2005) already described, no claim of consciousness can be made for such simple behaviour.

It seems easy to add complexity to the moth behaviour. An additional sense informing whether a 360° turn has recently been performed and additional instructions could be enough: if you have turned more than 360°, then reverse your flight. If the turning sensor informs you again of a turn of more than 360°, then ignore the light source, abandon turning and continue straight and return to your search of parallel light beams (the moon parallel light beams are the natural stimulus driving the behaviour of the moth). Thus, at a first view, the successive addition of sensors and program lines seems to end producing the multimodal primary consciousness.

The honeybee could be considered the summit in the progression of multimodality and program complexity in insects. Honeybees recognize their food by colour, odour, and location in complex and precise mental maps. They even have an extremely sophisticated language which can provide accurate reports on the direction, the distance and the nature of food sources to other members of the hive. They can substitute the sun-centred communication system by the polarized light pattern generated in the sky by the scattering of sunlight as a backup when the sun is obscured, but also landmarks constitute a third reference system and there is no reason to suppose that this exhausts the list of failsafe plans available for honeybees. In addition, they must convert the horizontal polar coordinates of the food source into vertical, gravity oriented angles,

Table 3.1 Effects of the differential stimulation in paired sensory organs in the control of a tropotactic movement. The behavioural output is in fact the result of applying an "exclusive or" logic.

Right eye input	Left eye input	Output
Dark (0)	Dark (0)	Continue straight
Dark (0)	Light (1)	Turn left
Light (1)	Dark (0)	Turn right
Light (1)	Light (1)	Continue straight

that is, they use symbols to signify physical concepts. Moreover, their symbolic language is in some way arbitrary, as shown by the existence of different "dialects" in different races of bees. In conclusion, social insects seem to be able to construct an extremely complex "Innenwelt" and seem to have a subjective sense of self, i.e., they seem to be conscious.

However, there is room for doubt. The complex behaviour of bees could be explained by extremely rigid mental rules. The following lines basically follow the account of Gould and Grant-Gould (1986, 1995). When a forager bee discovers a flower, an extremely mechanical and rigid sequence of processes is set in motion. The bee should learn many things about the food source, such as colour, shape, odour, location, nearby landmarks etc. It has been recognized, for instance that colour is learned only in the final three seconds after landing on the flower. In fact, the colour visible to the bee before the landing sequence is never registered. A naïve bee carried to a coloured feeder from the hive and placed over the food source will circle repeatedly after taking a load of sugary water as if "studying" the source, but when returning a few minutes later it is quite unable to choose the correct feeder colour. However, if the bee is interrupted while feeding, so that it must take off and land again on its own accord, the second landing will permit to choose the correct feeder colour on its next visit. Similarly, the landmarks are only learned after taking off and a bee, which is transported back to the hive while feeding, remains without the slightest memory of the landmarks that it must have seen on its arrival. Also, changing any component of the set, odour, for instance, forces the bee to relearn painstakingly all the other pieces of information even though they have not changed.

So, logical and impressive as the associative flower leaning of bees appears, it seems clear that these insects are simply well programmed learning machines, but, most important, the regulatory processes of behaviour are always specific for each behavioural stage. They attend only to the cues deemed salient by evolution in well defined contexts and often during precise critical periods. They file the information thus obtained in pre-existing arrays, i.e., they have independent mechanisms to control the consequences of every sensory input, but lack a central system capable of applying a single set of general rules for every situation, the basic system to elaborate the reality space already described (Merker 2005).

3.2 Consciousness in cephalopods

Apart from arthropods, large brained cephalopods are perhaps the logic non-vertebrate option to look for animal consciousness. They have complex multiple sensory systems well comparable to vertebrates and an important deal of behavioural repertory. Besides, their nervous system shows an increased centralization when compared with the ganglionar system of gastropods. They have also a unique chromatophoric effector system comprising an important part of nervous tissue and a high number of nerve-controlled chromaphoric organs, which allow the display of a variety of well coordinated colour flashes and tetanus. It has been stated that this system constitutes a language which serves for inter-specific communication (Moynihan and Rodaniche, 1977), but it is also indicative of behavioural states which in some way could be assimilated to "mind" states, as they can show excited (sexual or agonistic) cryptic and restful patterns of body coloration. Also, an Indo-Malayan octopus has been observed to display a repertoire of postures and body patterns, several of which are clearly impersonations of venomous animals co-occurring in the habitat, a behaviour that was called "dynamic mimicry". Interestingly, in these cases the octopus took decisions on the most appropriate form of mimicry to use according to the nature of the perceived threat (Norman *et al*, 2001). In addition, it has been asserted that the octopus is able to learn an operant behaviour just by watching the behaviour of a congener (Fiorito and Scotto, 1992) although this has been contested (Boal *et al.*, 2000). Consistent individual personality differences have been also observed in *Octopus rubescens*, with three behavioural dimensions, Activity, Reactivity and Avoidance (Mather and Anderson, 1993). In *Octopus bimaculoides* the number of observed personality dimensions is four, namely, Active engagement, Arousal/readiness, Aggression and Avoidance/disinterest. These dimensions have been identified as early as in three week old animals, with related individuals showing more similar personality than unrelated ones (Sinn *et al*, 2001).

Other observations show that *Octopus vulgaris* develops detailed navigational spatial maps for their home range, taking return shortcuts after complex trajectory hunting trips. This behaviour means that cephalopods are able to resolve the detour problem, which will be defined below. In addition, they cover different parts of their home range one after another on subsequent hunts and days suggesting memory for places where food might be found and working memory for places already visited (Mather, 1991). Also, *Octopus vulgaris* can manipulate their environment moving sand and rocks to adjust the size and the opening of their den, a behaviour that has been equalled to tool use (Mather 1994). They also use water as a tool to convert a suitable shelter into homes and to get rid of scavenging fish searching in the food remains in the vicinity of their shelter (Mather 1992).

To end with a description of the performances of cephalopods, play behaviour has also been observed. When a floating pill bottle was offered to several *Octopus dofleini,* some individuals directed water jets to move the bottle around the water tank, a behaviour that persisted more than 10 min. (Mather and Anderson, 1999; Kuba *et al.*, 2006). If this behaviour would have been performed by a human, it would have undoubtedly been described as bouncing a ball.

3.3 Consciousness in vertebrates

3.3.1 The evolution of land vertebrates

Looking at the vertebrate evolutionary tree, the land vertebrates evolved from the primitive tetrapod stock, being the coelacant (*Latimeria*) a living example of such trunk. Tetrapoda gave origin to amphibians and also to amniotes. Three main groups of amniota are recognized; (1) anapsids (turtles), (2) diapsids, which comprises tecodonts (birds and crocodiles) and lepidosauria (lizards, snakes and sphenodonts) and (3) synapsids, from which mammals were developed (Laurin and Reisz, 1995). In the search of vertebrate consciousness, we should look backwards, from conscious humans to their ancestors. Indeed, consciousness leaves no fossil remnants and therefore the study must be limited to the extant representatives of those earlier groups. Nobody can warrant the actual groups to be equivalent to those from which they appeared. In fact, extant reptiles, for instance, could be as different from their ancestors as current mammals are from their predecessors; all extant animals have evolved during the same amount of time and the assumption that reptiles would have remained stable during a time in which mammals suffered progressive changes is unsupported. However, there is a continuity between the brains of current fish, amphibians, reptiles, mammals and birds, which could give evolutionary value to comparative studies (Butler and Hodos, 1996).

3.3.2 Comparative neuroanatomical and neurophysiological roots of consciousness in vertebrates

Neuroanatomical complexity

Since the development of Darwinism, anthropocentrism plagued the evolutionary thinking. Evolution was considered to be a pathway to perfection and the concept of "scala natura" seemed the natural result of the evolution. Animals were higher (mammals, for instance) and lower (insects, fish), superior or inferior, evolved and non-evolved and so on. Nowadays, this way of thinking still pervades even among highly educated. However, it is beginning to be acknowledged that, for instance, evolutionary progress is extremely difficult to define (Ayala, 1988). Possibly, it can only be defined a posteriori, i.e., after observing success. As the long life of the Volkswagen beetle car shows, only bad designed models must be changed. This way, living fossils, like cockroaches or crocodiles, which have survived for millions of years, could be summits of the evolutionary progress, while the perfection of humans, with an antiquity of less than one hundred thousand years, remains to be demonstrated.

Increasing complexity is neither a tendency in evolution. In the first moments of evolution, life was undoubtedly simple. Then, complexity quickly increased. However, after a while, the acceleration of increasing complexity decayed and today, extremely complex organisms exist, man being a good example, but also extremely simple organisms, bacteria and protozoans, etc. which constitute the great majority of species as well as more than 99 % of living matter (Gell-Mann, 1994).

Not even the human brain could be considered as a summit. Comparing the brains of different animals is a difficult task. Small animals have small brains, big animals, big brains. Therefore, it would be unfair to compare the brain of a 3 g shrew, with that of a 30 ton whale. Instead, a seemingly sensible comparison would consist of comparing the total number of neurons in the brain of different animals. The neuronal density has been evaluated in the motor cortex of different animals, with relative values ranging from 142 in mice to 10.5 in humans and 6.8 in whales (Pribram and Meade, 1999) and from these figures an inverse relation to body size is evident. A decrease in neuron density could be perhaps compensated by an increase in neuronal branching and number of connexions which could provide a higher behavioural flexibility. However, no correlation has been found between the dendritic structure of pyramidal neurons in the parietal cortex of rats and their behavioural achievements (Anderson, 1995).

Another way to discard the effects of body size could consist of comparing the ratio between body size and brain size, assuming that a big body also needs to be controlled by a big brain (Butler and Hodos, 1996). Figure 3.1 shows the result of such a comparison. It should be a hard blow to the human pride to see that elephants and whales continue showing more brain per body weight unit than the self-crowned king of creation. In anatomical terms, the supposed intellectual advantage of human beings is difficult to support and the gap between the brain of humans and other mammals, at least with respect to size, does not exist.

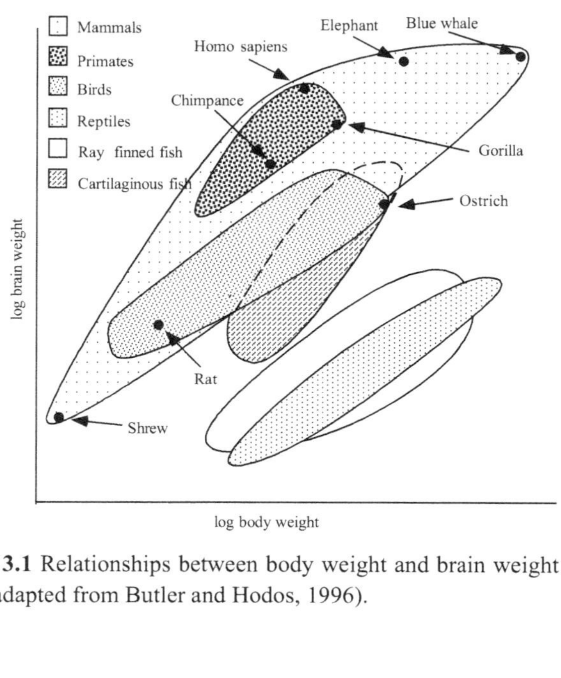

Fig. 3.1 Relationships between body weight and brain weight in mammals. (Figure adapted from Butler and Hodos, 1996).

The absence of evolutionary progress is more evident when looking at other details of Fig. 3.1. While reptiles and ray-finned fishes seem to be placed in a "correct" place, always having a smaller brain than mammals and birds, some cartilaginous fishes, rays for instance, have a brain size comparable to a medium sized mammal. It is not known what a ray does with its "brain excess"; one would have expected that these animals "should" be placed well under mammals (Butler and Hodos, 1996)

One feels tempted to consider other regions instead of the whole brain, although it should be kept in mind, that the use of a tailored criterion is always suspicious. One can remember how Descartes selected the pineal gland as the place for the human soul in the belief that it was unique and distinctive of humans. It is widely believed that the size of the isocortex constitutes a very distinctive human quality. However, the general impression is, once again, inaccurate. It has been demonstrated that the size of the primates isocortex is related to the size of the whole brain and in this respect, the human brain has an amount of cortex quite appropriate to its size, neither bigger, nor smaller than any other ape (Butler and Hodos, 1996).

In conclusion, no gross anatomic feature could be identified yet to support the superiority of humans over other animal species. Humans could be proud of our intelligence and our achievements, of being special. But every animal species is special and, most interestingly for the purposes of the present review, no anatomical gap exists between humans and other mammalian relatives that could explain the existence of consciousness in humans and its absence in animals. If we are an evolutionary summit because we have consciousness, this feature must be an evolutionary development and, given the continuity between our species and the rest of the animals, they must share, at least, some portion of consciousness (Medicus, 1987). Following the increase in brain complexity, consciousness should have emerged somewhere between the first vertebrates and *Homo*.

Neuroanatomical connectivity

Looking at the connectivity of the brain, the most salient difference between mammals, birds and reptiles, is the telencephalic overdevelopment of mammals and birds. Correspondingly, the thalamus has also experienced important modifications. The analysis of the relationships between telencephalon and thalamus is pertinent here because it is usually considered that the neurological activity of these two regions constitutes a landmark of consciousness in mammals (Seth et al., 2005).

The dorsal thalamus of tetrapods has two main divisions: collothalamus and lemnothalamus (Butler and Hodos, 1996). The distinction is based on differences in the input: the collothalamus receives collicular (mesencephalic) sensory input, while the lemnothalamus receives lemniscal sensory input. Regarding the outputs, both regions project contralaterally to the pallium, the striatum or both. These connections could be specific (somatotopic), but also unspecific (diffuse). In addition, the dorsal thalamus also receives telencephalic and mesencephalic feedback lines.

In primitive tetrapods, the lemnothalamus was constituted by a single anterior nuclear group. Instead, the collothalamus had visual, auditory and somatosensory divisions. The general division in lemnothalamus and collothalamus persisted in amniotes, but their structure was further elaborated. The main three groups remained in the collothalamus, but the lemnothalamus was divided in rostromedial and dorsal lateral (visual) groups.

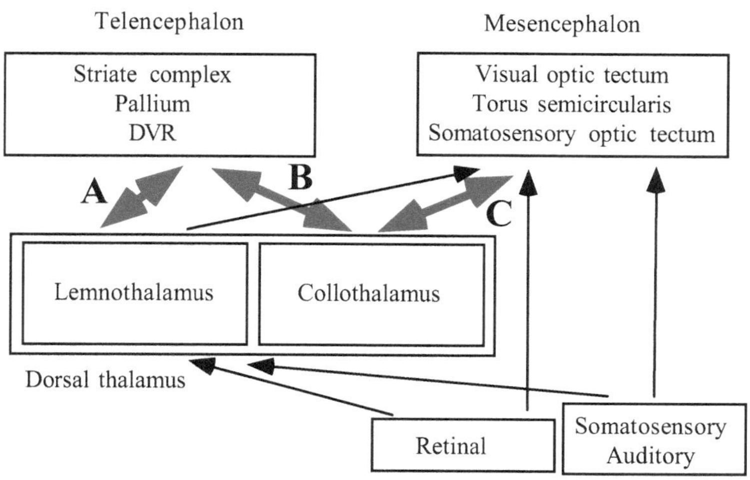

Fig. 3.2 Simplified relationships between dorsal thalamus, mesencephalic tectum and telencephalon in reptiles, mammals and birds. The connections are fundamentally similar in the three groups; the main changes lie in the relative strength of the re-entrant connections between dorsal thalamus, mesencephalon and telencephalon, represented by wide double arrows. In mammals, the connection A dominates over B and C. In Reptiles C dominates over A and B and in birds, B and C have similar weight dominating over A.

In diapsid reptiles the lemnothalamic rostromedial group was differentiated, with olfactory, spinal, somatosensory, retinal and auditory regions, while the dorsal lateral group received the bulk of visual input. In the collothalamus, the visual, auditory and somatosensory areas were enlarged and their telencephalic projections increased their importance. Especially, the Dorsal Ventricular Ridge (DVR), a region placed below the lateral ventricles, was highly developed and differentiated, with a specific topographic organization, i.e., with separated visual, auditory and somatosensory topographic projections. However, the most important input to the DVR and the striate complex proceeds from the collothalamus with visual (from the optic tectum) auditory (from the torus semicircularis) and somatosensory (also from the mesencephalic tectum) information. The collothalamus of diapsid reptiles is composed of (1) the nucleus rotundus, which receives visual tectal input, (2) the medial nucleus (nucleus reuniens pars compacta) which receives auditory input from the torus semicircularis, (3) the posterior medial nucleus (medialis complex) which receives somatosensory tectal input and (4) the nucleus reuniens pars diffusa which receives multisensory tectal input. These four nuclear groups project both to the striatum and the DVR. However, while the first three groups send specific topographic information to the telencephalon, the last one sends diffuse, non-specific projections.

These thalamo-telencephalic relations are summarised in Fig. 3.2. It should be remarked that the connexions between lemnothalamus and telencephalon are bidirectional, the same as between collothalamus and mesencephalon, that is, they show "reentrancy", a central property advocated for the development of consciousness (Edelman and Tononi, 2000).

Interestingly, the diagram is almost identical for diapsids (reptiles and birds) and synapsids, the group which gave rise to mammals. As a result, the thalamic relationships with mesencephalon and telencephalon are extremely old and represent the primitive organization of the thalamus. This is shown in Fig. 3.2 by the three arrows, marked A, B, and C. In reptiles, the arrow C dominates over arrows A and B. In mammals, the dominant connexions are represented by the arrow "A" while in birds, the weight of arrows "B" and "C" is similar. Thus, the differences between diapsids and synapsids lie only in the relative importance of the collothalamus and the lemnothalamus, the latter having expanded considerably in mammals, an expansion paralleled by a pallial overdevelopment (Fig. 3.2b).

It should be noted that reptiles are not a taxonomic category. It has already been mentioned that turtles are anapsids while crocodiles, birds (tecodonts), lizards and snakes (lepidosauria) are diapsid reptiles. In consequence no major differences exist between lepidosaurians and tecodonts in the connections between collothalamus, lemnothalamus, tectum and telencephalon. However, birds show a high telencephalic development. Thus, the telencephalon of birds and mammals seems to be in some way equivalent: mammals developed the pallium, while birds overdeveloped the striate complex.

Consciousness, arousal, waking and sleep

In spite of the general similarity, important differences between diapsid and synapsid thalamic relationships are reflected in the physiology. In mammals, the cortical arousal reaction, as recorded in the EEG, consists of a reduction in amplitude and synchrony (Fig. 3.3, upper record), while in reptiles, the reaction is the opposite (Fig. 3.3, lower record). Increased EEG amplitude and synchronization after sensory stimulation has also been found in mammals. This paradoxical activation has been recorded in (1) early brain developmental states (Villablanca, 1965), (2) as a result of cortical lesions (paradoxical comma reactivity), (3) in cortical atrophy, (4) in the oldest mammalian cortical regions, such as the olfactory bulb (olfactory spindles) or in the hippocampus (theta activation), and (5) most interestingly, during NREM sleep (where the brain is considered to be in a state of functional decortication) as shown by sensory evoked K-complexes and spindles. Hence, it can be deduced that the EEG increase in amplitude and synchrony following sensory activation is, in phylogenetic terms, the oldest arousal response (Bert and Godet, 1963) and the desynchronized response is a modern development only present in neocortical regions. It has been asserted that the desynchronized arousal response heralds the entrance in the conscious state while the slow cortical waves are the sign of consciousness loss (Seth *et al.*, 2005; Edelman *et al*, 2005). However, after considering the evidence of increased EEG synchrony in full waking poikilothermic animals, the generality of the assertion could be wrong.

It is likely that the opposite arousal response is the result of the switching between collicular and cortical processing modes. In diapsid reptiles the main sensory processing is mesencephalic. It has been observed that a snake only relies on vision to strike a prey. Then, other sensory modalities successively determine the localization of the captured prey, opening of the mouth and swallowing, respectively, but the whole set is still coordinated from the mesencephalic tectum. Composite visual + smell + taste + acoustic + tactile images seem not to existent in reptiles (Sjölander, 1994). On the contrary, multisensory processing should be the result of the improved associative capability of the mammalian cortex.

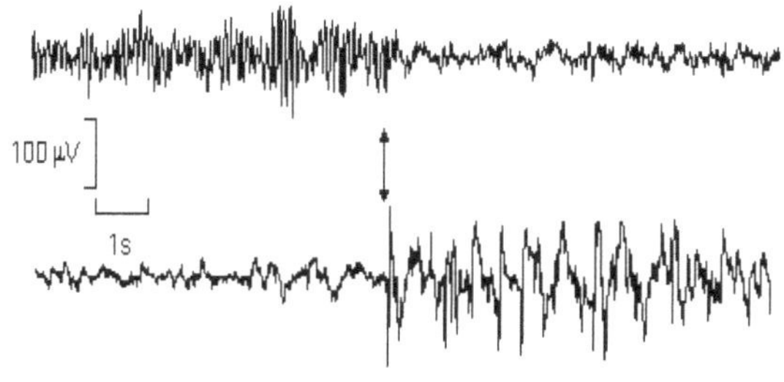

Fig. 3.3 The EEG arousal response to sensory stimulation (arrow) of mammals (upper record) and poikilothermic animals (lower record) is opposite.

However, the opposition between cortex and mesencephalic tectum is, most probably, only relative. This is evident after observing the collothalamic and lemnothalamic multisensory projections to the mesencephalic tectum, but also to the telencephalic striatum and DVR. Most probably, the absence of multisensory guided reptilian behaviour is no more than a consequence of limited observations in which the animals only showed a part of their true capabilities. On the other side, mammals can also produce mesencephalic goal directed behaviour as it is shown, for instance in blind vision cases and also in the Sprague effect (Sprague, 1996) in which the effects of a cortical lesion are recovered (in part) after further lesions in collicular commisures.

The cortical EEG desynchrony represents the fine sculpturing of activation and inhibition of different cortical areas necessary to perform the multisensory analysis of the environment (Steriade, 2000). On the contrary, the synchrony, with delta EEG or with sigma spindles is the sign of cortical inhibition. We propose here that the relative weight of telencephalic and mesencephalic processing suffers oscillations between waking and sleep in mammals, i.e., they show reciprocal inhibition. We also propose that the same switching occurred in phylogeny, in the transition from amniota to the ancestors of diapsids and synapsids as shown in the different responses of their descendants, extant reptiles and mammals.

The changes in consciousness observed during waking and sleep are due to variations in the functional dominance of cortical and collicular working modes in mammals. Cortical processing, with desynchronized EEG (= mammalian type) is dominant during waking. On the contrary, collicular processing, with cortical EEG synchrony (= reptilian type) dominates during NREM sleep. However, explained in this way, the switching between the two modes would produce a paradoxical behaviour: a mammal would be always active, with cortical (mammalian) waking during half of the activity-rest cycle and with collicular (reptilian) waking during the second half. It is obvious that this does not occur during normal conditions. Instead, the paradoxical alternation between the two behavioural modes is probably avoided thanks to two mechanisms. First, the collicular waking is severely atrophied in mammals, as it is shown by the low visual efficiency achieved after the destruction of the visual cortex. In addition, **a** powerful inhibitory activity which should have been developed to aid in the suppression of collicular controlled behaviour when the cortical function is depressed, i.e., during NREM (Nicolau *et al*, 2000). Most probably, sleepwalking represents a failure in the inhibitory mechanisms and in fact the behaviour observed during sleepwalking fits strongly reminds the behaviour observed in experimental studies of the already described Sprague effect.

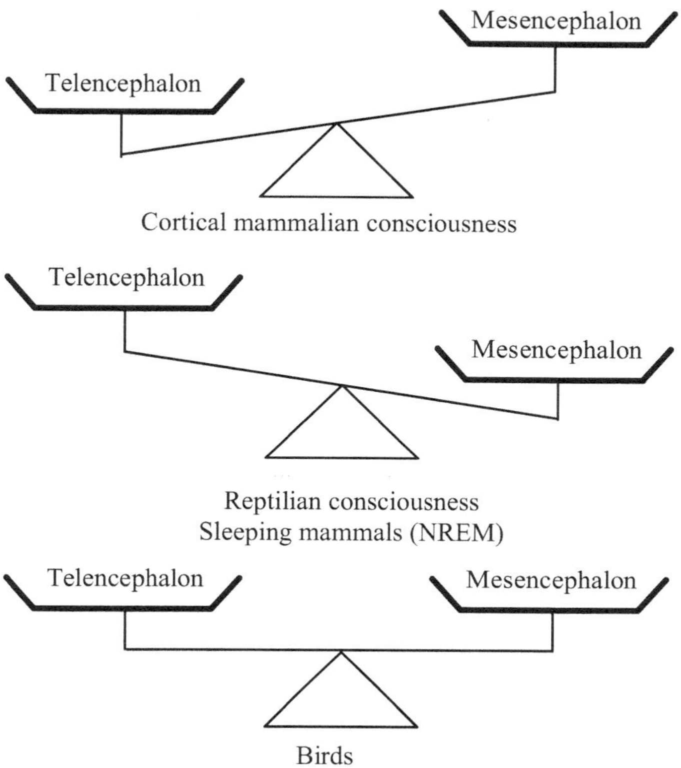

Fig. 3.4 The balance between telencephalic and mesencephalic consciousness. In waking mammals the telencephalic control dominates. In waking reptiles but also in NREM sleeping mammals, the main controlling regions are mesencephalic. In both cases the switching between them depends on thalamic structures. In birds, the weight of the striate complex (telencephalon) and that of the mesencephalon are similar, although both fundamentally rely on collothalamic (reptilian type) connections.

As a result of the inhibition/atrophy of the collicular activity, consciousness disappears during NREM sleep. However, the evident conscious experience of dreams poses further questions. The first reports of REM sleep stated that dreams only occurred during this state, but recent developments have shown that dreams also occur during NREM. The dreams produced during REM and NREM seem to have different qualities, as those of REM sleep seems to be vivid and reportable, while those recorded during NREM seem to be more thought-like. However this distinction is not too clear and vivid dreams have been recorded during stage 2 of NREM in the last part of a night sleep. These findings led to suppose that the dreaming machinery is independent from the NREM and REM machinery (Solms, 2000). At this point it seems appropriate to recall the scheme proposed by Parmeggiani showing the hierarchic changes in brain dominance along the different brain states (Table 3.2a). During waking, cortex is on top, controlling mesencephalic and rhombencephalic activity. This organization changes during NREM where the (reduced) mesencephalon takes the lead over the rhombencephalon and the cortex. Finally, during REM sleep, the rhombencephalon is in command controlling the descendent inhibition in motor activity (Parmeggiani, 1982). After this scheme, and Solms proposal too, the cortex remains isolated from environmental connections but free during non-waking states and dreaming is the result of such freedom.

Translating the scheme of the Table 3.2a to poikilothermic animals (without cortex), only two brain permutations are possible: mesencephalon in top, controlling collicular wakefulness, and rhombencephalon on top, blocking motor output (primitive REM) (Table 3.2b). Is dreaming possible under such a scheme? According to Solms (2000), the dream machinery is cortical and, lacking a developed cortex, poikilotherms would be unable to dream. However, considering the collicular consciousness defended in this review for poikilotherms, collicular activity could exist, not only during waking, but also during rest, with no motor output due to the descending inhibitory influence determined by rhombencephalic dominance. However, the eventual dreams produced by collicular activity during poikilothermic rest should be hindered by the unavoidable low body temperature attained during the poikilothermic nocturnal rest. It is difficult to speculate on the aspect of the eventual reptilian dreams.

Concerning birds, it is generally accepted that they share with mammals the same behavioural states as they show waking, NREM and REM. However, important differences also exist. First, the avian REM sleep episodes are extremely brief (Amlaner and Ball, 1994). In addition, birds show major differences in control as, for instance, they can support long sleepless periods when exposed to continuous light, without showing need of recovery when the light/dark cycles are restored (Berger and Phillips, 1994). Thus, the sleep of birds shows a much higher dependence on environmental conditions. Most probably, these differences reflect the different working modes of the avian and mammalian brain, the former working in diapsid mode, i.e., with dominance of collothalamic connections (as it is shown, for instance, by the extreme development of avian tectal vision). However, the relative telencephalic-mesencephalic weight should be more balanced in birds (Fig. 3.3).

Table 3.2a Permutations in hierarchical organization of the brain in different vigilance states. During waking, the cortical processing is at the top, while the mesencephalic optic tectum and the rhombencephalon are in a subordinate position. This changes in NREM and REM when the collicular and rhombencephalic processing, take the dominant position, respectively. The arrows represent the possible state changes from waking to NREM and from NREM to REM. The direct transition from waking to REM does not occur in normal conditions. The transitions from NREM and REM to wakefulness, though possible, are not shown. Modified after Parmeggiani (1982).

Waking	NREM	REM
Cortex	Mesencephalon	Rhombencephalon
Mesencephalon	Rhombencephalon	Cortex
Rhombencephalon	Cortex	Mesencephalon

Table 3.2b As poikilotherms lack cortex, only two different permutations are possible, determining only two vigilance states, waking and rest,, under collicular and rhombencephalic control, respectively. Comparing the two tables the homology between the mammalian NREM and the reptilian waking is evident.

Waking	REST
Mesencephalon	Rhombencephalon
Rhombencephalon	Mesencephalon

3.3.3 Empirical studies of consciousness in vertebrates

The neuroanatomical, neurophysiological and neurochemical studies outlined so far can provide circumstantial evidence on the presence or absence of consciousness in a given animal group. In addition, they provide hints on how the brain works to produce consciousness. However, it would be convenient to develop empirical tests to identify the actual presence of consciousness in a given animal group. It has been already mentioned that the capability of accurately reporting the properties of the environment is a strong indication of consciousness. Accordingly, humans and vervets, but not honeybees, have surpassed the threshold of consciousness. However, tests based on speech abilities have a reduced field of application. Therefore, more general tests for consciousness should be developed.

Anticipatory behaviour

The capability to produce anticipatory responses is considered a fundamental step for survival. Animals – and human beings – feel pain and react with reflective escape responses to painful stimuli, but this kind of behaviour needs no consciousness at all, in the same way that a photocell can open a door without consciousness. However, most animals can anticipate pain showing avoidance behaviour, for instance, the value of a particular stimulus, the sound of a bell, after a few repetitions, can change from indifferent, to pleasant (food time) or unpleasant (imminence of foot-shock). Every animal avoiding a dangerous stimulus is therefore reporting a change in the affective value of a given stimulus, i.e., the stimulus shows alliesthesia. In this case, the new value is currently called fear, an emotion, but also a change in the actual conscious state. Indeed, alliesthesia could also occur when a previously indifferent stimulus changes to tag the promise of a future pleasant stimulus, in which case, the conscious state changes to joy.

Examples of anticipatory responses are emotional fever and tachycardia. Mammals show increases in body temperature when gently held in the hands of the experimenter (Briese and Quijada, 1970, Briese and Cabanac, 1991, Cabanac and Briese, 1991). It is probably an autonomic manifestation of fear, an adapted behaviour anticipating a probable lesion. The emotional fever is also known in humans (Renbourn, 1960). As a consequence, it has been claimed that the emotional fever is a proof of consciousness. Interestingly, birds, which share homeothermy with mammals, also show emotional fever (Cabanac and Aizawa, 2000).

At a first sight, fever should only be observed in homeothermic animals. However, it is well known that active terrestrial poikilotherms maintain a rather accurate control of body temperature by behavioural means. The existence of variations in the preferred body temperature can be tested in poikilotherms. In fact, it has been shown that after gentle handling, lizards (Cabanac and Goselin, 1993, Balasko and Cabanac, 1998, Cabanac and Cabanac, 2000) and turtles (Cabanac and Bernieri, 2000) also show behavioural fever in the form of preference for a warmer environment. However, when the experiment was repeated with toads (Cabanac and Cabanac 2000, 2004, Myhre *et al*, 1977) or fish (Cabanac and Laberge, 1998) no emotional fever was observed despite behavioural fever was shown after they were injected with pyrogens (Cabanac and Laberge,1998). Similar results have been obtained with respect to tachycardia (Cabanac and Bernieri 2000, Cabanac and Cabanac, 2000). According to these results, fear (emotion) first appeared after the advent of true terrestrial animals.

Capability to rank different sensations

The capability to rank different sensations is also well acknowledged in mammals and it has been also tested in lizards. A palatable food should have a high hedonic value. However, when the taste of a palatable food is associated with some unpleasant experience the pleasantness of the food changes, that is, the quality of the sensory properties of the food are modifiable. Thus, it can be said that the animal is capable of ranking different sensations and produces a behavioural output in agreement with the resulting ranking. Green iguanas were tested in a conflict between exposition to a low temperature and a palatable food (Balasko and Cabanac, 1998). They responded decreasing the amount of ingested palatable bait and reducing the time spent in each visit in proportion to the environmental cooling of the bait's place, that is, the lizards traded off the pleasure of the palatable bait with the displeasure of the cold. This supports the capability of the lizards to develop a common currency for different sensory experiences.

Similarly, the taste aversion experiments performed in mammals and birds show that the hedonic value of a palatable food changes when it is associated to a gastrointestinal unpleasant after-effect (García *et al.*, 1955). Taste aversion learning has been extended to reptiles in which a new food was associated with visceral illness produced by injection of a subtoxic dose of lithium chloride. As a result, iguanas and skinks showed aversion to the new food lasting for several weeks. On the contrary, toads and newts showed no change in the amount of food ingested after lithium chloride injection, that is, they showed no conditioned taste aversion in spite of showing evident signs of visceral discomfort (Paradis and Cabanac, 2004). In conclusion, lizards can modify the hedonic value of sensations, but amphibians do not and the capability to rank different sensations seems to have appeared in amniotes.

The detour behaviour

The detour behaviour represents the ability of an animal to reach a goal by moving round an interposed obstacle with temporal loss of sensorial contact. It has been considered of particular interest for comparative and cognitive research, as a peculiar and quite naturalistic example of delayed response. The acquisition of "object constancy" in the human child, i.e., the ability to understand that an object temporally hidden is the same after being retrieved, has received considerable attention (Piaget, 1954). Similarly, the detour behaviour requires the maintenance of a memory of the location of a disappeared object, that is, an internal representation of the environment and the production of a "mental" experiment as the animal should construct a complex motor trajectory in advance to the final behavioural performance. This way, the detour behaviour is generally considered a form of "insightful learning" (Eibl-Eibesfeldt, 1975). Looking at comparative and phylogenetic studies on the detour behaviour, numerous examples have been described in mammals. In birds, it has been convincingly demonstrated in chickens, quails and in herring gulls, but not in canaries (Vallortigara, 2000). It was suggested that the failure of canaries was due to differences in the adaptation to terrestrial or aerial environment. Most likely, canaries in their natural habitat can always surpass barriers simply by flying (Vallortigara, 2000). Regarding to other vertebrates, impressive examples of detour are known for hunting chameleons for a long time (Fig. 3.5) (Von Frisch, 1962). As in birds, the superior performances of chameleons could be ascribed to the particular features of their arboreal habitat.

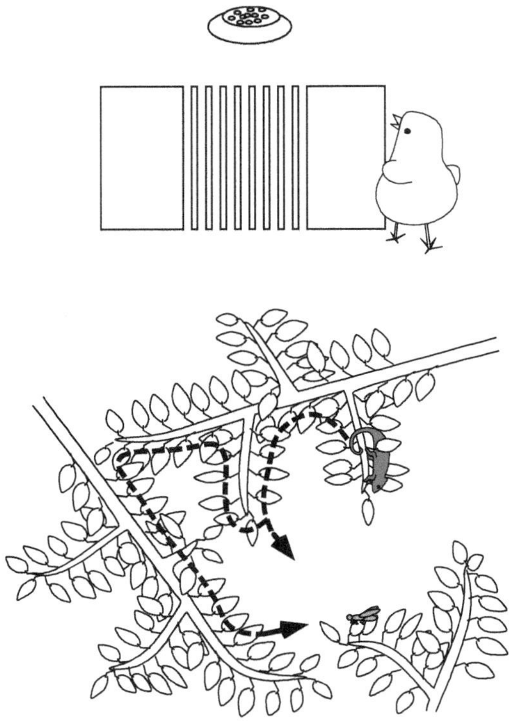

Fig. 3.5 The detour behaviour in chickens (upper panel; redrawn after Vallortigara, 2000) and in chameleons (lower panel). The complex pathway followed to reach the tongue flick position is shown with a dashed line (after Von Frisch, 1962).

Play Behaviour

The same way as it happens with consciousness, recognition of play is easy in humans. However, animal play is notoriously difficult to be defined and the problem increases when trying to extend the concept of play to non mammalian animals. One of the most accepted definitions of play is: "all motor activity performed postnatally that appears purposeless, in which motor patterns from other contexts may often be used in modified forms or altered sequencing (Bekoff, 1984).

Undoubtedly, the motor output is extremely costly in energetic terms. Therefore, the production of purposeless motor activity seems paradoxical and a logical principle of economy should forbid the production of playing behaviour. Several explanations have been put forward to justify the universality of play, not only in humans, but also in other mammals in which evident playful behaviour has been repeatedly observed. A description of the function(s) of play is out of the scope of this review. However, the inclusion of play at this point is relevant because play shows several traits indicative of consciousness. Besides of being an onerous activity, play seems to be always pleasant. The only explanation for the play paradox lies in considering that the expenditure of energy must have a wide variation in hedonic value, from rather unpleasant to extremely pleasurable, that is, it shows a wide range of alliesthesia. An animal confronted with the possibility of playing should rank the costs and the benefits of each alternative and its final decision will aim at maximizing pleasure. Therefore, the presence of play should be a sign of consciousness.

Play behaviour has been greatly neglected as an object of scientific study and the number of species in which play has been described is still extremely low. Obviously, most examples of play have been described in mammals. There are however, interesting descriptions of play in birds, mainly in psittaciforms and passeriforms. In particular, numerous examples of catching play, flight play, sliding, hanging and even playful interactions with other species have been described in ravens (Heinrich and Smolker, 1998).

Looking far in the phylogeny of vertebrates, no play behaviour has been described up to now in amphibians, but a number of reports have described sound examples of play in turtles. Object, locomotor and social play has been thoroughly described in a captive Nile soft shelled turtle, *Trionyx triunguis*, which even seemed to enjoy playing with his keeper with whom a relation was established. (Burghardt, 1998). The play behaviour has been extended to other chelonians, but also to lepidosaurians (Burghardt, 2005). In consequence, play seems to have been originated in amniotes.

4. Discussion

The basis of the eventual existence of consciousness in two invertebrate groups has been described. Regarding arthropods, although the empirical evidence shows a set of extremely sophisticated capacities in honeybees, the existence of consciousness in these animals seems to be improbable. Most likely, its behavioural performance depends on separate mechanisms specifically developed for each particular situation, in the form of a simple and linear program, that is, they seem to lack a central processing module to produce a unified reality space in the sense demanded by Merker (2005 and 2006).

Concerning cephalopods, the list of their behavioural performance is certainly impressive, suggesting that they have a simple form of primary consciousness (Edelman *et al.*, 2005). Between the recognized abilities of octopuses, the capability to perform detour behaviour and play are particularly interesting. Undoubtedly, this conclusion would be extremely robust if in addition (1) the interspecific language ability claimed by Moynihan and Rodaniche in 1977 and (2) if the capability of observational learning described by Fiorito and Scotto (1992) would be confirmed. However, these two claims have been disputed. On the other hand, it should be extremely interesting to perform tests on the capability of octopus to rate sensations and to develop a common currency for behavioural output, following the lead of Cabanac (1996).

In vertebrates, the basis for the eventual existence of consciousness, as well as its phylogenetic development has been described in anatomical and empirical terms. Concerning anatomy, the telencephalic cortex is generally believed to be the main structure associated with consciousness, although alternative brainstem locations have recently been proposed (Merker, 2007). However, it seems likely that the anatomy only is a permissive factor in the production of consciousness, and the anatomical development should be correlated with empirical demonstrations of enrichment in mental activity. In fact, the phylogenetic development of most anatomical factors seems to be quantitative. An important question concerning the development of consciousness is, therefore, whether it is a quantitative property, that is, whether consciousness appeared as the result of a continuous and slow process, or whether it is a discontinuous trait, attained by the brain after having reached a critical mass.

After reviewing the main anatomical traits and the evolution of the vertebrate brain, we found no good correlation between anatomical features and the production of high level mental operations. However, the regions with multisensorial and reentrant projections found in amniotes, seem to be likely candidates for being crucial in producing consciousness. Consciousness could have appeared in the first terrestrial vertebrates in parallel with the increase in metabolic activity achieved after having reached thermal independence and having got access to unlimited amounts of oxygen. Two main evolutionary lines regarding the development of the thalamus seem to exist in amniotes. Collothalamic connections between mesencephalon and the thalamus are well developed in extant reptiles and in birds. Lemnothalamic connections dominate in mammals. No obvious differences in efficiency between the two developmental lines are evident.

Assuming, in agreement with other authors (e.g. Seth *et al*, 2005), that mammals have the capacity for consciousness, we, therefore, suggest that consciousness also could exist in birds and reptiles.

With respect to empirical studies, a number of testable properties have been put forward to detect consciousness (Seth *et al*, 2005). However, precise experiments aiming at identifying whether a given animal fulfils a particular criterion, are scarce. Amongst them, the capability to rank different sensations, the capability to record either pleasure or displeasure and to feel joy and to be able to perform complex insight learning tasks have been applied to fish, amphibians, reptiles, mammals and birds.

Beginning with insight learning, the capability to perform detour behaviours seem to be an impressive proof of consciousness. When making a complex detour, the animal shows memory and capability to produce a detailed map of the environment. In addition, the animal seems to be able to distinguish the sensory changes produced by its own movements to compensate for the apparent movement of the goal. However, this capability could be achieved by simple procedures and, as a matter of fact, a Global Positioned System (GPS) such as those included in modern cars can perform such a task. What refrains us from attributing consciousness to the GPS of a car is, in the first place, the absence of emotions: if the advice of a talking GPS is repeatedly ignored and the driver does not take the pathway that the GPS is proposing, a new alternative is immediately calculated. This is equivalent to performing a detour around an obstacle, but what the GPS never does is showing frustration or annoyance after repeated disdain in following its instructions (unlike most accompanying passengers would do...!). In addition, if the GPS results would have been accomplished using multisensory input and achieved by a unique central control (e.g., according to the requirements proposed by Merker, 2005), the assumption of consciousness would be boosted. Most probably, a detour capability exist in many species, including fish, and amphibians, but also in insects with good visual acuity, such as honeybees and dragonflies (Sjölander, 1997). However, the detour behaviour is in these animals unaccompanied by emotions. Furthermore, they lack a central correction mechanism.

The complex detour capability observed in chameleons may be an indication of a complex mental life. It could be a particular development in chameleons, animals living in a complex environment and having an extremely good visual system. The mental life of chameleons could be an example of not only a quantitative, but a qualitative development, where a conscious capability is reached after having surpassed a certain threshold in sensory and motor control.

However, reptiles can also show other interesting performances. Due to the existence of emotions and the development of a limbic system (contradicting the assumptions of the Triune Brain hypothesis), reptiles have been called "walking limbic preparations". The main neuroanatomical difference between mammals and reptiles does not lie in the limbic structures, but in the relative weight of thalamocortical versus thalamocollicular connections. Therefore, reptiles could have a complex emotional life. Their capability to show tachycardia and emotional fever is a powerful proof of fear. The experiments of sensation ranking show that reptiles are quite able to experience taste pleasure, to modify the affective dimension of a sensory experience (alliesthesia) and in trading off the positive value of palatable food with the negative of cold exposure. The demonstration of play adds evidence in favour of a complex mental life. Indeed, the value of the capability to detour, to have emotions, to rank different sensations and to perform trade-offs is only relative. However, the simultaneous presence of these abilities in a single group seems to be rather strong evidence.

The failure of fish and amphibians to show the abilities observed in reptiles can be taken as an indication of a discontinuity in the development of consciousness in contrast to the continuity in the development of CNS complexity.

In conclusion, we suggest that consciousness appeared in amniotes, the first truly terrestrial animals. Up to now, the existence of consciousness has been generally accepted in mammals and birds, the two homeothermic vertebrate groups. However, all reptiles also control their body temperature, if not by physiological means, so by behavioural procedures. Thus, the idea of a reptilian consciousness is compatible with the idea of consciousness being tied to a high body temperature. It is not easy, however, to determine if consciousness in the two homeothermic vertebrate groups is mono- or polyphyletic due to the existence of the two divergent lines of brain development, one involving collothalamic and the other lemnothalamic re-entrant connections.

References

Amlaner Jr CJ, Ball NJ (1994) Avian Sleep. In: Kryger MH, Roth T, Dement WC (Eds) *Principles and Practice of Sleep Medicine*. Philadelphia: WB Saunders Co, pp 81-94.

Anderson B (1995) Dendrites and cognition: A negative pilot study in the rat. *Intelligence* **20**:291-308.

Ayala FJ (1988) Can “progress” be defined as a biological concept? In: Nitecki (Ed) *Evolutionary Progress*. Chicago: University of Chicago Press, pp 75-96.

Baars BJ (2005) ‘Subjective experience is probably not limited to humans: the evidence from neurobiology and behaviour’. *Conscious Cogn* **14**:7-21.

Balasko M, Cabanac M (1998) Behavior of juvenile lizards (*Iguana iguana)* in a conflict between temperature regulation and palatable food. *Brain Behav Evol* **52**:257-262.

Bekoff M. (1984) Social play behavior. *Bioscience* **34**:228-233.

Berger RJ, Phillips NH (1994) Constant light suppresses sleep and circadian rhythms in pigeons without consequent sleep rebound in darkness. *Am J Physiol* (Regulatory Integrative Comp Physiol) **36**:R945-R952.

Bert J, Godet R (1963) Reaction d’eveil telencephalique d’un dipneuste. *Soc Biol. Ouest Africain*, Seanc 12 juillet, **24A**:199.

Boal JG. Wittenberg KM. Hanlon RT (2000) Observational learning does not explain improvement in predation tactics by cuttlefish (*Mollusca: Cephalopoda*) *Behav Processes* **52**:141-153.

Briese E, de Quijada MG (1970) Colonic temperature of rats during handling. *Acta Physiol Latinoam* **20**:97-102.

Briese E, Cabanac M. (1991) Stress hyperthermia: physiological arguments that it is a fever. *Physiol Behav* **49**:1153-1157.

Burghardt GM (1998) The evolutionary origins of play revisited: lessons from turtles. In: M Bekoff, JA Byers (Eds) *Animal Play*. Cambridge: Cambridge University Press, pp 1-26.

Burghardt GM (2005) *The Genesis of Animal Play*. Cambridge MA: MIT Press.

Butler AB, Hodos W (1996) *Comparative vertebrate neuroanatomy. Evolution and adaptation*. New York: Wiley Liss.

Cabanac M (1996) On the origin of consciousness, a postulate and its corollary. *Neurosci Biobehav Rev* **20**:33-40.

Cabanac A, Briese E (1991) Handling elevates the colonic temperature of mice. *Physiol Behav* **51**:95-98.

Cabanac M, Aizawa S (2000) Fever and tachicardia in a bird (*Gallus domesticus)* after simple handling. *Physiol Behav* **69**:541-545.

Cabanac M Bernieri C (2000) Behavioral rise in body temperature and tachycardia by handling of a turtle (*Clemmys insculpta*) *Behavioural Processes* **49**:61-68.

Cabanac A, Cabanac M (2000) Heart rate response to gentle handling of frog and lizard. *Behavioral Processes* **52**:89-95.

Cabanac A, Cabanac M (2004) No emotional fever in toads. *J Therm Biol* **29**:669-673

Cabanac M, Laberge F (1998) Fever in goldfish: after pyrogens, yes, after handling, no. *Physiol Behav* **63**:377-379.

Cabanac M, Goselin F (1993) Emotional fever in the lizard *Callopistes maculatus* (Teiidae). *Anim Behav* **46**:200-2002.

Cannon CM, Bseikri MR (2004) Is dopamine required for natural reward? *Physiol Behav* **81**:741-748.

Chomsky N (1988) *Language and the Problems of Knowledge: The Managua Lectures.* Cambridge MA: MIT Press.

Cheney D, Seyfarth R (1990) *How Monkeys See the World.* Chicago: University of Chicago Press.

Dawkins R (1986) *The Blind Watchmaker.* London: Penguin.

Dennet D (1995) *Darwin's Dangerous Idea*. London: Penguin.

Diamond J (1992) *The Rise and Fall of the Third Chimpanzee.* London: Vintage.

Edelman GM (1989) *The Remembered Present*. New York: Basic Books

Edelman DB, Baars, BJ, Seth AK (2005) Identifying hallmarks of consciousness in non-mammalian species. *Conscious Cogn* **14**:169-187.

Edelman GM, Tononi G (2000) *Consciousness. How Matter Becomes Imagination.* London: Penguin.

Eibl-Eibesfeldt I (1975) *Ethology, the Study of Behavior.* New York: Holt Rinehart and Winston.

Fiorito G. Scotto P (1992) Observational Learning in *Octopus vulgaris Science* **256**:545-547.

García, J. Kimerdorf DJ, Koelling R (1955) Conditoned aversion to saccharin resulting from exposure to gamma radiation. *Science* **122**:157-158.

Gell-Mann M (1994) *The Quark and the Jaguar. Adventures in the Simple and the Complex*. New York: WH Freeman Co.

Gould JL, Grant-Gould C (1986) *Invertebrate Intelligence*. In: RJ Hoage, L Goldman (Eds) Animal intelligence. Washington: Smithsonian Inst Press, pp 21-36.

Gould JL, Grant-Gould C (1995) *The Honey Bee*. New York: Scientific American Library.

Heinrich B, Smolker R (1998) Play in common ravens (*Corvus corax*). In: M Bekoff and JA Byers (Eds) *Animal Play*. Cambridge: Cambridge University Press, pp 27-44.

Huber R, Smith K, Delago A, Isaksson K, Kravitz EA (1997) Serotonin and aggressive motivation in crustaceans: Altering the decision to retreat. *Proc Nat Acad Sci USA* **94**:5939-5942.

Iversen SD, Iversen LL (1981) *Behavioral Pharmacology*. New York: Oxford University Press.

Kelley AE (2005) Neurochemical networks encoding emotion and motivation. In: JM Fellous and MA Arbib (Eds) *Who Needs Emotions?* Oxford: Oxford University Press.

Kuba MJ,, Byrne RA, Meisel DV, Mather JA (2006) When Do Octopuses Play? Effects of Repeated Testing, Object Type, Age, and Food Deprivation on Object Play in Octopus vulgaris. *J Comp Psychol* **120**:184-190.

Laurin M, Reisz RR (1995) A reevaluation of early amniote phylogeny. *Zool J Linn Soc* **113**:165-223.

LeDoux JE (2000) Emotion circuits in the brain. *Annu Rev Neurosc* **23**:155-184.

MacLean PD (1990) *The Triune Brain in Evolution*. New York: Plenum Press.

Mather JA (1991) Navigation by spatial memory and use of visual landmarks in octopuses *J Comp Physiol A* **168**:491-497.

Mather JA (1992) Interactions of juvenile *Octopus vulgaris* with scavenging and territorial fishes. *Mar Behav Physiol* **19**:175-185.

Mather JA, Anderson RC (1993) Personalities of octopuses (Octopus rubescens) *J Comp Psychol* **107**:336-340.

Mather JA (1994) Home choice and modification by juvenile *Octopus vulgaris* (Mollusca: Cephalopoda): specialized intelligence and tool use? *J Zool London* **233**:359-368.

Mather JA, Anderson RC (1999) Exploration, play and habituation in octopuses (*Octopus dofleini*). *J Comp Psychol* **113**:333-338.

Medicus G. (1987) Toward an etho-psychobiology. A phylogenetic tree of behavioural capabilities proposed as a common basis for communication between current theories in psychology and psychiatry. *Ethol Sociobiol* **8**:131S-150S.

Merker B (2005) The liabilities of mobility: A selection pressure for the transition to consciousness in animal evolution. *Consc Cogn* **14**:89-114.

Merker B (2007) Consciousness without a cerebral cortex: A challenge for neuroscience and medicine. *Behav Brain Sci* **30**:63-81.
Moynihan MH, Rodaniche AF (1977) Communication, crypsis, and mimicry among cephalopods. In: T Sebeok (Ed) *How Animals Communicate*. Bloomington: Indiana University Press, pp 293-302.
Myhre K, Cabanac M, Myhre G (1977) Fever and behavioural temperature regulation in the frog *Rana esculenta*. *Acta Physiol Scand* **101**:219-29.
Nicolau MC, Akaârir, M, Gamundí, A, González J, Rial, RV (2000) Why we sleep: the evolutionary pathway to the mammalian sleep. *Prog Neurobiol* **62**:379-406.
Norman MD, Finn J, Tregenza T (2001) Dynamic mimicry in an Indo-Malayan octopus. *Proc R Soc Lond B* **268**:1755-1758.
Pani L, Gessa GL. (1997) Evolution of the dopaminergic system and its relationships with the psychopathology of pleasure. *Int J Clin Pharmacol Res* **17**:55-58.
Panksepp J (1998) *Affective Neuroscience*. New York: Oxford University Press.
Panksepp J (2005) Affective consciousness: Core emotional feelings in animals and humans. *Consc Cogn* **14**:30-80.
Paradis S, Cabanac M (2004) Flavor aversion learning induced by lithium chloride in reptiles but not in amphybians. *Behav Processes* **67**:11-8.
Parmeggiani PL. (1982) Regulation of physiological functions during sleep in mammals. *Experientia* **38**:1405-1408.
Peroutka SJ, Howell TA (1994) The molecular evolution of G protein-coupled receptors: Focus on 5-hydroxitryptamine receptors. *Neuropharmacol* **33**:319-324.
Piaget J (1954) *The Construction of Reality in the Child.* New York: Basic Books.
Pribram KH, Meade SD (1999) Conscious awareness: processing in the synaptodendritic web. *New Ideas Psychobiol* **17**:205-214.
Renbourn ET (1960) Body temperature and the emotions. *Lancet* **2**:475-478.
Rumbaugh D, Savage-Rumbaugh S (1986) Reasoning and language in chimpanzees. In: RJ Hoage and L Goldman (Eds) *Animal Intelligence*. Washington: Smithsoninan Inst Press, pp 57-75.
Seth AK, Baars, BJ, Edelman DB (2005) Criteria for consciousness in humans and other animals. *Consc Cogn* **14**:119-139.
Sinn DL, Penin NA, Mather JA, Anderson RC (2001) Early Temperamental Traits in an Octopus (*Octopus bimaculoides*) *J Comp Psychol* **115**:351-364.
Sjölander, S (1997) On the evolution of reality - Some biological perspectives and evolutionary stages. *J Theor Biol* **187**:595-600.
Solms M (2000) Dreaming and REM are controlled by different mechanisms. *Behav Brain Sci* **23**:843-850.
Sprague JM (1996) Neural mechanisms of the visual orienting response. *Progr Brain Res* **112**:1-15.
Steriade M (2000) Brain electrical activity and sensory processing during waking and sleep states. In: MH Kryger, T Roth and WC Dement (Eds) *Principles and Practice of Sleep Medicine*. Philadelphia: WB Saunders Co, pp 93-111.
Uexküll J. von (1909) *Umwelt und Innenwelt der Tiere.* Berlin: Springer.

Vallortigara G (2000) Comparative neuropsychology of the dual brain: A stroll through animals' left and right perceptual worlds. *Brain Lang* **73**:189-219.

van Ree JM, Niesnink RJ, van Wolfswinkel L, Ramsey NF, Kornet MM, van Furth WR, Vanderschuren LJ, Gerrits MA, van der Berg CL (2000) Endogenous opioids and reward. *Eur J Pharmacol* **405**:89-101.

Villablanca J (1965) The electrocorticogram of the chronic cerveau isolé cat. *Electroenceph Clin Neurophysiol* **19**:576-586.

von Frisch O (1962) Zur Biologie des Zwergchamäleons. *Z Tierpsychol* 19: 276-289.

Williams GC (1966) *Adaptation and Natural Selection. A Critique to Some Current Evolutionary Thought.* Princeton: Princeton Univ Press.

Chapter 4

On the Origin of Consciousness — Some Amniote Scenarios

Peter Århem, B.I.B. Lindahl, Paul R. Manger and Ann B. Butler

1. Introduction

The phylogenetic origin of consciousness is virtually unknown. One reason for this is that there is no consensus about usable empirical markers of consciousness. Starting from the only relatively safe point, that present-day humans are conscious, the search for such markers may be conducted along two basic lines: the first focusing on cognitive-behavioural characteristics, the second on physiological-anatomical characteristics. The outcome of inquiries conducted in accordance with the respective approach may vary considerably. Cognitive and behavioural criteria may range from simple stimulus-response characteristics to a capacity for language, and physiological and anatomical criteria from a primitive sensory apparatus to a six-layered cerebral cortex. Consequently the results can be as divergent as speculations about consciousness in insects (Griffin, 2001), and even in protozoans (Mayr, 2001), to the view that consciousness is limited to humans alone (Macphail, 1998).

What we will do here is to follow these two lines, the cognitive-behavioural and anatomical-physiological, from *Homo sapiens* to the origin of amniotes more generally, searching for steps or discontinuities, suggesting the emergence of traits that can be associated with consciousness. As we shall see when we come to the investigation of reptiles (in the following understood as non-avian, non-mammalian amniotes; see Tudge, 2000) in our search for consciousness markers we may have to rely on both physiological-anatomical and cognitive-behavioural characteristics, rather than risking a too narrow focus by utilizing either approach alone.

Consciousness Transitions: Phylogenetic, Ontogenetic, and Physiological Aspects
Edited by Hans Liljenström and Peter Århem

The results of this search will be analysed from the point of view of different scenarios, each suggesting different phylogenetic origins of consciousness. We take consciousness to be a biological phenomenon that is a result of neural activities, and we see the distinction between "consciousness" and "neural activities" as a division between *subjective* and *objective* factors regulating overt behaviour (Lindahl and Århem, 1994; Lindahl, 1997). We recognize various degrees and forms of consciousness, ranging from a state of experiencing perceptions, thoughts and feelings, without a sense of a self (a state corresponding to primary consciousness (Edelman, 1989) and feeling-consciousness (Macphail, 1998)) to a state of awareness of being aware, with a sense of a self (a state corresponding to higher-order consciousness (Edelman, 1989) and self-consciousness (Macphail, 1998)).

2. The basic consciousness criteria are anthropocentric

It may seem odd that most attempts to identify universal criteria of consciousness are based on parameters within an anthropocentric framework. Nevertheless, it is the only practical starting point we presently have at our disposal. The reason is of course that this is the only case in which verbal, descriptive reports are available, although cleverly constructed tests may circumvent this obstacle (Cowey and Stoerig, 1995; but see Macphail, 1998). Using this starting point we may try to develop cognitive-behavioural and/or physiological-anatomical criteria by correlating the tentative markers with the conscious states specified in such reports.

At present, the attempts to establish criteria seem more fruitful along the physiological-anatomical line than along the cognitive-behavioural line (although cognitive-behavioural criteria, based on, for instance, working memory and linguistic abilities may be valuable and have been used extensively). We will first consider some attempts to search for physiological-anatomical consciousness criteria within the anthropocentric framework. Such attempts have been made using various non-invasive brain imaging techniques as EEG, MEG, fMRI and PET, and have resulted in several sets of hypothetical criteria. The most general, and consequently most vague, assign critical roles for low-amplitude, high-frequency (gamma band) and irregular EEG patterns (see Seth *et al*, 2005; Edelman *et al*, 2005) and for wide-spread MEG activity patterns over cortex (Edelman and Tononi, 2000) in defining conscious states. (While the EEG, etc. level for the conscious state reflects large-scale patterns of asynchronous cortical activity, the phenomenon of consciousness also has been correlated at a finer scale with the synchronous activity (rhythmic oscillations) of selected populations of neurons, as measured by microelectrodes and also by neuromagnetic responses (Engel *et al*, 1999; Srinivasan *et al*, 1999; Llinás *et al*, 2005)). In contrast, at the EEG level, high-amplitude, synchronous, low-frequency and local cortical activity define unconscious brain states. These, mainly physiological, criteria have been further tested by applying them to pathological and pharmacological cases such as epilepsy and anaesthesia (see below). Thus, EEG patterns in epileptic and anaesthetized human brains display the expected high-amplitude, synchronized and low-frequency oscillatory activity.

Admittedly, these preliminary criteria are very general and may be used only approximately. The tentative physiological criteria can then be used to establish anatomical criteria. Which brain structures are causing the low amplitude, unsynchronized, high-frequency oscillations and consciousness, transforming the basal high amplitude, synchronized, low-frequency oscillations of the unconscious brain? A very general answer is that thalamo-cortical circuits play a key role in creating consciousness (Seth *et al*, 2005); but what parts of this system? Which thalamic nuclei? Which areas of cortex? We have to note here that there is a crucial difference between neural activity that *generates* conscious experiences and activity that only enables or supports such experiences. This distinction is seldom made in the current literature.

3. Sharpening the consciousness criteria — some principal consciousness-brain theories

There are a number of ambitious attempts to sharpen the anatomical criteria mentioned above. They may roughly be divided into those based on a bottom-up or a top-down approach, and into those emphasizing the sensory or the motor system (Butler *et al*, 2005). By "bottom-up theories" we mean those focusing on specific neurons and neuronal activity as generators of consciousness; by "top-down theories" we mean those focusing on large-scale neural structures and on activity of widespread brain regions as generators. Again, it should be stressed that these criteria are for the most part based on verbal reports and thus human brains.

The criteria or theory proposed by Crick and Koch (Crick, 1994; Crick and Koch, 1995; Rees *et al.*, 2002; Crick and Koch, 2003; Koch, 2004; Crick and Koch, 2005) represent a bottom-up sensory system approach. The theory concerns the visual system and asserts that visual awareness is associated with activity in higher order visual cortical areas that are in direct contact with prefrontal cortex. Although the "cortical system" covered by this theory includes most of the cerebral cortex and the dorsal thalamus, claustrum, dorsal striato-pallidal complex, cerebellum, and brainstem projection systems, we tentatively interpret the writings of Crick and Koch up to 2005 as stating that the generator neurons are limited to temporal, parietal and prefrontal cortex. Crick and Koch seem to limit the generator structure further by assuming that activity in a subpopulation of neurons in cortical layer V, characterized by firing in burst patterns, is crucial. A prominent feature in the theory of Crick and Koch is the insistence that the primary visual cortex is not a generator structure (Crick and Koch, 1995; Rees *et al.*, 2002). A further bold specification of this hypothesis is the recent addendum that the claustrum plays a crucial role in producing visual consciousness (Crick and Koch, 2005). Possibly, this specification should even be regarded as a new theory, replacing the previous one; viz. that the claustrum is the single generator structure, and that the cortical neurons are downgraded to merely supporting elements.

The theory of Edelman and Tononi (Edelman, 2003; Edelman and Tononi, 2000; Tononi and Edelman, 1998) appears to be an example of a top-down sensory approach. This theory focuses on the general features of consciousness — such as complexity and unity (Tononi and Edelman, 1998) — and asserts that consciousness is associated with activity in the temporal and frontal associative and motor regions of the cortex: a "dynamic core", characterized by "re-entrant" interactions within limited portions of the CNS. Structures supporting the dynamic core seem to be the septal region, amygdala, hippocampus, dorsal thalamus, hypothalamus, and the reticular activating system. Edelman and Tononi seem to assume a larger and more dynamic population of neurons associated with consciousness than Crick and Koch. But the critical issue whether the dynamic core is equivalent to the generator structure is not explicitly dealt with in the writings of Edelman and Tononi.

Making a wider survey of the neuroanatomy proposed to underlie (but not necessarily generate) conscious experiences we find that most investigators include the majority of the brain in their proposals. When it comes to the suggested neural generators of conscious experiences, the situation is different. The proposals show great variation with respect to their extent and dynamics, with relatively limited neuronal populations in the theory of Crick and Koch, to the almost global population of neurons in the theory of Edelman and Tononi. Nevertheless, the cortico-thalamic system plays a central role in these and other theories (for a discussion see Århem *et al*, 2003 and Butler *et al*, 2005). This means that the generation of conscious experiences seems to require a six-layered cortex and a specialized dorsal thalamus with both specific sensory relay nuclei and generalized projections (Jones, 2001). Except for the interesting suggestion by Merker (2007, this volume) no proposal to date regarding the neural basis of consciousness has specifically excluded the cerebral cortex.

Within the cerebral cortex, the theories focus on different regions of the cortex, with the theory of Crick and Koch ruling out primary visual cortex as contributing to the conscious experience (Crick and Koch, 1995; Rees *et al*, 2002) whereas that of Edelman and Tononi specifically include this cortical area (Edelman, 2003). Emphasis has been placed on components of the temporal and parietal association areas and the anterior cingulate, premotor, and prefrontal cortical areas (Crick and Koch, 1995; Tononi and Edelman, 1998; Edelman and Tononi, 2000; Cotterill, 2001; John, 2001; Rees *et al*, 2002).

4. Testing the principal criteria — the case of general anaesthesia

These principal theories or sets of criteria can, as indicated above, best be critically assessed within an anthropocentric framework. Do the principal theories described above provide criteria for trying to trace conscious experiences in other species? A way to test this is to explore the canonical consciousness transitions in sleep and general anaesthesia. These phenomena are clearly related, although the precise nature of the relation is still poorly understood.

Let us here briefly review how the brain reacts when general anaesthetics are applied. General anaesthetics form a large group of chemically unrelated compounds that reversibly induce unconsciousness. The group includes compounds such as the intravenously used agents propofol and ketamine and the volatiles halothane and ether. "General anaesthesia" is basically an umbrella term for a complex state comprising the endpoints amnesia, immobility, analgesia and hypnosis, each depending on different mechanisms and different molecular and anatomical targets (Rudolph and Antkowiak, 2004; Grasshoff *et al*, 2006). Of these substates hypnosis is the most interesting for us, associated as it is with loss of consciousness. Using general anaesthetics, we are thus allowed to explore brain activity at different stages in the transitions between consciousness and unconsciousness.

What do such studies reveal about which structures of the brain are critically affected by general anaesthetics and which structures are critically associated with the loss of consciousness? Most hypotheses of critical target structures for general anaesthesia aim at explaining the characteristics of cortical activity during the successive periods of general anaesthesia as reflected by EEG; thus during the *sedation period* there is a paradoxical increase in power at all frequencies, during the *hypnosis period* there is the occurrence of spindles of relatively high frequency waves and a decrease in power at high frequencies, and during the *anaesthesia period* there is the emergence of burst suppression, i.e. irregular low power activity interrupted by periods of silence (Antkowiak, 2002). The changes in cortical activity are paralleled by changes in the activity of thalamic relay neurons; at the transition between sedation and hypnosis tonic firing is transformed into burst activity, and at the transition between hypnosis and anaesthesia the burst activity is temporarily blocked.

These changes in EEG during the period of general anaesthesia are not uniform, the anterior and posterior parts of the cortex showing separate modifications. John (2001) summarizes the results from a long series of quantitative EEG studies of anaesthetic effects in three points: (i) a general lowering of frequency over the whole cortex; (ii) a relative increase in the activity of anterior cortical regions compared with posterior regions; and (iii) an increased coherence between low frequency activity in frontal areas and an uncoupling between activity in anterior and posterior regions. Both increased and decreased activity as well as increased and decreased coherence thus characterize anaesthetic-induced unconsciousness. From this we may conclude that both *suppression* and *disruption-of-coherence* mechanisms seem simultaneously at work during the states of general anaesthesia.

From such EEG oriented studies three families of hypotheses about the target structures for general anaesthesia have emerged. The *cortical hypotheses* suggest that the observed effects are caused by a direct effect on cortical neurons. This is supported by neocortical-slice studies as well as simulation studies (Steyn-Ross *et al*, 2004). A major problem with these studies is that delta waves have so far not been possible to record from isolated neocortical slices. The *thalamic hypotheses* suggest that the observed effects are caused by direct actions on thalamic intralaminar neurons. This implies that general anaesthetics cause the observed tonic-to-bursting transition, in turn causing the observed cortical effects. The thalamic hypotheses are supported by the fact that experiments on neocortical slices, including thalamic neurons, show delta waves (Antkowiak, 2002). The *reticular-system hypotheses* have at present little experimental support. Very little is known about direct effects on reticular-activating-system neurons. The reticular-system hypotheses mainly derive their plausibility from the similarity between the neural processes associated with general anaesthesia and sleep.

Remembering the fact that the general anaesthetic target structure is not necessarily the consciousness generator structure, we may conclude that these three families of general anaesthesia hypotheses might be compatible with the very general thalamo-cortical consciousness hypothesis discussed in section 2 (Seth *et al*, 2005). Recent metabolic and magnetic imaging studies take the issue further. A comparative study by Alkire *et al* (2005) points again to the thalamus and cortex as critical for consciousness, but also to the parietal cortex, and possibly the underlying claustrum as crucially involved. This would be compatible with both the Crick-Koch and the Edelman-Tononi criteria discussed above. However, we are still far from identifying generator neurons and generator structures.

It should be noted that all anaesthetics do not suppress global cerebral metabolism — the measured parameter in the imaging studies discussed — and cause a regionally-specific effect on thalamic activity. Nevertheless, if such agents disrupt signals associated with normal neuronal network reverberant activity, it is still conceivable that those few anaesthetics not associated with a global metabolic suppression effect have their consciousness-effects mediated at the level of the thalamocortical interactions.

To some extent the three families of hypotheses — the cortical, thalamic and reticular-system hypotheses — display the difference between *process specific* theories and *neuron specific* theories. In the cortical hypotheses the general anaesthetics are often assumed to simultaneously attack a large dynamic network of cells while in the reticular-system hypotheses they are assumed to attack a smaller and more specified neuronal target. Within the frames of all these theories we also find two competing ideas: that of direct *suppression* and that of *disrupting coherence* (Cariani, 2000). Presumably, the final theory of general anaesthesia will make use of a combination of these ideas.

5. The phylogenetic origin of consciousness

Using the principal consciousness-brain theories or sets of criteria examined above, we are now in the position to discuss the phylogenetic origin of consciousness. This will be done in

Table 4.1 Four scenarios of the phylogenetic origin of consciousness within amniotes

Scenario	Phylogenetic origin	Explanatory strength	Explanatory weakness
I	*Homo sapiens*	Unique language	No unique brain structure
II	*Mammalia*	Six-layered cortex, claustrum/amygdala	No unique cognitive characteristic
III	*Mammalia* and *Aves*	Behavioural-cognitive similarities	Dual emergence
IV	*Reptilia*	Single emergence	Relatively poor cognitive performance

terms of four different scenarios, each suggesting different phylogenetic origins of consciousness.

5.1 Scenario I — Consciousness emerged exclusively with the first humans

The tests of the consciousness-brain theories examined above were based on a relatively safe assumption: humans are conscious. To derive criteria from the two principal theories for trying to trace the neural basis of consciousness in other species is far more problematic. However, it is of course possible to envisage a totally anthropocentric scenario that consciousness emerged with *Homo sapiens*. We will call this *scenario I* (see Table 4.1). This scenario has been forcefully defended by Macphail (1998). A merit of this scenario is that it takes the uniqueness of human language into account. Human language is the only natural communication system that comprises both a descriptive and an argumentative function (for a discussion, see Popper and Eccles, 1977, Ch. P3:17). All natural communication systems comprise signal and expressive functions. Some systems comprise spatially descriptive functions. Bees can communicate distance and direction to food sources (von Frisch, 1967). Black birds (*Turdus merula*) warn differently depending on whether a predator is approaching from above or from the ground. Crows warn differently when they provoke a sparrow hawk and a goshawk. However, no animal except humans are known to communicate temporal information.

5.2 Scenario II — Consciousness emerged exclusively with the first mammals

A major problem with scenario I is that no decisive structural difference between human and other mammalian brains has yet been found (Preuss, 2000; Hof and Van der Gucht, 2006). This problem is taken into account in a more general mammalian scenario, here called *scenario II* — the hypothesis that all mammals, but only mammals, are conscious (Table 4.1). A merit of this scenario is that it *prima facie* offers tentative neural markers of consciousness: a six-layered cerebral cortex, perhaps the claustrum, a massive number of pallial commissural connections, and reciprocal projections from pallial areas to dorsal thalamic nuclei. These criteria seem closely related to the sharper criteria discussed above and are suggested by Crick and Koch (2003) and by Edelman and Tononi (2000). A problem with scenario II, however, is that, despite lacking an overtly laminated neocortex and possibly some of its neuronal components (Karten, 1997; Medina and Reiner, 2000), the avian brain has recently been shown to be structurally more similar to the mammalian brain than previously thought, with numerous homologous regions and many shared neural features (Karten, 1997; Reiner *et al.*, 2004, 2005).

5.3 Scenario III — Consciousness emerged independently in mammals and birds

The problem of the mammalian-avian homologues is taken into account by a scenario envisaging consciousness in both birds and mammals, here called *scenario III* (Table 11.1). In this case the differentiation is dual, separating mammals and reptiles, and birds and reptiles. A large number of traits, cognitive and behavioural as well as anatomical and physiological, support this view — traits that are shared by some mammals (mostly primates) and some birds (Pepperberg, 1999, 2002; Pepperberg and Lynn, 2000; Clayton *et al.*, 2003; Hunt and Gray, 2003; Emery and Clayton, 2004; see also Butler *et al.*, 2005, and Butler and Cotterill, 2006), but are not found in reptiles; nor, for that matter, in amphibians. These traits thus represent true qualitative differences between mammals/birds and other tetrapods and thus strongly indicate a mammalian-avian versus reptilian-amphibian dichotomy. The shared mammalian versus avian physiological and behavioural traits include: homeothermy, extended parental care of the young, and grasping ability for the manipulation of objects. Although more difficult to assess and therefore more controversial are shared cognitive traits that suggest a mammalian-avian/reptilian dichotomy. These include, for the mammalian-avian category: social complexity, in terms of group and intraspecific interactions; tool manufacture and use; cultural transmission of tool design; and production of long and complex vocalization sequences.

Table 4.2 Brain characteristics of three of the origin-of-consciousness scenarios discussed in the text. +++ denotes robust presence of trait, + denotes minor presence or presence in some taxa, and - denotes its absence.

Scenario	Brain charctersistic	Mammalia	Aves	Reptilia	Amphibia
II	Neocortical cytoarchitecture	+++	-	-	-
	Massive number of pallial commissural connections	+++	+	+	+
	Claustrum or homologue	+++	?	?	-
	Reciprocal pallium-dorsal thalamus projections	+++	+	+	-
III	High brain-body ratio	+++	+++	-	-
	High number of glial cells	+++	+++	?	-
	Enlarged pallium, neuronal complexity	+++	+++	+	-
	Robust corticospinal tract	+++	+++	-	-
	Thalamic reticular nucleus	+++	+++	+	-
	Multiple basal ganglia-thalamic loops	+++	+++	+	-
	Enlarged cerebellum	+++	+++	+	-
IV	Pallial elaboration, including multimodal association areas	+++	+++	++	-
	Elaboration of hippocampus, double pyramidal neurons	+++	+++	++	-
	Ascending sensory pathways via dorsal thalamic nuclei	+++	+++	++	-
	Intralaminar nuclei with projections to pallium	+++	+++	++	-

As we have noted, neural traits provide a promising basis for the identification of markers of consciousness. Some brain characteristics shared by birds and mammals in contrast to other tetrapods are listed in Table 4.2.

The pallium in mammals is dominated by the expansive cerebral cortex, which, in most of its areas, exhibits a distinctive cytoarchitecture with six layers and specific populations of glutamatergic pyramidal and spiny stellate cells and GABAergic aspiny stellate, basket, and chandelier cells. In contrast, the avian pallium has a relatively small region that is pseudolaminar in architecture — the Wulst or hyperpallium — and lacks the specific six-layered architecture of cerebral cortex (Medina and Reiner, 2000). Most of the pallium in birds consists of regions within the dorsal ventricular ridge, now most frequently called mesopallium and nidopallium, which contain some pyramidal-like neurons and numerous, highly interconnected, multipolar neurons with densely spiny dendrites (Tömböl, 1995a,b), but they are generally nuclear in their cytoarchitecture. The Wulst receives lemnothalamic inputs (Butler, 1994a,b), like the mammalian frontal and striate cortices (Manger *et al.*, 2002a), while the dorsal ventricular ridge, receives collothalamic inputs (Butler, 1994a,b), like the mammalian parietal and temporal lobes, the extrastriate parts of the occipital lobe, and the basolateral part of the amygdala. Identification of the mammalian homologue(s) of the dorsal ventricular ridge is a matter of continuing debate (see Butler *et al*, 2005). One interesting possibility in the context of consciousness is that the dorsal ventricular ridge corresponds to the mammalian amygdala-claustral complex (Bruce and Neary, 1995; Puelles *et al.*, 2000). Whatever the resolution of this question, the bird-reptile dorsal ventricular ridge is clearly a major site of ascending collothalamic sensory projections, and it contains multiple regions, including multisensory association areas (Martinez-García *et al*, 2002; Manger *et al.*, 2002b).

In summary, in whatever way the specific mammalian-avian brain homology possibilities are resolved, the high number of shared pallial and other forebrain traits makes scenario III an attractive one. Except for differences in terms of the large, glutamatergic projection neurons being specifically pyramidal in morphology, an overtly six-layered cortex, and some cortical neuronal components, the many shared traits are consistent with the hypotheses advocated by a considerable number of proponents such as Crick and Koch (1995), Edelman and Tononi (2000) and Sjölander (1997).

The many common traits between mammalian and avian brains may seem surprising, considering the traditional view of a fundamental structural difference between the two brain types, reflected in the traditional nomenclature. In a previous study (Butler *et al*, 2005) we tested the consciousness-brain theories mentioned above (Crick and Koch, 1995; Edelman and Tononi, 2000) on the avian brain, assuming that birds are conscious. The general conclusion was that these theories appear to be valid for the avian brain. The critical structures assumed to be necessary for consciousness in mammalian brains have their homologous counterparts in avian brains, perhaps even the claustrum, a combined claustrum/amygdala formation, or the latter in addition to collothalamic-recipient neocortex as a field homology within the dorsal ventricular ridge (see Butler and Molnár, 2002). Thus, the macrostructure portion of the theory of Crick and Koch (Crick and Koch, 1995), assuming that the higher collopallial visual areas and the lemnopallial prefrontal cortex are necessary for visual experience, is by and large compatible with the hypothesis of avian brain consciousness. Likewise, the theory of Edelman and Tononi (2000), assuming a special role for multimodal areas and the prefrontal cortex, is compatible with the hypothesis that birds are conscious. In summary, the *macrostructure* portion of both the theory proposed by Crick and Koch and the one proposed by Edelman and Tonini seem compatible with scenario III. The situation is less evident for the *microstructure* portion of the theory of Crick and Koch; a critical role for layer 5 neurons is at present difficult to verify for avian brains, although layer 5-like and 6-like neuronal components are indeed present in birds (Medina and Reiner, 2000).

A problem with scenario III is that it implies that consciousness has emerged twice. Such an idea of a parallel emergence of consciousness might be derived from the fact (see Tudge, 2000) that reptiles and birds (sauropsids) form a monophyletic group that originates from stem amniotes independently of the other amniote monophyletic group, the mammals. In fact, the earliest mammals appear in the fossil record 10 million years before the earliest reptiles. Birds are most closely related to crocodiles, these two groups together comprising the theocodonts, and the thecodonts are more closely related to turtles and the tuatara *Sphenodon* (Rhynchocephalia) than any of these taxa are to the squamates (lizards and snakes). The squamates thus constitute the sister group to the taxon that includes crocodiles, birds, turtles, and the tuatara. This is not an insurmountable obstacle, however, as consciousness seems to coincide with the other convergent mammalian-avian adaptations listed above.

Another problem with scenario III is the clear evolutionary continuity between birds and reptiles in terms of brain cytoarchitecture, which is more evident than the corresponding stem amniote-mammalian evolutionary transition (see Butler and Hodos, 2005). This makes it difficult to entirely exclude reptiles from the consciousness-expressing species, as will be discussed next.

5.4 Scenario IV — Consciousness emerged with the first reptiles

A simple solution to the problems encountered when assuming scenario III is a scenario where the cut-off point for consciousness is assumed to be between reptiles and amphibians. This is *scenario IV*. This would make the continuous reptilian-avian brain evolution less problematic, but it also introduces the difficulty of finding a crucial consciousness generating brain structure lacking in the *amphibian* brain but not in the reptilian one. Scenario IV must also take into account the conclusions from the scanty cognitive and behavioural evidence available suggesting that reptiles are much less capable of information processing than mammals and birds (but see Macphail and Bolhuis, 2001). We will discuss both issues below.

In terms of neuronal cytoarchitecture and organization, the reptile-bird brain evolutionary transition shows more continuity than the corresponding stem amniote-mammalian evolutionary transition (see Butler and Hodos, 2005; Manger *et al*, 2002b). Although the brain-body ratios of birds and mammals substantially overlap each other and mostly exceed the brain-body ratios of reptiles, the neural features of avian brains are qualitatively similar to those of reptiles. In contrast to the mammalian neocortex, the reptilian pallium has a relatively small region that is cortical in architecture, and in birds the homologous structure, the Wulst or hyperpallium, is pseudolaminar in structure. Rather, both the avian and reptilian pallia are dominated by the dorsal ventricular ridge (the mesopallium and nidopallium of birds), that is nuclear in cytoarchitecture. In contrast, in similarity with the mammalian pallium, the reptilian and avian pallia show: (i) some elaboration, including the presence of at least a few multimodal association areas; and comprise (ii) elaborated hippocampal formations, including presence of double pyramidal neurons; (iii) ascending sensory pathways via dorsal thalamic nuclei; and (iv) wide-spread projections from thalamic intralaminar nuclei (Table 11.2).

These structural characteristics, more quantitative than qualitative, form distinguishing features of scenario IV, structural discontinuities separating scenario IV from one assuming consciousness in amphibians. The amphibian telencephalon comprises a medial pallial area that is believed to be homologous to the hippocampal formation of amniotes, a very small dorsal pallial area that probably is homologous to the dorsal cortex (hyperpallium) of sauropsids, and some amygdalar areas. However, of importance in the question of homology of the dorsal ventricular ridge to mammalian pallial structures, is the recent discovery that amphibians entirely lack the relevant basolateral component of the amygdala (Laberge *et al*, 2006; Dicke and Roth, 2007). The medial pallium is the only part of the pallium that shows any degree of elaboration in amphibians, and it is not quantitatively or qualitatively comparable to the medial pallial, hippocampal formations of amniotes. One crucial difference is the lack of double pyramidal cells that are uniquely present in the hippocampal formation of amniotes (Westhoff and Roth, 2002). Another important difference is that in amphibians the medial pallium is the predominant target of ascending lemnothalamic information (Dicke and Roth, 2007), but the dorsal thalamic nucleus that relays to the medial pallium does not receive direct sensory inputs. Instead, it receives mostly or all GABAergic inputs relayed to it from the ventral thalamus (Roth and Grunwald, 2000; Dicke and Roth, 2007) — a situation in marked contrast to the direct retino-lemnothalamic-pallial and the direct somatosensory-lemnothalamic-pallial projections present in amniotes. Likewise, collothalamic relays to the telencephalon in amphibians differ markedly from those in amniotes. In the latter, these inputs are predominantly to the pallium, with collateral inputs to the striatum. In amphibians, the collothalamus relays information predominantly to the striatum, giving rise to only very sparse pallial projections. For the present, however, it is important to note that we cannot make the assumption that consciousness is absent in amphibians. More species-sensitive behavioural testing needs to be carried out to assess amphibian capabilities, akin to the recent work that has revealed much greater cognitive capacities in birds than previously appreciated.

6. Reconsidering reptilian cognitive behaviour

Assuming a reptilian consciousness scenario (scenario IV), we have to reconsider the reports on reptilian cognitive behaviour. As mentioned, relatively little is known about the potential for reptilian cognitive behaviour relevant to the consciousness question. The general behavioural repertoire of reptiles seems less complex than that of mammals and birds (MacLean, 1990; Stein and Meredith, 1994; Sjölander, 1997). Reptiles seem to go through a circadian sleep-wake cycle, although the states involved are poorly defined and at least four different scenarios have been forwarded to describe the phenomenology.

Thus reptile sleep has been suggested to comprise: (i) both slow wave (SW) and rapid eye-movement states, equivalent to the corresponding states in humans; (ii) exclusively a single slow-wave state (Garau *et al*, 2005; Eiland *et al*, 2001); (iii) exclusively a single rapid-eye-movement state (Rial *et al*, 1993; Nicolau *et al*, 2000); and (iv) exclusively a single sleep state, but a state deviating from both slow-wave and rapid-eye-movement states (Rattenborg, 2006). The presence of a sleep-wake cycle in reptiles does, of course, not necessarily imply that reptiles go through a consciousness-altering cycle, but only that they go through an active-inactive behaviour cycle. However, if the states involved show phenomenological similarities to the human sleep-wake states, which have "accurate report" support, the probability of an occurrence of conscious states in reptiles gains credibility.

Concerning cognitive behaviour, most studies that address aspects of such behaviour use either mammals or birds as subjects, with occasional forays into the invertebrate realm. Reptiles can be trained for relatively simple discrimination paradigms, such as a visual discrimination task (Bass *et al*, 1973; Bass, 1977), but studies that involve more complex cognitive tasks for either reptiles or amphibians have not been published.

In extensive studies of the behavioural repertoires of various reptiles in semi-natural settings, MacLean and Greenberg (Greenberg, 2002; MacLean, 1990) observed the daily routines and component behaviours of *Anolis* lizards (and some other reptilian species) associated with territorial and reproductive issues. MacLean and Greenberg documented a list of specific behaviours in various contexts, including establishment of an individual's home-site and home-range, patrolling territory, ritualistic defence displays, intraspecific fighting, courtship displays, mating and breeding, flocking, and migration. The *Anolis* lizards, under the semi-natural vivarium conditions of the laboratory, exhibit social displays that include not only an "assertion" display that indicates a general arousal but also challenge displays that are specifically addressed to a conspecific male and courtship displays that are addressed to a conspecific female. Such context-specific responses might be interpreted as indicating some level of cognition (and consciousness), but their stereotypy indicates that they are more probably in the category of simpler stimulus-response behaviours.

Nonetheless, we must be cautious in accepting negative evidence for complex cognitive capacities in reptiles and even amphibians. An important caveat derives from studies of behaviour in goldfish. Although a number of previous studies of goldfish behaviour following complete destruction of the telencephalon indicated that these animals could behave in a virtually normal manner with only their brainstem intact, more recent studies by Salas, Rodríguez and colleagues, which were designed to investigate behavioural questions in a species-relevant manner, have revealed that both hippocampal-like spatial mapping and memory functions and amygdala-like fear conditioning functions are carried out in localized regions of the telencephalic pallium (Broglio *et al*, 2005). These functional capacities of the pallium had been erroneously discredited by the earlier work. It should also be noted that among the reptiles, crocodiles have the most avian-like brains and also the highest brain-body ratios (Jerison, 1973) and yet have been little studied in terms of their behavioural repertoires and cognitive capacities. They may represent an important taxon in this regard and worthy of further investigation.

7. Conclusions

In our search for the phylogenetic origin of consciousness we have discussed the prospects of identifying markers of consciousness along two lines of evidence, the one cognitive-behavioural, the other physiological-anatomical. We conclude that at present the physiological-anatomical approach tends to yield the most concrete results. We have tested some principal consciousness-brain theories (represented by the theory of Crick and Koch, and that of Edelman and Tononi, respectively) as candidates by applying them to what is known about human brain physiology, under the constraints of general anaesthesia, and about the brain physiology of other phyla.

In summary, we are presented with a number of scenarios, each in and of itself consistent, and it is at present not possible to conclusively single out one as the most probable. All scenarios are built on chains of fragmentary evidence, and which scenario will prevail depends on when and how its weakest links are challenged. Scenario I suggests that only humans are conscious. The merit of this scenario is that it takes the uniqueness of human language into account. The problem with this scenario is that no qualitative structural differences between human and other mammalian brains have yet been found. This problem is taken into account in Scenario II, the hypothesis that all mammals, but only mammals, are conscious. The critical differentiation is then between mammals and other species, most importantly birds. A merit of this scenario is that it offers tentative, but clear, structural markers of consciousness, the six-layered cerebral cortex, and perhaps the claustral-basolateral amygdalar formation. A problem is that the avian brain has recently been shown to be structurally more similar to the mammalian brain than previously thought, including a claustrum and/or amygdalar homologue either as a discrete or field homology, and consequently this scenario has few adherents. Scenario III takes certain recent comparative anatomical and cognitive data into account. The critical cut-off point here is dual, separating mammals and birds, respectively, from reptiles. The problem with this scenario is that consciousness must be assumed to have emerged twice within amniotes. This is not an insurmountable obstacle, as other such convergent features are well known, homeothermy being one such case. Also there are many quantitative mammalian-avian/reptilian dichotomies supporting this scenario, and it has consequently been advocated by many. A problem with this scenario is the clear evolutionary continuity between birds and reptiles in terms of brain structure. Scenario IV solves this problem. The critical cut-off point here is assumed to be between reptilians and amphibians, thus assigning consciousness exclusively to amniotes. This would take the continuity of reptilian-bird brain evolution into account, but it also introduces the problem of finding the crucial consciousness-generating brain structure lacking in the amphibian brain but not in the reptilian one. The clear candidates for such a generator structure are consistent with most current consciousness hypotheses—the dorsal, lateral, and ventral pallial derivatives and their dorsal thalamic circuitry. Interestingly, some of these structural features that distinguish the amniote brain from the amphibian brain are also present in a number of taxa of fishes, including many teleosts and some cartilaginous fishes. Consistent with the possibility of more widespread instances of consciousness across the entire vertebrate taxon, is the well established position that amphibian brains are actually regressive, exhibiting paedomorphy and a resulting secondary simplification of their neural organization (Roth *et al.*, 1993). While the brain-body ratios of amphibians fall considerably lower than those of amniotes, many species of fishes have brain-body ratios that overlap amniotes. Further cognitive-behavioural and physiological-anatomical studies on representatives of these taxa might provide a fruitful test of current views of the evolution of consciousness.

Comparing the different amniote scenarios discussed here, the question forced upon us seems to be whether there is a dichotomy between the mammalian+avian and the reptilian brain (Scenario III) or between the mammalian and the avian+reptilian brain (Scenario II). While the qualitative and many of the quantitative neural traits, as well as the behavioural and cognitive traits, suggest the former, a case can also be made for the latter, particularly in regard to neural cytoarchitecture, phylogenetic relationships, and some of the quantitative neural traits listed above. Of the neural comparisons being examined here, the most compelling dichotomy clearly exists between amphibians and amniotes, while that between birds and reptiles in terms of neural architecture is in large part one of degree. That quantitative difference is clearly a crucial one, since the *cognitive* dichotomy between mammalian+avian taxa and reptilian+amphibian taxa is also compelling.

The picture is thus complex. Other scenarios are of course also possible and have been suggested; ranging from various scenarios assuming consciousness in vertebrates in general, with a focus on cognitive-level behavioural studies on fish, to more global ones, assuming a gradual evolution of consciousness from a diminutive start, as posited by Griffin (2001). A major challenge, however, is to answer the fundamental question: What, if any, is the *function* of consciousness?

References

Alkire MT, Miller J (2005) General anesthesia and the neural correlates of consciousness. *Prog Brain Res* **150**:229-244.

Antkowiak B (2002) In vitro networks: cortical mechanisms of anaesthetic action. *Br J Anaesth* **89**:102-111.

Århem P, Liljenström H, Lindahl BIB (2003) Consciousness and comparative neuroanatomy. Report on the Agora Workshop in Sigtuna, Sweden, on 21 August, 2002. *J Consciousness Studies* **10**:85-88.

Bass AH (1977) Effects of lesions of the optic tectum on the ability of turtles to locate food stimuli. *Brain Behav Evol* **14**:251-260.

Bass AH, Pritz MB, Northcutt RG (1973) Effects of tectal and telencephalic ablations on visual behavior in the side-necked turtle, *Podocnemis unifilis*. *Brain Res.* **55**:455-460.

Broglio C, Gómez A, Durán E, Ocaña FM, Jiménez-Moya, Rodríguez F, Salas C (2005) Hallmarks of a common forebrain vertebrate plan: specialized pallial areas for spatial, temporal and emotional memory in actinopterygian fish. *Brain Res Bull* **66**:277-281.

Bruce LL, Neary TJ (1995) The limbic system of tetrapods: a comparative analysis of cortical and amygdalar populations. *Brain Behav Evol* **46**:224-234.

Butler AB (1994a) The evolution of the dorsal thalamus of jawed vertebrates, including mammals: cladistic analysis and a new hypothesis. *Brain Res Rev* **19**:29-65.

Butler AB (1994b) The evolution of the dorsal pallium in the telencephalon of amniotes: cladistic analysis and a new hypothesis. *Brain Res Rev* **19**:66-101.

Butler AB, Manger PR, Lindahl BIB, Århem P (2005). Evolution of the neural basis of consciousness: a bird-mammal comparison. *BioEssays* **27**:923-936.

Butler AB, Cotterill R (2006) Mammalian and avian neuroanatomy and the question of consciousness in birds. *Biol Bull* **211**:106-127.

Butler AB, Molnár Z (2002) Development and evolution of the collopallium in amniotes: a new hypothesis of field homology. *Brain Res Bull* **57**:475-479.

Butler AB, Hodos W (2005) *Comparative Vertebrate Neuroanatomy: Evolution and Adaptation*, 2nd Edition. New York: Wiley-Liss.

Cariani P (2000) Anesthesia, neural information processing, and conscious awareness. *Consc Cogn* **9**:387-95.

Clayton NS, Bussey TJ, Dickinson A (2003) Can animals recall the past and plan for the future? *Nat Rev Neurosci* **4**:685-691.

Cowey A, Stoerig P (1995) Blindsight in monkeys. *Nature* **373**:247-249.

Cotterill RMJ (2001) Cooperation of the basal ganglia, cerebellum, sensory cerebrum and hippocampus: possible implications for cognition, consciousness, intelligence and creativity. *Prog Neurobiol* **64**:1-33.

Crick F (1994) *The Astonishing Hypothesis: The Scientific Search for the Soul.* New York: Simon & Schuster.

Crick F, Koch C (1995) Are we aware of neural activity in primary visual cortex? *Nature* **375**:121-123.

Crick F, Koch C (2003) A framework for consciousness. *Nat Neurosci* **6**:119-126.

Crick FC, Koch, C (2005) What is the function of the claustrum? *Phil Trans R Soc B.* **360**:1271-1279.

Dicke U, Roth G (2007) Evolution of the amphibian nervous system. In: JH Kaas (Ed) Evolution of Nervous Systems, Vol 2. Oxford UK: Academic Press, pp 61-124.

Edelman, GM (1989) *The Remembered Present: A Biological Theory of Consciousness.* New York: Basic Books.

Edelman GM (2003) Naturalizing consciousness: A theoretical framework. *Proc Natl Acad Sci USA* **100**:5520-5524.

Edelman GM, Tononi G (2000) *A Universe of Consciousness: How Matter Becomes Imagination.* New York: Basic Books.

Edelman DB, Baars BJ, Seth AK (2005) Identifying hallmarks of consciousness in non-mammalian species. *Consc Cogn* **14**:169-87.

Eiland MM, Lyamin OI, Siegel JM (2001) State-related discharge of neurons in the brainstem of freely moving box turtles, *Terrapene carolina major. Arch Ital Biol.* **139**:23-36.

Emery NJ, Clayton NS (2004) The mentality of crows: convergent evolution of intelligence in corvids and apes. *Science* **306**:1903-1907.

Engel AK, Fries P, Konig P, Brecht M, Singer W (1999) Temporal binding, binocular rivalary, and consciousness. *Consc Cogn* **8**:128-151.

Grasshoff C, Drexler B, Rudolph U, Antkowiak B (2006) Anaesthetic drugs: linking molecular actions to clinical effects. *Curr Pharm Des* **12**:3665-3679.

Greenberg N (2002) Adaptive functions of the corpus striatum: the past and future of the R-complex. In: GA Cory and R Gardner Jr (Eds) *The Evolutionary Neuroethology of Paul MacLean.* Westport CT: Praeger, pp 45-81.

Griffin DR (2001) *Animal Minds: Beyond Cognition to Consciousness.* Chicago: The University of Chicago Press.

Garau C, Aparicio S, Rial RV, Esteban S (2005) The evolution of sleep: birds at the crossroads between mammals and reptiles *Rev Neurol* **40**:423-430.

Hof PR, Van der Gucht E (2006) Structure of the cerebral cortex of the humpback whale, Megaptera novaeangliae (Cetacea, Mysticeti, Balaenopteridae). *Anat Rec* **290**:1-31.

Hunt GR, Gray RD (2003) Diversification and cumulative evolution in New Caledonian crow tool manufacture. *Proc Roy Soc Lond B* **270**:867-874.

Jerison H (1973) *The Evolution of the Brain and Intelligence*. New York: Academic Press.

John ER (2001) A field theory of consciousness. *Consc Cogn* **10**:184-213.

Jones EG (2001) The thalamic matrix and thalamocortical synchrony. *Trends Neurosci* **24**:595-601.

Karten HJ (1997) Evolutionary developmental biology meets the brain: the origins of mammalian cortex. *Proc Natl Acad Sci USA* **94**:2800-2804.

Koch C (2004) *The Quest for Consciousness: A Neurobiological Approach*. Eaglewood: Roberts and Company Publishers.

Laberge F, Mühlenbrock-Lenter S, Grunwald W, Roth G (2006) Evolution of the amygdala: new insights from studies in amphibians. *Brain Behav Evol* **67**:177-187.

Llinás R, Urbano FJ, Leznik E, Ramirez RR, van Marle HJF (2005) Rhythmic and dysrhythmic thalamocortical dynamics: GABA systems and the edge effect. *Trends Neurosci* **28**:325-333.

Lindahl BIB (1997) Consciousness and biological evolution. *J Theor Biol* **187**:613-629.

Lindahl BIB, Århem P (1994) Mind as a force field: comments on a new interactionistic hypothesis. *J Theor Biol* **171**:111-122.

MacLean PD (1990) *The Triune Brain in Evolution: Role in Paleocerebral Functions*. New York: Plenum.

Macphail, EM (1998) *The Evolution of Consciousness*. Oxford: Oxford University Press.

Macphail EM, Bolhuis JJ (2001) The evolution of intelligence: adaptive specializations versus general process. *Biol Rev Camb Philos Soc* **76**:341-364.

Manger PR, Elston GN, Pettigrew JD (2002a). Multiple maps and activity dependent representational plasticity in the anterior Wulst of the barn owl (*Tyto alba*). *Eur J Neurosci* **16**:743-750.

Manger PR, Slutsky DA, Molnar Z (2002b) Visual subdivisions of the telencephalon of the iguana, *Iguana iguana*. *J Comp Neurol* **453**:226-248.

Martínez-García F, Martínez-Marcos A, Lanuza E. (2002) The pallial amygdala of amniote vertebrates: evolution of the concept, evolution of the structure. *Brain Res Bull* **57**:463-469.

Mayr, E (2001) *What Evolution Is*. New York: Basic Books.

Medina L, Reiner A (2000) Do birds possess homologues of mammalian primary visual, somatosensory and motor cortices? *Trends Neurosci* **23**:1-12.

Merker B (2007) Consciousness without a cerebral cortex: A challenge for neuroscience and medicine. *Behav Brain Sci* **30**:63-81.

Nicolau MC, Akaârir M, Gamundi A, González J, Rial RV (2000) Why we sleep: the evolutionary pathway to the mammalian sleep. *Progress in Neurobiology* **62**:379-406.

Pepperberg IM (1999) *The Alex Studies: Cognitive and Communicative Abilities of Grey Parrots*. Cambridge, MA: Harvard University Press.
Pepperberg IM (2002) In search of King Solomon's ring: cognitive and communicative studies of grey parrots (*Psittacus erithacus*). *Brain Behav Evol* **59**:54-67.
Pepperberg IM, Lynn SK (2000) Possible levels of animal consciousness with reference to grey parrots (*Psittacus erithacus*). *Amer Zool* **40**:893-901.
Popper KR, Eccles JC (1977) *The Self and Its Brain: An Argument for Interactionism*. Berlin: Springer-Verlag.
Preuss T (2000) What's human about the human brain? In MS Gazzaniga (Ed) *The New Cognitive Neurosciences*. Cambridge MA: MIT Press, pp 1219-1234.
Puelles L, Kuwana E, Puelles E, Bulfone A, Shimamura K, Keleher J, Smiga S, Rubenstein JLR (2000) Pallial and subpallial derivatives in the embryonic chick and mouse telencephalon, traced by the expression of the genes Dlx-2, Emx-1, Nkx-2.1, Pax-6, and Tbr-1. *J Comp Neurol* **424**:409-438.
Rattenborg (2006) Evolution of slow-wave sleep and palliopallial connectivity in mammals and birds: A hypothesis. *Brain Res Bull* **69**:20-29.
Rees G, Kreiman G, Koch C (2002) Neural correlates of consciousness in human. *Nature Reviews* **3**:261-270.
Reiner A, Bruce L, Butler AB, Csillag A, Kuenzel W *et al* (2004) Revised nomenclature for avian telencephalon and some related brainstem nuclei. *J Comp Neurol* **473**:377–414.
Reiner A, Yamamoto K, Karten HJ (2005) Organization and evolution of the avian forebrain. *Anat Rec* **287A**:1080-1102.
Rial R, Nicolau MC, Lopez-Garcia JA, Almirall H (1993) On the evolution of waking and sleeping. *Comp Biochem Physiol Comp Physiol* **104**:189-193.
Roth G, Nishikawa KC, Naujoks-Manteuffel C, Schmidt A, Wake DB (1993). Paedomorphosis and simplification in the nervous system of salamanders. *Brain Behav Evol* **42**:137-170.
Roth G, Grunwald W (2000). Morphology, axonal projection pattern, and responses to optic nerve stimulation of thalamic neurons in the salamander Plethodon jordani. *J Comp Neurol* **428**:543-557.
Rudolph U, Antkowiak B (2004) Molecular and neuronal substrates for general anaesthetics. *Nat Rev Neurosci* **5**:709-720.
Seth AK, Baars BJ, Edelman DB (2005) Criteria for consciousness in humans and other mammals. *Consc Cogn* **14**:119-139.
Sjölander S (1997) On the evolution of reality — some biological prerequisites and evolutionary stages. *J Theor Biol* **187**:595-600.
Srinivasan R, Russell DP, Edelman GM, Tononi G (1999) Increased synchronization of neuromagnetic responses during conscious perception. *J Neurosci* **19**:5435-5448.
Stein BE, Meredith MA (1994) *The Merging of the Senses*. Cambridge MA: MIT Press.
Steyn-Ross ML, Steyn-Ross DA, Sleigh JW (2004) Modelling general anaesthesia as a first-order phase transition in the cortex. *Prog Biophys Mol Biol* **85**:369-85.

Tömböl T (1995a) *Golgi Structure of Telencephalon of Chicken (Gallus domesticus)*. Budapest: Semmelweis University Medical School.

Tömböl T (1995b) *Golgi Structure of Telencephalon of Chicken (Gallus domesticus). II.* Budapest: Semmelweis University Medical School.

Tononi G, Edelman GM (1998) Consciousness and complexity. *Science* **282**:1846-1851.

Tudge, C (2000) *The Variety of Life*. Oxford: Oxford University Press.

von Frisch K (1967) *The Dance Language and Orientation of Bees*. Cambridge MA: Harvard University Press.

Westhoff G, Roth G (2002). Morphology and projection pattern of medial and dorsal pallial neurons in the frog *Discoglossus pictus* and the salamander *Plethodon jordani*. *J Comp Neurol* **445**:97-121.

Chapter 5

A Bird's Eye View of Consciousness

Euan M. Macphail

1. Introduction

The major contending ideas concerning the nature of consciousness have been current for some hundreds, if not thousands, of years, and the ideas that I shall advance here are no exception. It might be reasonable to be somewhat discouraged by the apparent lack of progress, but we do have the advantage of far greater knowledge about living organisms, knowledge that does have a bearing on the plausibility of old ideas. The most important knowledge that we have is that living species evolved and that the mechanism of evolution is natural selection. It is because we are related to nonhuman species (for brevity, I shall henceforth refer to nonhumans as animals) that a better understanding of animals may throw up insights of direct relevance to humans. My own experimental work has focused largely on pigeons, and I shall use avian studies to illustrate the basic claims that underlie my approach. I shall begin, however, by outlining what appear to me to be the central issues in consciousness.

2. Which organisms are conscious?

We human beings are conscious living organisms. Not all living beings are conscious – most of us suppose that only animals, not plants or fungi or protozoa, can achieve consciousness. Biologists have by now achieved a reasonable grasp of the way in which nonliving constituents can be assembled to produce a living, reproducing, organism: but it remains a mystery how living cells, supposedly nonconscious entities, can be assembled to produce a conscious being.

Consciousness Transitions: Phylogenetic, Ontogenetic, and Physiological Aspects
Edited by Hans Liljenström and Peter Århem

2.1 Definition of consciousness

Our lack of understanding means that we cannot easily provide a definition of consciousness, and to focus the discussion here, I shall suppose that the essence of consciousness is the ability to feel pleasure and/or pain, to prefer some states to others. This approach affords a natural link with our ethical systems, as we do not suppose that one can do wrong to nonconscious entities: to damage a nonconscious object, such as a clock or a painting, may well be morally wrong, but if so, it is because of the effect on humans, not on the object. Conscious beings are the only entities towards which we may act in an ethically wrong way; nonconscious beings are wholly indifferent to their treatment. It is, then, important to be able to establish whether an organism is conscious or not. How is this to be done? The approach adopted here argues that this is equivalent to asking how we should establish whether an organism feels pleasure or pain.

2.2 Criteria

At first sight, it seems a relatively easy matter to establish that at least some animals feel pain. A dog that stands on a thorn squeals, and withdraws his paw. Just as we would in a comparable situation – and we, of course, would experience pain. Are these outward signs of discomfort sufficient to demonstrate conscious experience? We would hardly want to use vocalisation as a criterion for feeling, as there seems no obvious rationale for excluding animals – such as fish – from the ability to feel simply because they cannot vocalise. But there are two problems facing straightforward acceptance of withdrawal as an index of feeling.

The first problem is that the behaviour shown in reaction to potentially damaging stimuli may also be shown in circumstances in which we know or strongly suspect that conscious experience is not occurring. The case has been well made out in the case of the earthworm by Julian Jaynes. Worms do not, of course, squeal, but they do show vigorous withdrawal when jabbed with a sharp object. "Most people will identify with a struggling worm. But as every boy who has baited a fish hook knows, if a worm is cut in two, the front half with its primitive brain seems not to mind as much as the back half, which writhes in 'agony'. But surely if the worm felt pain as we do, surely it would be the part with the brain that would do the agonising. The agony of the tail end is our agony, not the worm's; its writhing is a mechanical release phenomenon, the motor nerves in the tail end firing in volleys at being disconnected from their normal inhibition by the cephalic ganglion." (Jaynes, 1993, p. 6)

It is also the case that spinal animals – vertebrate preparations in which the spinal cord has been severed from the brain – withdraw their limbs in response to sharp pricks. Sherrington observed that when the limb of a spinal dog is lowered onto a level surface the limb reflexly extends, and when lowered onto a sharp point, the limb withdraws

(Sherrington, 1906). In case it might be thought that these spinal preparations are conscious, it should be mentioned that these same reflexes are seen in spinal humans, who are entirely unconscious of the stimulation; it seems hard to believe that there could be two conscious beings in one human body.

3. Function of feelings

There is a second, quite different, problem facing the assumption that a dog withdrawing a paw with a thorn in it feels pain. This is, the difficulty of seeing how the capacity to feel pain could have evolved. We can approach the problem by considering some entirely hypothetical and relatively simple organism, a predecessor of more complex animals. This animal lives in the sea, ingests nutrients, and is capable of independent movement. It can sense aspects of the chemical composition of the matter that drifts past it, and temperature gradients. It accepts or rejects potential nutrients according to their chemical signature, and moves so as to maintain a safe internal temperature. Perhaps, like us, it accepts matter rich in carbohydrates, and swims up temperature gradients so as to remain at about 21 degrees centigrade. Does it have a sweet tooth? Does it find the water pleasantly warm? There seems absolutely no reason to suppose that it would – the entire system could function perfectly adequately in an entirely 'mechanical' way.

Now consider a slightly more complex animal, one capable of rapid withdrawal from a discrete potentially noxious stimulus. What is needed for such a response? A sensory system that can discriminate between noxious and neutral stimuli, and a link between that system and the motor system so that detection of noxious stimuli engages withdrawal. Once again, a mechanical process, with no evident role for feeling. This, we may think, is a plausible description of the state of affairs in the worm.

What environmental demand could lead to the superimposition of feelings on such a mechanical system, so that the detection of noxious stimuli now gives rise to pain? It may seem that the function of pain is to engender an urgency in withdrawal, so that self-preservation gains priority over other contending behaviour. But it is hard to see that any such effect of feeling could not be matched by an alteration in the priority of the signal from the sensory system: we know from experience that in fact rapid withdrawal from, say, a hot surface, typically precedes the conscious experience of heat.

Similar arguments apply to all aspects of the behavioural responses to noxious (and beneficial) stimuli. Consider the squeal of the dog in response to the thorn. There is no *necessary* link between any feeling and squealing. Vocalization has evolved as a response to aversive stimuli because it serves a function – such as, perhaps, to attract the attention of potentially helpful conspecifics, or to warn off potential attackers, or to warn offspring of possible danger – but whatever that function, it is equally well served whether or not accompanied by feeling. It is the response, not the feeling, that contributes to survival, and it is the response that will be selected.

The argument advanced here is, then, that there is no plausible function for feeling, and so, no way to see how sentient experience could have evolved through natural selection. Before turning to the awkward fact that, nevertheless, feeling *has* evolved (we humans certainly do feel things), there is one further aspect of this armchair exploration of the evolution of feeling that should be considered.

4. Emergence of consciousness

Suppose that a function for feeling can be imagined, so that animals that do feel will have a selective advantage over those that do not. What heritable changes in the structure of the organism could mediate the transition from a non-feeling organism (such as, perhaps, a worm) to a feeling organism? One traditional answer to this question has, of course, been to suppose that feeling is experienced by a non-physical entity, a soul or a mind, that is associated with the physical body in some way. I shall not discuss this possibility in any depth here, as I confess to difficulty in presenting an even remotely coherent account of it (I can make sense neither of the idea that the physical and the non-physical interact nor of the idea that there is a miraculous parallelism between a body's history and the history of its associated mind). What should, however, be observed here is that positing a non-physical account does not help in deciding which organisms feel, and which do not. Some metaphysical systems suppose that worms have souls, some do not, and a decision about which animals achieve sentience will have to be based on some other approach.

Given that sentient beings have evolved, the failure to establish any function for feeling suggests that experience may be a functionless epiphenomenon, the inevitable consequence of some other change whose function may be more apparent. It makes sense to begin the search for this change by comparing human beings, of whose consciousness there is no doubt, with animals, some of which, it seems possible, are not conscious. Two potentially relevant differences are readily apparent. First, we have relatively large brains; second, human cognition is more complex than that of any animal species: we are, for example, more intelligent, and we use language.

Could it be that our large brain size is responsible for consciousness? Two points can be made about brain size as an index of consciousness. First, our brains, although larger in relation to body size than those of any nonhuman, are not the largest vertebrate brains: the brains of elephants and blue whales are bigger than those of man (Jerison, 1973). To use large size alone as a criterion would imply that elephants, say, were conscious, but that monkeys might not be. One might, of course, argue that all mammals have relatively large brains, and that all are thereby conscious; fish and reptiles, whose brains are ten times smaller (taking body size into account) might (along with all invertebrates) be deemed not

to be conscious because of their small brain size. Any such proposal would, however, be patently arbitrary, and would require some further supporting rationale. It is hard to see what this might be, since the rationale underlying the use of brain size as a criterion must be that increases in numbers of neurons and their interconnections automatically induce consciousness. But the human spinal cord possesses many more neurons than the brains of many small mammals and I have already argued that the spinal cord is not a conscious entity.

The human brain differs, of course, not only in size, but in anatomical organization, and changes in organization of the nervous system might seem plausible candidates as causes of the emergence of consciousness. But changes in anatomical organisation – the emergence of the neocortex in mammals, for example – will be relevant only in so far as they are reflected in changes in function. If neocortex is in some way associated with consciousness, it is surely because of novel cognitive functions mediated by the neocortex. We should, then, explore the other aspect of human uniqueness, our cognitive capacities. And if the origin of consciousness is to be found in some aspect of cognitive processing, then in trying to decide which animals are conscious and which not, we should seek some step-change in cognition between one group of animals and another. The sections that follow will review efforts to establish differences in intelligence between different animal groups, in the expectation that analysis of those differences could yield insights into specific differences in cognitive processes amongst them. Work in this area has been concentrated on vertebrates, and that emphasis will be reflected here. I shall not explore the possibility that consciousness is universal in vertebrates, but absent in some (or all) invertebrates for a number of reasons: first, not enough is known of cognitive function across the huge range of invertebrate groups to be able to judge whether there might be some cognitive leap made somewhere amongst them; second, the work that has been done on honey bees has shown a perhaps surprising degree of parallels between their learning and vertebrate learning (Bitterman, 1996), so evidence of an abrupt shift in cognition between this invertebrate species and vertebrates in general is lacking; and, finally, because the case that I shall go on to make out is opposed most strenuously by those who suppose that consciousness is more widespread amongst vertebrates than I do.

5. Vertebrate intelligence

In 1982 I published the results of a survey of work on learning across the various classes of vertebrates, and came to the conclusion that, excluding humans, there was no conclusive evidence for any differences in intelligence either between or within those classes (Macphail, 1982). Because of my particular interest in birds, because birds are so distantly related to humans, and because phrases like 'bird-brained' have reflected a fairly widespread low opinion of the bird's intelligence, I shall illustrate my case by contrasting the supposedly stupid birds with supposedly intelligent mammals.

5.1 Birds versus mammals

Our understanding of both the behaviour and the brains of birds has advanced quite dramatically over recent decades. Birds are one of the seven classes of vertebrates (three classes of fish, amphibians, reptiles, birds, and mammals). Birds evolved from a group of reptiles, the archosaurs, some 140 million years ago, and may well be direct descendants of a type of dinosaur. Mammals evolved from a quite different group of reptiles, and to find a common ancestor of birds and mammals, we have to go back some 300 million years. Mammals are, then, very distantly related to birds, and my reason for emphasising this fact is precisely that recent decades have shown that birds and mammals have very much more in common in both brain and behaviour than had been thought. It will help set the context for the discussion of cognitive function to consider first the size and organization of the avian brain

5.2 The avian brain

The size of an animal's brain varies, naturally, with its body size (larger heads tending to hold larger brains), and the relative size of a bird's brain can be summarized very briefly: the brains of birds are approximately the same size as the brains of mammals of comparable body weight (and about ten times as large as those of the cold-blooded vertebrate classes) (Jerison, 1973).

The general appearance of the avian forebrain – the region of most interest here – is strikingly different from that of its mammalian counterpart. Whereas the mammalian telencephalon is clothed with a thin 6-layered structure – the neocortex – no comparably layered structure is found in the avian telencephalon, which consists of relatively featureless nuclear regions separated by fibre laminas. This difference in appearance, through which the avian forebrain comes to resemble the reptilian forebrain more than the mammalian, led early comparative neuroanatomists (Ariëns Kappers *et al*, 1936) to suppose that the major structures of the avian telencephalon found their mammalian counterparts in the basal ganglia - the corpora striata – and so labelled the avian structures as paleostriatum, archistriatum, neostriatum, augmented by regions of uncertain homology, the ectostriatum and the hyperstriatum.

The implicit assumption of these anatomists, that there was a major difference in organisation between the avian and mammalian forebrains, was supported by neuroanatomical work that attempted to establish the courses and destinations of fibre connections between the diencephalon and the telencephalon. Much of the mammalian neocortex is devoted to sensory areas that receive organised projections from specific sensory regions of the thalamus, and to motor regions, that project down to the spinal cord. But workers on the avian forebrain concluded that there were no comparable organised sensory projections from the thalamus to telencephalic structures (and so, no functional equivalent of the neocortex). This view held sway until surprisingly recently (Powell and Cowan, 1961), and, as subsequent work has shown, it was entirely erroneous, a consequence of reliance upon on inadequate anatomical techniques.

In fact, it has become clear that the avian homologues of the mammalian basal ganglia are to be found in the paleostriatum and associated structures (Medina & Reiner, 1995), and that homologues of regions within the neocortex are to be found within other 'striatal' regions.

There are organised projections from sensory nuclei within the avian thalamus to areas within the 'striatal' regions of the telencephalon: visual projections run to the ectostriatum and to an area in the hyperstriatal complex now known as the 'visual Wulst' (Karten *et al*, 1973), auditory projections, to the neostriatum (Campbell and Boord, 1974), and somatosensory projections, to regions within both the hyperstriatal complex and the neostriatum (Wild, 1987). And although there are no layers to be seen in these regions, the functional organisation exhibited by the visual system shows a number of striking parallels: for example, cells in the visual Wulst, like cells in the mammalian striate cortex, have a columnar organization, relatively small receptive fields, and are responsive to such features as narrow bars having a specific orientation and (in some species) binocular disparity (Macphail, 2001).

The other major 'striatal' region of the avian telencephalon, the archistriatal complex, shows many parallels to the mammalian amygdalar complex (Reiner and Karten, 1985; Zeier and Karten, 1971), and includes a region, the posteromedial 'limbic' archistriatum damage to which, as in mammals, disrupts fear motivation (Dafters, 1975, 1976).

Two further parallels between avian and mammalian forebrain are worth noting. First, there is an area within the avian telencephalon that corresponds, by a number of anatomical criteria, to the primate frontal cortex. What is of more interest is that damage to this area, in birds, as in mammals, brings about a selective deficit in short-term memory for spatial information (Güntürkün and Durstezitz, 2001). Second, there is an avian homologue of the mammalian hippocampal complex, damage to which, as in mammals, disrupts spatial memory (and aspects of homing in pigeons) and short-term memory for nonspatial information (Bingman *et al*, 1995; Good and Macphail, 1994).

In summary, the avian brain is (taking body size into account) roughly the same size as the mammalian brain, and there are substantial parallels in the organization of the avian and the mammalian forebrain. As a consequence of these findings, the nomenclature of avian forebrain regions has now been modified substantially to reflect the current view of the homologues between avian and mammalian regions (Jarvis *et al*, 2005). If we assume that the forebrain is closely associated with cognitive processing – with intelligence – the implication seems to be that these brains have evolved to meet comparable environmental demands in comparable ways. Nothing that we know about the avian brain entails a difference in cognition between birds and mammals, and we would not be surprised to find parallels also in their general cognitive functioning.

5.3 Avian intelligence

There is, of course, no generally accepted definition of animal intelligence, and I use the term to refer to an animal's capacity to learn and to solve problems. Since we do not know how animals solve problems, exploration of their abilities on both simple and (apparently) more complex learning tasks are equally relevant: animals might solve complex problems by using exactly the mechanisms they use for simple problems.

The basic 'simple' tasks used by comparative psychologists involve conditioning procedures, in which two events are consistently paired together. Two types of procedure are usually distinguished. In classical, or Pavlovian conditioning, two stimuli are paired, and typically one of them, the unconditioned stimulus (UCS), elicits a reflex which, after pairing, is now elicited by the other, initially neutral, stimulus (the CS). So, a series of pairings of a bell CS with a food UCS elicits, from a hungry dog, salivation to the bell. In instrumental conditioning, the experimenter arranges a contingency between some response of the animal (pressing a lever, for example) and a rewarding or punishing event; the rate of response of the animal alters accordingly. Conditioning procedures are used to demonstrate and to investigate associative learning.

In birds as in mammals, learning is readily demonstrated using both classical and instrumental training procedures (Macphail, 1982), and, there is no reason to suppose that the efficiency of associative learning varies between birds and mammals; it can be shown, for example, that associative learning in pigeons may occur in a single trial (Macphail *et al*, 1992). Similar demonstrations of successful rapid associative learning have been found in fish, amphibians and reptiles (Macphail, 1982), and it is generally accepted now that efficient conditioning is seen throughout vertebrates (and I shall refer at a later stage to conditioning in invertebrates also).

The fact that the rate of acquisition in conditioning procedures is equally rapid in two groups of animals does not of itself show that the same associative processes are engaged, and because association-formation contributes significantly to my account of animal intelligence and to my interpretation of consciousness, some discussion of the nature of the associations formed is necessary.

Over the first half of the twentieth century, the dominant view of animal psychologists was that, although both classical and instrumental procedures were successful, the only contingency that was in fact effective was that between a stimulus and a response; the behaviourist view, which enjoyed more than 50 years' dominance, was that animals could form only stimulus-response (S-R) bonds. According to this account, not only could animals not detect contingencies between stimuli, they could not detect contingencies between responses and stimuli. Animals did not, that is, 'know' (in Pavlovian paradigms) that the bell signalled food, nor (in instrumental paradigms) that pressing the lever obtained food. Where learning occurred, it occurred because a response by the animal in the presence of a stimulus was followed by a reward. This was, at least in retrospect, an astoundingly narrow view of animal cognition, even more so because it was supposed to be equally true of humans. And although it was largely through advances in our accounts of human cognition that behaviourism became generally rejected (Macphail, 1998), one important contribution to its final downfall was an experiment that used pigeons.

This critical study (Brown and Jenkins, 1968) introduced a procedure known as autoshaping. The procedure is simple: a small aperture in the front wall of the test cage gives access to a perspex key which can be illuminated from behind. Each trial consists of the illumination of the key for, say, 8 seconds, followed by the delivery of food in a hopper beneath the key. A series of keylight-food pairings takes at an average rate of once a minute, so that the keylight comes to act as a signal that food is about to be delivered. The arrival of food is entirely independent of the behaviour of the pigeon, but after a few pairings in this procedure, pigeons come to peck the illuminated key light. If animals could form only S-R bonds and there is no contingency involving the keypeck response, how could this come about? A diehard behaviourist could argue that, perhaps, the initial key peck was a chance 'spontaneous' event, but that because it would shortly be followed by the food reward, the bond between the lit key and the peck response would be strengthened. This possibility, unlikely as it is (pigeons in fact very rarely 'spontaneously' peck lit keys), was ruled out by the demonstration that if the procedure is now altered so that on all trials on which the pigeon does peck the key, no food is delivered, pigeons nevertheless begin to keypeck (and in fact persist in keypecking (Williams and Williams, 1969). The response of pecking the lit key cannot now be 'strengthened' by the food reward, as it has never been followed by food reward. The inescapable conclusion is, that the pigeon begins to peck because it has detected the contingency between the keylight and the food.

It might seem that the persistence of the pigeon's keypeck despite the fact that it prevents the arrival of the food suggests a low level of intelligence. But we may sympathise with the bird's problem if we imagine a comparable procedure using hungry humans, in which a light signals the arrival of some much-desired morsel. Suppose we introduce the rule that if the unfortunate person salivates, the morsel will be withheld, how easy would it be to stop salivating? The pigeon's problem may be, then, one of the 'reflexiveness' of the peck response, and the important issue here is that the response provides a clear demonstration that the bird detects the contingency between the two stimuli.

There is now general acceptance of the notion that in conditioning procedures animals do detect contingencies between events, whether they be stimuli or responses, and that they do form stimulus-stimulus associations, stimulus-response associations, and response-stimulus associations. This may not seem great progress, as it in fact represents a return to precisely the views on association formation of, amongst others, William James (James,

1890/1950). What is, however, now true is that we have available the analytic procedures that allow us to demonstrate the essential nature of the associations formed.

6. Complex learning

The ubiquity of efficient associative learning in vertebrates may not, perhaps, be surprising, as it appears to be a simple, basic, form of learning. Surely there are other, more complex, forms of learning that are found in some animals (dolphins or chimpanzees, for example) but not in others? My 1982 survey could find no such example. There are, of course, many reports of achievements by supposedly intelligent animals that have not been matched by supposedly less intelligent animals. But in many cases, the absence of demonstrations in the supposedly less intelligent could be attributed, not to their failure to meet the cognitive demand, but to their lack of the sensory or motor capacities needed for the task. To put it crudely, no goldfish has been seen piling crates one on top of the other in order to reach a banana otherwise out of reach. But the success of chimps in this task (Köhler, 1925/1957) does not demonstrate their intellectual superiority to goldfish. The only way to compare species is to provide tasks that both equate cognitive demands and are physically no more demanding for either species. In fact, when analogues of tasks mastered by such animals as chimpanzees and dolphins have been devised for supposedly less talented animals, the analogues have been successfully mastered (Macphail, 1987) (and this includes an ingenious analogue of Köhler's crate task devised for pigeons (Epstein *et al*, , 1984)). The difficulty in finding convincing demonstrations of differences between vertebrate species encourages the belief that some vertebrate species are more intelligent, and others less, than we had supposed. Both these conclusions may be illustrated by an experiment that compares pigeons with dolphins.

6.1 Spontaneous communication

Bastian reported a study (described in (Wood, 1973)) that appeared to show a highly sophisticated form of communication that developed spontaneously between two bottlenosed dolphins (*Tursiops*) in a tank with a dividing partition. One dolphin (the sender) was shown one of two stimuli, not visible to the second dolphin. The second dolphin (the receiver), which could hear but not see the first, had to choose to push one of two paddles, and if it chose correctly, both dolphins were rewarded. Which paddle was correct depended on which visual stimulus had been shown to the sender dolphin. Neither dolphin had received specific training before beginning the task, but the receiver was able to choose appropriately, so that it must have used the auditory information provided by the sender (as though the sender was 'telling' the receiver which stimulus it could see).

Boakes and Gaertner (1977) devised an analogous experiment using pigeons, the only significant change being that the birds could see each other through the transparent partition (although, again, the receiver, which had to peck one of two white keys on either side of the front wall of its section of the cage, could not see the stimulus – a red or a green light on a central key – shown to the sender in the other section of the cage). The pigeons, like the dolphins, solved the problem, and Boakes and Gaertner were able to show that the mode of solution was less complicated than might at first sight appear. In essence, what happens is as follows. Initially the receiver pigeon adopts a position habit - that is, it consistently responds to one of the two keys, say the left. Given that there is at that stage no way in which it could predict which key was correct, this is a sensible strategy, and position habits are adopted by a wide range of mammals when faced with an essentially insoluble choice. This has an interesting consequence, since the colour of the stimulus shown to the sender pigeon is now correlated with the outcome of the trial. Suppose the red key indicates that the left (receiver's) key is correct, then for the receiver, red now consistently predicts reward, and green, nonreward. As a consequence of the phenomenon of autoshaping (described above), the sender pigeon now comes to peck the red (but not the green) light. The receiver now detects that when the sender approaches the front wall, food follows its response and that when the sender retreats (as pigeons do from a stimulus that is negatively correlated with reward), no food is delivered. The receiver now avoids the left key when the sender retreats, and so pecks the (correct) right key.

The problem 'solved' by the dolphins is also 'solved' by the pigeons – but one would not describe the outcome as communication in any meaningful sense. In fact the solution is mediated by associative learning, and this is one of a number of examples of supposedly complex tasks that are actually mastered through associative processes (Macphail, 1996).

7. Association formation and causality

The preceding section gives a flavour of the evidence used to support my contention that there are currently no convincing demonstrations of differences in cognitive processes among the various vertebrate groups. I have also indicated that the apparently simple process of association formation may be the basic cognitive process involved in the solution of what at first sight may appear to be more complex problems. In fact it is my contention that association formation is at the heart of *all* problem-solving in animals. Whether other 'higher' processes are involved remains contentious. But what can be said is that no proposal of a specific higher cognitive process has yet gained universal – or even widespread – acceptance. We can, then, find no convincing evidence for a cognitive leap that might be responsible for the transition from non-feeling to feeling organisms. Before considering the implications of this conclusion, a little should be added about the humble process of association formation.

Rats and pigeons have been the most common subjects of psychologists' experiments for many decades now, and the results obtained have been used in the search to establish laws of animal learning as though it was obvious that they could safely be generalized across species. This has seemed a safe assumption given the now very large body of data showing that their performance is indeed similar.

It is natural, however, to ask why it is that, despite following quite independent evolutionary paths from an early stage in reptilian evolution, birds and mammals have evolved the same basic cognitive learning process of association formation?

The procedures used by experimental psychologists in investigating learning in animals typically arrange an arbitrary contingency between two events, so that pairings of those events occur. Association formation evolved in the real world, raising the question – under what circumstances will animals experience consistent pairings of events? The obvious answer is: when there is a causal link between the events. Association formation allows an animal to detect that one event is a signal for another, or that its own response causes a predictable change in the world about it. The ability to form associations will allow an animal to uncover what Antony Dickinson has called 'the general causal structure of its world' (Dickinson, 1980, p. 166).

7.1 Blocking

This 'causal' interpretation of association formation gains support from some aspects of association formation that are not intuitively obvious. The example I shall consider here, known as blocking, is one that shows that pairing of one event with another – even when both are salient events – is not necessarily sufficient to obtain association formation. In the blocking paradigm, a control group of animals is given a number of exposures to a compound stimulus, such as the simultaneous presentation of two stimuli, a light and a tone, say, followed by a shock. The treatment of a second, experimental, group is identical, except that in a preliminary phase, one of the stimuli, the light, say, is presented singly, paired with the shock. The amount of fear generated by the tone when presented singly is then assessed in each group. The animals in the experimental group show little if any fear of the tone, whereas those in the control group show significant fear of it. The animals in the experimental group had learned in the preliminary phase to fear the light, and this blocked the acquisition of fear to the tone when that stimulus, presented along with the light, was followed by shock. Thus, despite the fact that each group has experienced the same number of pairings of the tone with the shock, the control group now fears the tone, and the experimental group does not.

What is the significance of blocking? There are in fact alternative theoretical interpretations of the phenomenon. One approach supposes that the experimental group learn in the preliminary phase to attend, on account of its learned significance, to the light, and so ignores the tone; a second account supposes that learning only occurs when an event is unexpected and that, since the preliminary phase trains the animals in the experimental group to expect a shock when shown the light, no further learning takes place when the tone is presented along with the light. The proper explanation of blocking is, however, not pertinent here. Our interest lies in the question why blocking should have evolved as a feature of association formation. A ready explanation can be found given the supposition that the function of association formation is to detect causes. If we put ourselves in the position of a rat in the experimental group, then we have learned, in the preliminary phase, that light causes shock; now when we experience light and tone followed by shock, we would naturally attribute the shock to the light – there would be no reason to suppose that the tone caused the shock. The phenomenon of blocking – which, incidentally, could not be explained by the earlier behaviourist theories – serves to strengthen confidence in the idea that association formation evolved to exploit the causal links between events in the world.

This 'causal' interpretation of blocking evidently provides a plausible account of the ubiquity of association formation in vertebrates (and blocking has been demonstrated not only in mammals, but in birds and fishes). And if association formation allows animals to unravel the causal structure of their worlds, it is not surprising that it is the dominant cognitive process employed in their solution of problems. Finally, it is important to note that association formation appears to be equally ubiquitous in the (not extensively explored) learning of invertebrates; as blocking has been demonstrated in both slugs (Sahley *et al*, 1981) and bees (Cheng and Spetch, 2001; Couvillon *et al*, 1997), it seems that the same causal interpretation is justified.

8. Language

Although no convincing contrasts have yet been demonstrated between the cognitive processes of nonhuman vertebrates, the fact remains that we humans are both more intelligent than animals, and possess language. In considering the possible sources of these contrasts, an initial question is: do we talk because we are more intelligent, or are we more intelligent because we talk? I have argued (Macphail, 1982) that it is not possible to answer this question directly because, first, we use language either overtly or covertly to solve basically all the cognitive problems that we tackle, and second, because we cannot in practice assess the intelligence of humans who, despite otherwise normal development, do not talk.

An alternative approach is to ask whether any nonhuman (having inferior intelligence) could achieve at least a minimal grasp of language. As is well known, there have been numerous efforts to teach language to animals, and it is probably equally well known that the outcomes of these efforts have been highly contentious. One of the sources of contention lies in reaching agreement on what an animal must do to demonstrate its use of language. Many of the findings reported in the literature – for example, a chimpanzee making appropriate signs with her fingers in response to a variety of stimuli – appear readily amenable to associative accounts. The ability to 'label' stimuli in some arbitrary way, or to respond appropriately to some arbitrary 'request' is too easily accounted for as simply an example of association formation, a process whose existence is not in dispute. I share the belief of many commentators that in order to demonstrate significant use of language, it is critical to show that animals can use or understand grammar. At the centre of grasping grammar is the notion of understanding a sentence, consisting of a subject and a predicate. Although it is possible to train animals to respond differentially to two (or more) symbols according to the order in which they are delivered, it has not proved possible (and this seems the minimum to prove acquisition of grammar) to train them so that the significance (the 'meaning') of the order generalizes to other symbols in a way that could convince us that grammar is used.

This is not the place to attempt a review of the animal language controversy, but in what follows I shall assume that no nonhuman animal is capable of showing even the rudiments of language. The plausibility of this claim may for some be strengthened by noting that the achievements of various primates in language acquisition programs are in general no more impressive that of a bird – Alex, an African grey parrot (Pepperberg, 1981). Given the parallels outlined in previous sections between avian and mammalian brain size, brain organization and intelligence, this should not be surprising; but it may discourage those who suppose that we should see in our closest relatives signs of the cognitive predecessors of language.

The argument of this section has been that the cognitive capacities of animals are insufficient to support acquisition of grammar. This in turn suggests that association formation is inadequate, at least without support from other processes, for language acquisition. There was, of course, one well-known effort to produce a behaviouristic account of language (Skinner, 1957); that book was, however, the subject of a critical review (Chomsky, 1959) that was a central contribution to the downfall of behaviourism. Chomsky's own interpretation of language supposes instead that humans possess a species-specific language acquisition device, an innate universal grammar. This proposal draws indirect support from two features of human language: first, there are striking similarities between the grammars of different human languages; second, despite large variations in the size of vocabularies in different languages, the grammar of no human language is either more primitive or more sophisticated than that of any other.

It is not necessary here to attempt a detailed assessment of Chomsky's position: it will be sufficient to conclude that there is good reason to suppose that our possession of language does mark a major contrast between human and animal cognition. If it is the case, first, that animal cognition is dominated by association formation, and human cognition, by language, and second, that language is a capacity mediated by processes other than association formation, then we should consider what role, if any, is played by associations in human cognition. Has, for example, the evolution of language resulted in the loss of the associative system available to animals?

9. Association formation in humans

When people have been used as participants in conditioning experiments – in which, for example, a tone is paired with an airpuff – they do, unsurprisingly, detect the contingency (manifested by an eyeblink to the tone, in the present example). But it has been suggested (Brewer, 1974) that the learning shown is not the consequence of some incremental associative process such as that used by animals, but of the formation of conscious hypotheses and expectations. It is true (and unsurprising) that humans in conditioning studies can invariably report the operative contingency, and that 'conditioned' responses to a tone can be established simply by telling participants that tones signal airpuffs (and abolished by letting them know that airpuffs will no longer be delivered). It seems that in order to demonstrate an associative system, it will be necessary to find circumstances in which the participants' behaviour plausibly reflects the operation of associative processes, but is not accompanied by explicit hypotheses that could explain their performance.

9.1 Artificial grammars

Support for the notion that humans do use an associative system comes, oddly, from experiments using fairly complicated designs (Macphail, 1998). One set of experiments (Reber, 1993) explored the acquisition of 'artificial grammars'. In these studies, people are asked to recall lists of letters (consonants), such as PVV, PTTV, TSXXTVPS. The lists ('sentences') are in fact generated according to an arbitrary set of Markovian rules (the artificial grammar) that limit the legal transitions between the consonants. After memorizing 20 of these legal sentences, the participants were told that they had been constructed according to complex grammatical rules, and were then asked to classify novel strings as either legal sentences or illegal strings. The participants achieved the correct classification for about 70 percent of the strings, showing that they had learned something about the grammar. They were, however, very surprised to learn of the existence of the grammar, and were quite unable to articulate any of its rules, leading Reber to conclude that they had not consciously attempted to find any rules.

Other experiments have shown that participants memorize sets of 'grammatical' sentences more rapidly than sets of arbitrary lists, although, once again, they are unaware of any rules and cannot articulate any. In another effort to see whether some active process of hypothesis formation might nevertheless be involved, participants were told that the lists that they were to be asked to memorize were constructed according to formal rules. These participants showed, however, poorer recall than controls who were not given any information about the use of rules, and it seemed that the impairment in performance came about because the experimental group had devised rules that were in fact wrong. It is difficult to detect the rules used, and "looking for rules won't work if you cannot find them" (Reber, 1993, p. 48).

If the participants in studies using artificial grammars cannot detect the rules in use, what is it that they have learned, and that they profit by in their recall? Reber argues that they have used an implicit learning system that detects covariations or contingencies between events – noting also that this is precisely the function attributed to associative processes in theories of animal learning. They have, in other words, formed associations, but because the contingencies detected are not simple, they have not allowed the emergence of accurate conscious hypotheses. Reber goes on to argue that this implicit associative system evolved before the explicit, hypothesis-forming system that dominates where relatively simple rule-based solutions are available.

If humans use an associative system, and if, as I have argued above, the associative system evolved to detect causal links, then it makes sense to ask whether we actually use that system when asked to specify causes. Once again, a positive answer to this question will require that an alternative account, in terms of articulated hypotheses, can be ruled out. Some ingenious experiments on causality judgments have achieved this.

9.2 Causality judgments

Experiments on causality judgments expose participants to a series of presentations of two events, and ask them to judge the strength of the causal link between them (on a scale of 0 to 1). Typically, the potential 'cause' is an action, like pressing a button, and the 'effect' is an event like a light coming on. The effectiveness of generating explicit hypotheses is minimized by complicating the relationship between the action and the outcome. Instead of arranging the task so that the action is invariably followed by the outcome, and that the outcome is invariably preceded by the action, the contingency between the events is weakened, either by having presentations of the action that are not followed by the outcome, or by presentations of the outcome that are not preceded by the action. Formally, the strength of the contingency between two events is the difference between the probability of occurrence of the outcome following the action, and the probability of the outcome in the absence of the action. The strength can vary from −1 to +1. So, if the outcome is just as likely to occur in the absence of the action as following the action, the strength of the contingency between them is zero; and if the outcome always follows the action and never occurs in its absence, the contingency strength is +1.

Participants' estimates of the effectiveness of their actions in such studies are reasonably good (see Shanks, 1993, for a review). Accurate judgments could be the product of either an implicit associative process, or the generation of explicit hypotheses. To be accurate, these hypotheses would require assessments of the rates of occurrence of the outcome given the action and in the absence of the action. But in experiments in which participants have been asked to give their estimates of, first, the effectiveness of their action and, second, of the rates of occurrence of the outcome, the accuracy of their effectiveness judgment has been substantially better than could have been predicted on the basis of their estimates of the rates of occurrence of the outcome in the presence and absence of the action. The accuracy of causality judgments is, then, congruent with the notion that we use an implicit associative system in making them.

Work on causality judgments has also found features of the processes involved that show parallels with phenomena shown in animal studies. One such study (Dickinson, Shanks & Evenden, 1984) used a computer game in which participants were asked to rate the effectiveness of their action (pressing a space bar) in blowing up tanks (which could also be blown up by mines). In a preliminary phase in which the space bar was not available an experimental group were exposed to trials in which tanks were frequently blown up by mines. After a training phase in which there was a positive contingency between bar-presses and tanks blowing up, these participants rated the effectiveness of their actions significantly lower than the ratings of controls, who experienced identical conditions in the training phase, but had experienced a low rate of tanks being blown up by mines in the preliminary phase. Humans, like slugs, bees, and other vertebrates, show blocking in tasks that tap associative learning processes.

The conclusions from the work surveyed in this section may be summarized as follows: humans possess both the associative system found throughout vertebrates and a species-specific language–acquisition device. The human associative system is implicit, in that its activities are not open to conscious inspection. The fact that this system is operates unconsciously in ourselves does not, of course, prove that its operation is unconscious in animals also. But it does mean that we cannot use the efficient learning of animals as evidence that they are conscious. It also suggests that we should explore the question whether the emergence of language – the only major cognitive contrast found – might he responsible for the emergence of consciousness.

10. Infantile amnesia

The notion that language and consciousness enjoy a close association is not novel. Descartes' cogito argument is, in essence, the claim that rational thought – inconceivable without language – is both the proof and the core of consciousness. It was this link that led Descartes to argue that our minds continue to exist when asleep only because when asleep, we dream continuously (and that animals are not conscious). We are now happy to accept that our minds survive dream-free sleep and other forms of unconsciousness, so that if we are to find a link between language and consciousness, we must seek some other, slightly less rigid association.

One possibility is that the cognitive mechanisms used in the acquisition of language result, independently, in human consciousness; another possibility is that the use of language is necessary for the development of consciousness. These are issues that might be resolved if we could track the ontogeny of consciousness in humans. At what stage do babies become conscious – before or after they learn to talk? Most people probably suppose that embryos consisting of a handful of cells are not conscious, and that some degree of maturation of the central nervous system is a prerequisite of consciousness. And most of us would certainly suppose that the responses of young babies to, for example, pinpricks, show that they, like us, feel pain. But the same arguments advanced against the use of such responses in animals as proof of their consciousness apply to babies, although in the case of babies, those arguments may seem less forceful because we know that adult humans do feel pain. Surely the parallels between the infant's nervous system and the adult's are sufficient to guarantee the existence of feelings in infants? We could, of course, answer these questions with total confidence if we could simply ask infants what, if anything, they feel: an assurance in words that something is experienced is as good evidence as we shall ever have (it is, after all, what ultimately convinces us that all other humans experience feelings). Given that infants do not talk, the next step would seem to be, to wait until a child does acquire language, and then to ask him about his early experiences. At his point, we meet an interesting problem: despite the barrage of striking novel events that befall us as infants, we do not recall anything of this early preverbal life. Our earliest childhood memories tend to come from when we are 3 or 4 years old (Howe and Courage, 1993), some time after we have learned to talk. The phenomenon of infantile amnesia is fascinating in itself, and worth probing here as it may throw some light on the origins of consciousness.

It might be thought that infantile amnesia was the result of the immature nervous system being insufficiently stable to store new information over long periods. There is, however, a substantial body of evidence showing that infants can learn and retain successfully (Macphail, 1998). Experiments using the deferred imitation technique make the point. In these studies, infants watch an adult performing a particular action with a toy. In the test, the infants are given the toy and observed to see whether they perform the action modelled by the adult. One study used a toy car and a hinged board that could be unfolded to make an incline, and the experimenters modelled two actions: unfolding the board, and placing the car on the top of the incline so that it would run down the board. Children as young as 11 months old perform successfully in this task, and reproduce correctly the order of the two actions seen. In similar studies in which three actions were modelled in a given sequence, 21-month old babies showed significant recall of the actions and their sequence when tested after an 8-month delay (Bauer, 1996).

Why are the apparently stable memories formed by infants not recalled in later life? Does the advent of language somehow block access to them? There is evidence that counters this possibility from a study (Myers *et al*, 1987) that explored retention by five 'highly verbal' toddlers approaching three years old of events that had occurred when they were between 6 and 40 weeks old. The children, when infants, had been given auditory and visual tests in a laboratory, and the same tests were repeated, in the same laboratory, when they were verbal toddlers. Their performance in the re-test differed from that of control toddlers who had not been tested as infants (the difference being shown largely in their readiness to reach for and play with the objects used in the tests).

Although infants can learn and retain information over the transition to verbal life, they do not appear, when verbal, to be aware that they have remembered anything. The children in the study described above were asked questions about their previous visits, but showed no evidence of remembering them; none of the five children, for example, remembered having been in the laboratory previously. Another, perhaps more striking, example of this lack of conscious recall comes from a report of a child, 18 months old, who caught a fish bone in her throat, and was taken, in great distress, to a hospital emergency department, where the bone was removed. Six months later, this child, now talking, could recall nothing of the incident: but she did pick out a photograph of the emergency-room doctor, and has refused to eat fish since the incident (Howe, Courage, & Peterson, 1994). The failure of conscious recall, coupled with the inability of adults to access the associative system, suggests that infants may use the implicit, associative system that is found in animals also.

The work reviewed above suggests that conscious recall emerges with the development of language, and raises again the possibility that language in some way gives rise to consciousness. Some insight into how this might come about comes from an influential theory of infantile amnesia, that argues that a concept of the self is an essential prerequisite of explicit, autobiographical memory (Howe and Courage, 1993). In essence, the notion is that without a self-concept there is no framework available that could incorporate experiences into an explicit time-related autobiographical memory. If I have no concept of 'I' as being myself, how could I establish memories of 'my' life events?

The emergence in children of a 'self-concept' is, of course, not an abrupt event, and is generally taken to be marked by the use of words like 'I', 'me' and (a little later) 'you', along with obvious recognition of themselves in mirrors, photographs and mirrors. The self-concept emerges somewhere between 18 and 24 months – roughly the time at which language begins to develop, and a little before the time at which memories that can be recalled in later life start to form. The emergence of language, the self-concept, and explicit memories across a short phase of development encourages the notion that they may be causally-linked, and I have argued that language allows the development of self, which in turn is the basis of both conscious recall and consciousness itself.

Before outlining the steps in the argument that make the leap from language to consciousness, a summary of the conclusions drawn this far may be helpful. I have argued that the outward behavioural manifestations of 'feeling' cannot be taken as proof of experience, that no plausible adaptive advantage for feelings has been made out, and that no plausible physical mechanism for the generation of feelings is currently available. Given that we humans, at least, certainly do have feelings, I have argued that feelings must be associated with some shift in cognitive function at some stage in (presumably) vertebrate history. A search for contrasts in cognitive capacities between different groups of nonhuman vertebrates, illustrated by comparisons of the brains and intelligence of birds and mammals, finds only striking parallels, along with a clear dependence on associative learning in all groups. Humans, however, are distinguished by their possession of one species-specific capacity, the acquisition of language. Language dominates human thought, and when we do use associative mechanisms, we cannot access them consciously. Infants' early learning is dominated by associative activity, and does not generate explicit memories; the transition to the life of conscious memories occurs as language and the self-concept are acquired.

Up to this point, the claims made can be backed up by reasonable experimental support: although there is probably not universal agreement about any single claim, none would be universally rejected. But at this stage I have to stray further into speculation than before, and cannot expect to carry the same conviction. All that can be said in anticipatory defence is, that this is, I believe, the fate of all current (and previous) accounts of consciousness – all seem to have a dire and generally unsuccessful struggle with credibility.

My account ends, then, with an attempt to answer two questions: why is language a prerequisite for a self-concept, and what has the self to do with the capacity to experience feelings – with consciousness in its most critical sense?

11. Language and self

The basic feature of language is that sentences are produced and understood. This is the way in which we make statements about the world, and use language to convey information about the world. A sentence consists of a subject and a predicate; the speaker identifies a subject, and says something *about* it. The mental capacity to conceive of some property as being 'about' some other property underlies our capacity to acquire language. It is difficult to conceive a way in which the formation of associations could embrace the generation of sentences. Most theorists of learning in animals suppose that associations are formed between 'internal representations' of external events. A pigeon that has been trained in a procedure in which a red light regularly precedes food would, for example, form an association between representations of the red light and of the food. But that cognitive process is very different from the formation of a sentence involving those representations. The pigeon is not entertaining the proposition that the red light *is* food: it is simply the case that activation of the representation of the light leads (by association) to activation of the representation of the food (and so, as we have seen, to pecking).

The development of the concept of self and the emergence of language are roughly contemporaneous, and this makes sense if we assume that the ability to conceive of some property as being *about* some subject underlies the growth of a clear distinction between conceptions about oneself and conceptions about the world outside oneself. The development of the self-concept in turn plausibly leads to the construction of a cognitive structure that monitors and controls those cognitive processes that are 'explicit', while leaving other cognitive processes, association formation, for example, to run automatically – without interference – in the background. Once this 'self' structure is in place, autobiographical memory becomes possible, and infantile amnesia is overcome.

The notion that language is essential for the emergence of 'self' would, on the assumption that language is specific to humans, be countered by evidence that non-humans develop a self concept. There have in fact been claims of experimental demonstrations of the self in animals, many of which have explored the question whether animals can recognise themselves in mirrors. The issue of self in animals has been associated by some with the wider claim that the great apes alone have a 'theory of mind', and that they act upon their beliefs about the beliefs of their fellow apes. Much of the evidence concerning this claim has focused on attempts to determine whether apes are capable of deliberate deception. As was the case with claims about language in animals, these have been (and remain) highly contentious issues (see, for example, (Heyes, 1995, 1998), and it will be sufficient here to note, first, that many commentators (myself included) remain unconvinced by the evidence from the ape studies and, second, that many of the phenomena ascribed to apes are well matched by comparable work on birds (Emery, 2006).

The position taken here is, then, that the concept of self is peculiar to humans, and that it develops alongside the acquisition of language, both achievements being dependent upon the capacity of comprehending a predicate as being about a subject.

12. Self and consciousness

The final step of my argument, and frankly the most speculative, is to move from self-consciousness to feeling-consciousness. What is the relationship between these two aspects of consciousness, and does it make any sense to suppose that feeling-consciousness is dependent upon self-consciousness?

My basic reason for supposing that there may indeed be a dependence of feelings on a self relates to the question where feelings might occur, if not in a self. If we experience pain in a finger we do not suppose that the experience actually occurs in the finger; we suppose that it occurs in some sense centrally, in the brain, for example. But again we do not suppose that we could excise some particular set of neurons and equate their pattern of activity with the experience of pain. When pain occurs, a person feels it; it is 'I' who experience the pain, and no other answer seems admissible. So, because we do not believe that an isolated human spinal cord is a person, or has a concept of self, we do not believe that the cord can feel anything.

An armchair experiment may help bring out the flavour of this argument. Suppose you had to undergo some surgical procedure, involving an incision perhaps, that would be painful, and are given a choice between two drugs. The first is a general anaesthetic, but you may be allergic to it. The second is a drug that is entirely safe and has the effect of preventing access of any experience to conscious memory, but has no apparent effect on any of the pathways believed to mediate noxious inputs. When this drug is administered, the overt responses to noxious stimulation are the same as those of someone with no drug treatment, except that no verbal account of it is given at any stage during or after the operation. Its effects are, in other words, the opposite of those of curare, which prevents the overt response, but not the ability to form explicit memories which can be retrieved after treatment. Should you take the amnestic drug or the anaesthetic?

A person having chosen the amnestic drug would, after the procedure, deny having had any painful experience – he would have no recollection of it, and so no belief that it could have been *his* experience: and if not his experience, whose? The utility (if any) of this argument is that it clarifies the proposal that the ability to form explicit memories (in autobiographical memory, dependent upon the existence of a self) is a prerequisite of the possibility of experience. Organisms that have not developed a self cannot, then, be sentient beings; and if animals and infants have no self, then they do not feel anything. They do, of course, learn, using the implicit associative system: the participant in the armchair experiment outlined above might, once the amnestic drug wears off, be expected to withdraw from stimuli – the surgeon, perhaps – associated with the incision, but would have no notion of the source of the aversion.

13. Concluding comments

This review has emphasized what I see as those aspects that are of most general interest, whatever one's approach to the nature of consciousness might be. So, I have highlighted the difficulty of establishing which animals might have feelings; I have introduced data indicating that similarities among vertebrates in brains and intelligence are more marked than was at one time supposed (with association formation playing a critical role in animal cognition); I have discussed the implicit nature of association formation in humans; and I have argued that the phenomenon of infantile amnesia may be related to the absence of language and a self-concept in babies. I have, however, devoted little attention to rival accounts of animal cognition, preferring to note that the existence in animals of language, a self-concept, or a theory of mind all remain controversial issues. Similarly, I have provided only a brief sketch of the way in which I believe that the possession of language paves the way to consciousness.

A position that supposes that neither any animal nor human infants enjoy any form of conscious experience is unlikely to attract wide support, but I hope that those who reject it may nevertheless find something to mull over in the evidence used. As I have suggested previously, there is probably no theory of consciousness that currently enjoys widespread confident support, and this lack of confidence in any particular position serves to emphasize what, perhaps, hardly requires saying: whatever belief one may have, we should continue to treat animals and infants in the same way as we would have done in the absence of any philosophical excursions. Not only should we, but I suspect that even a near certainty would have little effect on our natural sympathetic anthropomorphism. Many of us can make no sense of the notion of free will, but this has virtually no effect on our behaviour – we continue to praise and blame exactly as though we do believe in free will, and only a psychopath would maltreat animals or infants on the excuse that he has concluded they feel nothing.

References

Ariëns Kappers CU, Huber GC, Crosby EC (1936) *The Comparative Anatomy of the Nervous System of Vertebrates, Including Man.* New York: Hafner.

Bauer PJ (1996) What do infants recall of their lives - memory for specific events by one-year-olds to 2-year-olds. *American Psychologist* **51**:29-41.

Bingman VP, Jones T J, Strasser R, Gagliardo A, Ioalè P (1995) Homing pigeons, hippocampus and spatial cognition. In: E Alleva, A Fasolo, H-P Lipp, L Nadel, L Ricceri (Eds) *Behavioural Brain Research in Naturalistic and Semi-Naturalistic Settings*. Dordrecht: Kluwer, pp 207-223.

Bitterman ME (1996) Comparative analysis of learning in honeybees. *Anim Learn Behav* **24**:123-141.

Boakes RA, Gaertner I (1977) The development of a simple form of communication. *Q J Exp Psychol* **29**:561-575.

Brewer WF (1974) There is no convincing evidence for operant and classical conditioning in human beings. In: WB Weimer and DJ Palermo (Eds) *Cognition and the Symbolic Processes*. Hillsdale, NJ: Erlbaum, pp 1-42.

Brown PL, Jenkins HM (1968) Auto-shaping of the pigeon's key-peck. *J Exp Anal Behav* **11**:1-8.

Campbell CBG, Boord RL (1974) Central auditory pathways of non-mammalian vertebrates. In: WD Keidel and WD Neff (Eds) *Handbook of Sensory Physiology,* Vol. 5, Part 1, Auditory system. Berlin: Springer, pp 337-362.

Cheng K, Spetch ML (2001) Blocking in landmark-based search in honeybees. *Anim Learn Behav* **29**:1-9.

Chomsky N (1959) Review of B. F. Skinner's Verbal Behavior. *Language* **35**: 26-58.

Couvillon PA, Arakaki L, Bitterman ME (1997) Intramodal blocking in honeybees. *Anim Learn Behav* **25**:277-282.

Dafters R (1975) Active avoidance behavior following archistriatal lesions in pigeons. *J Comp Physiol Psychol* **89**:1169-1179.

Dafters R (1976) Effect of medial archistriatal lesions on the conditioned emotional response and on auditory discrimination performance of the pigeon. *Physiol Behav* **17**:659-665.

Dickinson A (1980) *Contemporary Animal Learning Theory*. Cambridge: Cambridge University Press.

Dickinson A, Shanks DR, Evenden J (1984) Judgment of act-outcome contingency: the role of selective attribution. *Q J Exp Psychol* **36A**:29-50.

Emery NJ (2006) Cognitive ornithology: the evolution of avian intelligence. *Philos Trans R Soc B Biol Sci* **361**:23-43.

Epstein R, Kirschnit CE, Lanza RP, Rubin LC (1984) 'Insight' in a pigeon: Antecedents and determinants of an intelligent performance. *Nature* **308**:61-62.

Good M, Macphail EM (1994) The avian hippocampus and short-term memory for spatial and non-spatial information. *Q J Exp Psychol B* **47**:293-317.

Güntürkün O, Durstezitz D (2001) Multimodal areas of the avian forebrain - blueprints for cognition? In: G Roth and MF Wulliman (Eds) *Brain Cognition and Evolution*. New York/Heidelberg: Wiley/Spektrum, pp 431-450.

Heyes CM (1995) Self-recognition in primates - further reflections create a hall of mirrors. *Anim Behav* **50**:1533-1542.

Heyes CM (1998) Theory of mind in nonhuman primates. *Behav Brain Sci* **21**:101-134.

Howe ML, Courage ML (1993) On resolving the enigma of infantile amnesia. *Psychol Bull* **113**:305-326.

Howe ML, Courage ML, Peterson C (1994) How can I remember when "I" wasn't there: long-term retention of traumatic experiences and emergence of the cognitive self. *Consc Cogn* **3**:327-355.

James W (1890/1950) *The Principles of Psychology*, Volume 1. New York: Dover Publications, Inc. (reprint of 1890 edition).

Jarvis E, Gunturkun O, Bruce L, Csillag A, Karten H, Kuenzel W, et al. (2005) Avian brains and a new understanding of vertebrate brain evolution. *Nat Rev Neurosci* **6**:151-159.

Jaynes J (1993) *The Origin of Consciousness in the Breakdown of the Bicameral Mind*. London: Penguin Books.

Jerison HJ (1973) *Evolution of the Brain and Intelligence*. New York: Academic Press.

Karten HJ, Hodos W, Nauta WJH, Revzin AM (1973) Neural connections of the "visual Wulst" of the avian telencephalon. Experimental studies in the pigeon (*Columba livia*) and owl (*Speotyto cunicularia*). *J Comp Neurol* **150**:253-278.

Köhler W (1925/1957) *The Mentality of Apes*. Harmondsworth: Penguin Books.

Macphail EM (1982) *Brain and Intelligence in Vertebrates*. Oxford: Clarendon Press.

Macphail EM (1987) The comparative psychology of intelligence. *Behav Brain Sci* **10**:645-696.

Macphail EM (1996) Cognitive function in mammals: The evolutionary perspective. *Cognitive Brain Res* **3**:279-290.

Macphail EM (1998) *The Evolution of Consciousness*. Oxford: Oxford University Press.

Macphail EM (2001) Conservation in the neurology and psychology of cognition in vertebrates. In: G Roth, MF Wulliman (Eds) *Brain Evolution and Cognition.* New York/Heidelberg: Wiley/Spektrum, pp 401-430.

Macphail EM, Reilly S, Good M (1992) Long-term-memory of pigeons for stimulus outcome associations involving multiple stimuli, each seen in a single brief trial. *Q J Exp Psychol B* **45B**:81-98.

Medina L, Reiner A (1995) Neurotransmitter organization and connectivity of the basal ganglia in vertebrates - implications for the evolution of basal ganglia. *Brain Behav Evol* **46**:235-258.

Myers NA, Clifton RK, Clarkson MG (1987) When they were very young: Almost-threes remember two years ago. *Infant Behav Develop* **10**:123-132.

Pepperberg IM (1981) Functional vocalizations by an african grey parrot (psittacus-erithacus). *Zeitschrift für Tierpsychologie - Journal of Comparative Ethology* **55**:139-160.

Powell TPS, Cowan WM (1961) The thalamic projection upon the telencephalon in the pigeon (*Columba livia*). *J Anat* **95**:78-109.

Reber AS (1993) *Implicit Learning and Tacit Knowledge: An Essay on the Cognitive Unconscious.* Oxford: Clarendon Press.

Reiner A, Karten HJ (1985) Comparison of olfactory bulb projections in pigeons and turtles. Brain. *Behav Evol* **27**:11-27.

Sahley C, Rudy JW, Gelperin A (1981) An analysis of associative learning in a terrestrial mollusc. 1. Higher-order conditioning, blocking and a transient US pre-exposure effect. *J Comp Physiol A* **144**:1-8.

Shanks DR (1993) Human instrumental learning: A critical review of data and theory. *British J Psychol* **84**:319-354.

Sherrington CS (1906) *The Integrative Action of the Nervous System.* New Haven: Yale University Press.

Skinner BF (1957) *Verbal Behavior.* New York: Appleton-Century- Crofts.

Wild JM (1987) The avian somatosensory system: connections of regions of body representation in the forebrain of the pigeon. *Brain Res* **412**:205-223.

Williams DR, Williams H (1969) Auto-maintenance in the pigeon: sustained pecking despite contingent non-reinforcement. *J Exp Anal Behav* **12**:511-520.

Wood FG (1973) *Marine Mammals and Man: The Navy's Porpoises and Sea Lions.* Washington: R. B. Luce.

Zeier H, Karten HJ (1971) The archistriatum of the pigeon: organization of afferent and efferent connections. *Brain Res* **31**:313-326.

Chapter 6

The Molecular Biology of Consciousness†

Jean-Pierre Changeux

1. Introduction

David Ferrier in his 1878 book on "The localisation of cerebral diseases" gives a brief report of the experiments he carried out on the ablation of the prefrontal lobes in the monkey. He mentions notable alterations in the "character and manners of the animals" and notes that "instead of examining with interest and curiosity all what was happening in their field of observation, they remained apathic and weak, they were sleepy, responding only to sensations and impressions at the time"... "They had lost the capacity of intelligent observation and attention". In the same book Ferrier also reports the first observation of a frontal lobe lesion in humans, the famous case of Phineas Gage (see Damasio, 1995). He notes that following his accident, changes in his emotional reactions and intellectual faculties took place: "... he became nervous, irreverent... impatient... with an excessive obstination... capricious and undecided". Ferrier further states that "what is true for the monkey is true for man", thus introducing the important issue of the possible validity of animal models of human brain diseases. In the following years, the Italian neurologist, Luigi Bianchi (1921), went a step further. In his book "The mechanics of the brain: the function of the frontal lobes", he states that "among the phenomenal factors of the activities of living organisms" arises a "bond of coherence" which "progresses with the development and complexity of living organisms and of their nervous system". He refers to it as conciousness and further specifies that "the dawn of higher consciousness coincides with the apparition of the frontal lobes in the evolution of the brain" also noting their "inhibitory power" and their capacity for "intel-

† "Review. The Ferrier Lecture 1998: The molecular biology of consciousness investigated with genetically modified mice" by Jean-Pierre Changeux appeared in *Philosophical Transactions of the Royal Society B: Biological Sciences* 361:2239 – 2259 (2006). Reprinted with kind permission of The Royal Society.

Consciousness Transitions: Phylogenetic, Ontogenetic, and Physiological Aspects
Edited by Hans Liljenström and Peter Århem

lectual syntheses".

Since these early times the notion that non humans may possess cousciousness gave rise to contrasting attitudes. Following Thomas Huxley's provocative statement of his Belfast lecture that "we are conscious automata", that "brutes" share consciousness with humans (1874) and that "all states of consciousness in us, as in them, are immediately caused by molecular changes of brain substance", non human species have served as convenient experimental models for the scientific investigation of consciousness (Thorndike, 1898; Yerkes, 1916; Jasper *et al*, 1998; Koch, 2004; Dehaene and Changeux, 2004). In particular mice are easily accessible to pharmacological or genetic manipulations and carry out tasks that in humans involve awareness of the stimuli (e.g. Picciotto *et al,* 1995, 1998; Maskos *et al*, 2005). Yet, it may be argued that mice lack the most characteristic features of consciousness and "what it is like to be a mouse?" cannot compare to "what it is like to be a human being ?" (see Nagel, 1974). Accordingly, only humans would feel and introspect. Subjective – first person – experience would be the "hard problem" to explain, which requires the invention of far-fetched bridging principles (Chalmers, 1998).

The issue then becomes to what extent studying mice may help to understand consciousness in humans? Can the laboratory mouse become a relevant animal model to investigate the molecular biology of consciousness or at least some of its aspects?

A first point to make is that looking at the broad spectrum of living species including humans, and the developing infants, "we must question the monitoring of this mental activity". As already stated by Preyer in his 1894 book "Mental development in the child", there is not the least reason for assuming in advance that every human being comes into the world endowed with complete consciousness of self" and "that at the moment of birth an immortal soul is in waiting, as it were, for the new and as yet unconscious citizen of the world in order to take possession of him forever". Preyer also noted that in the newborn the different senses gradually develop, in a non-concomittant manner, the feeling of selfhood that he refers to as "cortical ego" developing later. He further states that, as the last step, "out of knowledge (science) (he) has developed his conscience". Unambiguously, "there are several grades of consciousness". The development of these grades, or *levels of consciousness* has been reinvestigated recently, in humans from the newborn to the adult (Bartocci *et al*, 2006; Johnson, 2005; Lagercrantz *et al,* 2002; Lagercrantz 2005; Lee *et al,* 2005) as in the course of biological evolution (see Boakes 1984; Changeux, 2002; Zéki, 2003; Zhou *et al,* 2005).

Regrouping the evolutionary analysis of Barresi and Moore (1996) primarily based on social interactions (see also Trivers 1985) and the developmental data of Zelazo (1996, 2004), Lagercrantz (Lagercrantz *et al,* 2002; Bartocci *et al,* 2006) and others (Johnson, 2005; Lee *et al,* 2005) on the human newborn, one may propose a breakdown of "consciousness" into multiple nested hierarchical levels. Beyond Edelman's (1989) cleavage between what he refers to as "primary" and "higher order" consciousness primarily on the basis of language use, one may suggest the following distinctions:

(i) A lowest level of "*minimal consciousness*" for simple organisms, like rats or mice, is characterized by the capacity to display spontaneous motor activity and to create representations, for instance, from visual and auditory experience, to store them in long-term memory and use them, for instance, for approach and avoidance behavior and for what is referred to as exploratory behavior; these organisms are more present oriented and do not make reference to an *explicit* sense of self. The comparability between the action of self and others is reduced. The 25-30 week preterm fetus, according to Lagercrantz (2005) would have reached a stage of brain maturation analogous (though not identical) to a newborn rat/mouse; he/she, may process tactile and painful stimuli in the sensory cortex (Bartocci *et al,* 2006), would percieve pain and thus would show signs of minimal consciousness;

(ii) "*recursive consciousness*" present, for instance, in vervet monkeys (possibly also in some birds), manifests itself by functional use of objects, proto-declarative pointing and searching for hidden objects; organisms at this level may display elaborate social interactions, defined imitations, social referencing and joint attention; they possess the capacity to hold several mental representations in memory simultaneously, and are able to evaluate relations of self and other in a common conceptual framework, yet, without mutual understanding; along these lines, the new born infant exhibits sensory awareness, ability to express emotions and processes mental representations (i.e. of a pacifier); he/she would already differentiate between self and non-self touch (Rochat, 2003) and imitate the tongue protrusion of an adult (Meltzoff, 1990); using the distinction of Gärdenfors (2003), he/she would be able to contrast perceptions of an "*inner world*" and perceptions of the "*outer world*" and would have reached the stage of recursive consciousness (see Barresi and Moore, 1996);

(iii) explicit "*self-consciousness*" develops in infants at the end of the second year, together with working and episodic memory and language; it is characterized by self-recognition in mirror tests and by the use of single arbitrary rules with knowledge of one's own behavioral potential and self-other distinction; to some extent chimpanzees might reach this level (see Boakes, 1984);

(iv) "*reflective consciousness*", theory of mind and full conscious experience, with first person ontology and explicit report, is unique to humans and develops following 3-5 years in children.

The case of mice and rats is simple and unambiguous. These species do not go far beyond the level of minimal consciousness. The issue of self- or reflective consciousness is irrelevant. Nevertheless, the mouse may be useful in providing some elementary "building blocks" (Searle, 2000) of consciousness. As mentioned, it offers also the possibility of joint investigations at the behavioral, physiological and molecular levels, together with an easy access to a large spectrum of techniques (from molecular genetics to brain imaging), making possible the investigation of what I may refer to as the *fundamentals* of minimal consciousness. These fundamentals, which are anticipated to be universally shared by higher organisms (from birds to humans but also possibly some invertebrates) (see Baars, 2001) may include: (i) the access to multiple states of consciousness, like wakefulness, sleep, general anaesthesia…; (ii) the capacity for global integration of several sensory and cognitive functions in goal directed behaviour (see Hobson, 1999), Bianchi's "bond of coherence", together with behavioural flexibility and thus what may be referred to as a minimal form of "intentionality" (see Discussion). In his recent book, the "Quest for consciousness", Christof Koch (2004) has reduced the issue of consciousness in animals to what he refers to as a "Turing test" for consciousness. "Take your sensor-motor routine in some species and enforce a waiting period of a few seconds between the sensory input and the execution of an action. If the organism's performance is only marginally affected by the delay, then the input must have been stored in some sort of *intermediate buffer,* implying some measures of consciousness" (Koch, 2004). This definition, by its simplicity, is of significant interest; yet, as we shall see, might not be sufficient, even in the case of the mouse, to specify the fundamentals of consciouness.

2. Modelling consciousness: the neuronal workspace hypothesis

In an attempt to implement in neural terms the "intermediate buffer" (Koch, 2004) or «global integration" (Hobson, 1999; Changeux, 1983) aspects of consciousness, neuro-computational models referred to under the generic name of Neuronal workspace hypothesis (Dehaene *et al*, 1998, 2003; Dehaene and Changeux, 2004, 2005) were proposed. These models, which extend earlier formal modelling of delayed-response, card-sorting tasks and the Tower of London test (Dehaene and Changeux, 1989, 1991, 1997), offer, in particular, neuronal implementations of Baars (1998) psychological global workspace. The proposed architectures separate, in a first minimal description (see also Norman and Shallice, 1980), two distinct computational spaces, each characterized by particular patterns of connections. Subcortical networks and most of the cortex would be viewed as a mosaic of specialized processors, each attuned to a particular type of information processing such as sensory (*present*), motor (*future*), long-term memory (*past*), evaluation (*value*, *emotions*), or attention focusing (*selective amplification or inhibition*). In spite of their diversity, processors share specialization, automaticity and fast feed-forward and feed-back processing (Dehaene *et al,* 2003) via a limited number of local and medium range connections. On top of this automatic non conscious level, a distinct set of cortical "workspace" neurons was postulated and characterized by their ability to send and receive projections to many distant areas through long range excitatory axons. These neurons, mainly pyramidal neurons from cortical layers II and III, particularly dense in prefrontal, cingulate and parietal regions would break the modularity of the cortex through the recruitment of sets of neurons in the underlying processors; they would allow the broadcasting of information, in a spontaneous and sudden manner – or "ignition" – to multiple neural targets, thus creating a global availability (or *integration*) that would be experienced in consciousness and would give rise to reportability. Other computational theories on consciousness, several of them consistent with the present proposal, have been recently reviewed (Dehaene and Changeux, 2005; Maia and Cleeremans, 2005).

An important feature of these neuronal models of cognitive functions (Dehaene and Changeux, 1989, 1991; Dehaene *et al*, 1998) has been the deliberate incorporation into their architecture of systems of evaluation through reward neurons (Dehaene and Changeux, 1989) and even of auto-evaluation (Dehaene and Changeux, 1997) (see also the algorithms of Barto *et al,* 1983; Friston *et al,* 1994) implemented in terms of limbic/mesencephalic aminergic neurons, with which the prefrontal cortex is densely interconnected, and to which dopaminergic and cholinergic, nicotinic systems may directly, or indirectly, contribute. More elaborate reward learning algorithms based on predictive Hebbian learning were also proposed (Montague *et al,* 1995, 1996) and their implementation studied both *in silico* and *in vivo* in the monkey (Schultz *et al,* 1997). In contrast with other views (Crick and Koch, 2003; Shallice, 1988; Koch, 2004), a close link was thus proposed between the neural basis of "making a conscious mental effort" and the modelisation of reward neurons (Dehaene *et al,* 1998; Dehaene and Changeux, 2000).

Simulations performed with the proposed neural architectures account for cognitive tasks, which tax the prefrontal cortex, such as delayed response tasks, the Stroop task, the attentional blink (Dehaene *et al,* 2003) or inattentional blindness, taking into account the ongoing spontaneous activity of the network (Dehaene and Changeux, 2005). Yet, at this stage, many features which characterize the higher levels of consciousness, such as the recursive and reflective aspects of consciousness, have not been dealt with. On the other hand, a simple connectionist circuit initially proposed for self-evaluation (Dehaene and Changeux, 1991, 1995) may serve as a basic ingredient to begin modelizing features of explicit self-consciousness.

The mouse lies at a low level of the evolutionary scale with a rather reduced prefrontal cortex. Yet, in this species long range frontal and callosal connections are abundant, though with the distinctive feature (at variance with primate and cat) of large-scale maintenance of simultaneous projection by frontal cortical and callosal projection neurons in the adult (Mitchell and Mackli, 2005). Also, in mice and rats, strong evidence exists for a relationship between working memory and prefrontal function and interaction (Thinus-Blanc, 1996; Jones, 2002). Moreover, in mice and rats, all the neuronal modulatory systems (cholinergic, dopaminergic, serotoninergic, adrenergic ...) are present and functional in the regulation of states of wakefulness, attention but also in reward. The mouse may thus possess a sufficient number of building blocks to make the scientific investigation of the fundamentals of minimal consciousness of relevant interest.

3. The neuronal nicotinic receptors: allosteric membrane proteins that modulate higher brain function

Physiological, psychopharmacological and pathological evidence supports the concept of an implication of cholinergic systems in cognitive functions and conscious awareness (Woolf, 1997; Perry *et al,* 1999; Perry and Perry, 2004) to the extent that Koch (2004) states that "if a single neurotransmitter is critical for consciousness, then it must be acetylcholine" (p. 90). Of particular interest in the context of consciousness, cholinergic neurons form discrete nuclei that provide widely divergent projections throughout the brain (Semba, 2004):

i. The basal forebrain neurons (including the nucleus basalis and medial septal nucleus) constitute the main afferent system to the cerebral cortex and thalamus and its functions include arousal, selective attention and REM sleep;
ii. The pedunculopontine neurons and lateral dorsal tegmental nucleus widely project to the forebrain regions, such as the thalamus, but also to the brain stem reticular formation; this system also contributes to the control of sleep and wakefulness in mammals (Semba, 1999); moreover, both nuclei provide the cholinergic innervation of the dopamine cells of the *ventral tegmental area* and *subtantia nigra* and play a role in the acquisition of drug-rewarded behaviour and drug addiction in particular nicotine self-administration (DiChiara, 2000; Balfour, 2002; Robbins 2000; Gutkin *et al*, 2005), together with rewarded cognitive tasks performance (Kobayashi and Isa, 2002).

Last, intrinsic cholinergic neurons are also distributed in the cerebral cortex and striatum, together with a variety of cholinergic cells dispersed in several brain regions (Semba, 2004).

Acetylcholine in the brain interacts with two distinct categories of pharmacological receptors: muscarinic receptors which belong to the broad family of G-protein coupled receptors with 7 transmembrane helices and nicotinic acetylcholine receptors (nAChR), which are pentameric, transmembrane ligand-gated ion channels. Muscarinic receptors control working memory, declarative memories, sustained visual attention and psychomotor speed (see Ellis *et al*, 2005), and drugs that antagonize muscarinic receptors cause hallucinations and reduce the level of consciousness (Perry *et al*, 1999; Perry and Perry, 2004). On the other hand, nicotine acting on brain nAChRs enhances arousal, attention and working memory, causes EEG desynchronisation and higher amounts of REM sleep (Levin 2002). Moreover, nAChRs have been implicated in the mechanisms of general anaesthesia. Volatile anaesthetics and ketamine have been reported to inhibit nicotinic receptors ($\alpha 4\beta 2$ and $\alpha 3\beta 4$) at clinically relevant doses (Tassonyi *et al*, 2002; Franks and Lieb, 1994) (see Discussion).

The nAChRs are among the best characterized pharmacological receptors, and their contribution to brain functions is actively investigated (Changeux and Edelstein, 2005 a,b). Their potential role in the control of consciousness will thus exclusively be considered in the subsequent chapters of this lecture.

Muscle nAChR was initially identified from fish electric organ (Changeux, 1981, 1990; Changeux and Edelstein, 2005a,b) as a protein of about 300 000 MW made up of five subunits of several different types. Extensive biochemical studies including reconstitution experiments and expression studies from cloned subunit genes have shown that this transmembrane oligomer carries several categories of binding sites : two distinct acetylcholine (nicotine and snake venom α-toxins) sites located at the interface between subunits, a unique ion channel located in the axis of pseudosymmetry perpendicular to the membrane plane and all the structural elements that mediate the signal transduction process elicited by the neurotransmitter acetylcholine : the activation and desensitization of the ion channel (Changeux and Edelstein, 1998, 2005a,b).

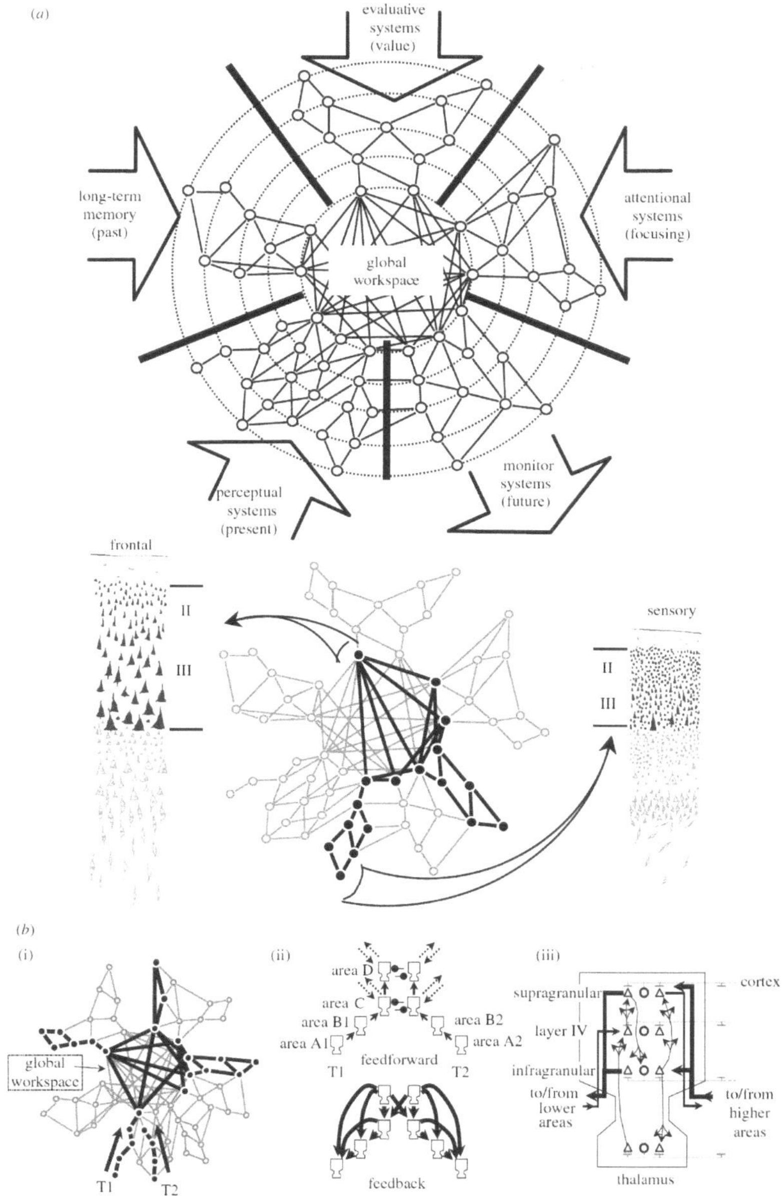

Fig. 6.1 The Neuronal Workspace Hypothesis: schematic representation of brain networks comprising multiple specialised processors and a central network of high level areas interconnecting them by neurons with long range axons. (From Dehaene *et al,* 1998, 2003).

Biochemical equilibrium studies, *in vitro* rapid kinetics and *in vivo* electrophysiological recordings *in situ* or from functional receptors reconstituted in frog oocytes from genetically engineered subunits (Changeux and Edelstein, 2005a,b), support the extension to the nAChR of the scheme proposed by Monod, Wyman and Changeux (1965) (MWC) for regulatory enzymes (Changeux, 1980; Changeux and Edelstein, 1998, 2005a,b). In short, the signal transduction processes of activation and desensitization are viewed as resulting from: (1) the spontaneous occurrence of the receptor protein under discrete conformations in reversible equilibrium; (2) the differential stabilization of one of these few conformations by the ligand for which it exhibits the highest affinity; (3) the conservation of the molecular symmetry of the receptor oligomer in the course of these transitions; (4) the differences in the states of activity of the discrete conformations involved: the active state is the only conformation where the channel is open, the desensitized state exhibits the highest affinity for agonists (and some antagonists); (5) the presence of multiple allosteric sites for physiological and/or pharmacological ligands (e.g. for local and general anaesthetics) through which ligand binding may differentially stabilize the conformations of the protein.

Determination of the X-ray crystal structure of the snail acetylcholine binding protein (an homolog of the extracellular domain of the nAChR (Brejc *et al*, 2001; Celi *et al*, 2004)) together with the cryo-electron microscopy data of the membrane domain collected on *Torpedo marmorata* muscle type nAChR (Unwin, 2000, 2005) have recently led to refined three-dimensional models of the receptor molecule (Taly *et al*, 2005; Unwin, 2005; Paas *et al*, 2005). These studies demonstrate that: (1) the ACh binding sites are far distant from each other (ca. 30Å), and (2) the signal transduction mechanism that links the neurotransmitter sites and the ion channel involves a global structural change of the quaternary and tertiary structure of the protein that unites the two distinct domains of each subunit. In the absence of X-ray structural data, the conformational transitions were explored *in silico* by a computer method referred to as normal mode analysis. The first mode (over the first 10) was found to produce a structural reorganisation of the model molecule compatible with channel gating. A wide opening of the channel pore resulted from the concerted symmetrical quaternary twist motion of the protein (Taly *et al*, 2005; Paas *et al*, 2005; Cheng *et al*, 2006). The data were found consistent, in first approximation, with the MWC two-state model of allosteric transition

Among the 17 genes encoding nAChR subunits that have been identified and cloned in mammals, nine (α2-α7 and β2-β4) are expressed in the brain. These subunits may co-assemble to form a wide variety of functional pentamers that are distributed throughout the brain with different pharmacological and electrophysiological properties. The most abundant combinations of subunits are: *α4β2 containing nAChRs* (α4β2*), which exhibit high affinity for acetylcholine and nicotine, do not bind α-bungarotoxin and are abundant in cerebral cortex and thalamus; *α3β4*, which is primarily present in the peripheral nervous system; and α7 homopentamers, which display a low affinity for ACh, tightly bind α-bungarotoxin, desensitize very quickly and are abundant in the limbic systems and cerebral cortex (Le Novère *et al,* 2002).

The existence of overlapping patterns of expression of these subunits has thus hampered the elucidation of the role of a particular subunit (or subunits) in brain functions until mice deleted for specific nAChR genes became available (Picciotto *et al*, 1995, Champtiaux and Changeux, 2004). In parallel, gene targeting and homologous recombination experiments were carried out to create mice carrying mutations in nAChR genes associated with different diseases in humans. Several of them are relevant to this review, since they cause idiopathic epilepsies (Steinlein *et al*, 1995) and, as we shall see, may affect states of consciousness (Steinlein, 2004; Champtiaux and Changeux, 2004).

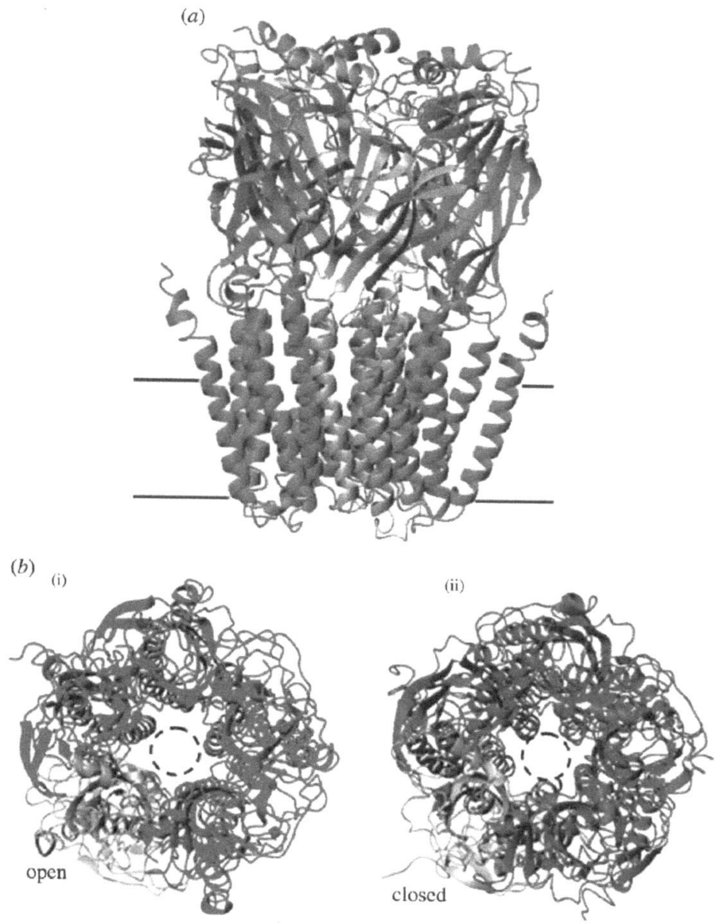

Fig. 6.2 The nicotinic acetylcholine receptor: a typical allosteric membrane protein. Reconstituted atomic model of the alpha 7 receptor in the resting and open channel states. (From Taly *et al*, 2005).

4. The integration of neuronal nicotinic receptors to the neuronal workspace architecture

The prefrontal areas of the neocortex are known to play a key role in higher cognitive functions (see Levin, 1992, 2002; Sacco *et al*, 2004) and were postulated to be essential components of the conscious neuronal workspace architecture (Dehaene *et al*, 1998; Dehaene and Changeux, 2004; Changeux, 2004; see Bianchi 1921). An eventual contribution of nAChRs to this network was investigated at the cellular and synaptic levels.

Early work on synaptosomal preparations identified presynaptic nAChRs *in vitro* (rev. Wonnacott *et al*, 1995). Yet, their cellular distribution and function in neuronal networks had not been established. In a first electrophysiological approach (Vidal and Changeux, 1989), extracellular recordings of field potentials and unit discharges were taken from the superficial layers of slice preparations of prelimbic area of the prefrontal cortex. They revealed excitatory effects of nicotine that were blocked by neuronal bungarotoxin and thus possibly involved $\alpha3\beta2$ or most likely $\alpha4\beta2$ nAChRs. In subsequent experiments, still on *in vitro* slices (Vidal and Changeux, 1993), intracellular recordings were made from pyramidal neurons located predominantly in layers II-III (where nicotine binding sites were found concentrated (Clarke, 1993)) and which belong to the long-axon neurons, subsequently postulated by the neuronal workspace hypothesis (Dehaene *et al*, 1998; Dehaene and Changeux, 2004). Iontophoretic application of nicotine near the recording site increased the amplitude of the postsynaptic potentials evoked by electrical stimulation of the superficial cortical layers. In contrast, muscarinic agonists decreased the amplitude of the postsynaptic response. In all instances, the early postsynaptic potentials were blocked by AMPA glutamate receptors antagonists (CNQX), but not by NMDA glutamate receptors blockers (APV). The facilitation of glutamatergic excitatory synapses by presynaptic nAChR has since then been confirmed in several laboratories (see McGehee and Role, 1996). Many neurons in the brain have also been found to generate fast inward currents upon application of nicotine to the somatodendritic compartment, in particular cells from the *locus coeruleus* (Léna *et al*, 1999), ventral tegmental area (VTA) (Klink *et al*, 2001; Champtiaux and Changeux, 2004), medial habenula and interpeduncular nucleus (Mulle and Changeux, 1990; Mulle *et al*, 1991). Last, inhibitory GABAergic neurons have also been found to express nAChR, in particular at the postsynaptic level, where they enhance synaptic inhibition (Klink *et al*, 2001; Maggi *et al*, 2003, 2004).

In conclusion, nAChRs are abundant throughout the brain, in the prefrontal cortex and dopaminergic neurons in particular, where they differentially enhance both excitatory and inhibitory transmission. nAChRs are thus present at the level of workspace neurons and reward neurons where they may possibly control the access to the workspace and the processing of conscious information.

In relation with this hypothesis, another unexpected target of nAChR was found on myelinated axons of the white matter. The discovery was casually made through whole cell recordings of acutely isolated neurons from the interpeduncular nucleus that had retained synaptic contacts from myelinated fibbers still attached to their cell body. As anticipated, nicotine dramatically increased the frequency of GABAergic postsynaptic currents on the IPN neurons. Yet, rather unexpectedly, the Na channel blocker tetrodotoxin (TTX) was found to block these effects. The specificity of TTX is strictly limited to voltage-sensitive Na channels. These channels are thus mobilised in the coupling between nAChR and the terminal GABA sites release. The nAChR involved, thus are *not* located on the nerve ending. The term "preterminal" was coined to qualify the occurrence of these nAChRs along the axon and the observation was confirmed by other groups (McMahon *et al*, 1994a,b; McGehee and Role, 1996). The only plausible localizations of these axonal nAChRs are the "nœud de Ranvier". Their precise role at this level is still not understood. As previously noted, the expansion of the neocortex and in particular of the prefrontal cortex in high mammals is accompanied by a massive increase of the white matter. A possible - but still highly speculative function - would be that they are the targets of as yet unidentified cholinergic terminals that would exert global control of the state of activity of the long axons that compose the white matter and regulate in a "paracrine" or "volume" manner the states of consciousness mediated by the global workspace!

5. Nicotinic receptors and states of consciousness in the mouse

In mammals, including mice and men, two cholinergic systems exert activating effects on thalamocortical and cortical neurons: first, the pedunculopontine and laterodorsal tegmental nuclei, that project to all thalamic nuclei but have no direct projection to the cortex and second, the nucleus basalis, that projects to the cortex and thalamic reticular nucleus. Acetylcholine release from both systems brings thalamic and cortical neurons

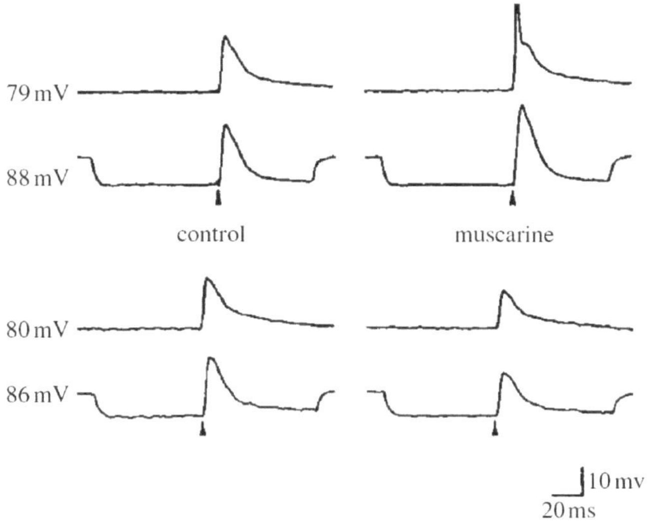

Fig. 6.3 Presence of functional nicotinic receptors in prefrontal cortex slices. Opposite effects of nicotine and muscarine on post-synaptic potentials in layer II/III neurons. (From Vidal and Changeux, 1993).

closer to their firing threshold ensuring fast responses to external stimuli during waking or to inner ones during REM sleep (Steriade, 2004). Acetylcholine excites pyramidal neurons through a prolonged depolarisation mediated by muscarinic receptors but also produces fast depolarisation of interneurons through both muscarinic and nicotinic receptors (ref. Jones, 2004). Moreover, nicotine intake has long been known to cause EEG desynchronisation (Yamamoto and Domino, 1965) and to increase arousal and attention (Lawrence *et al*, 2002; Levin, 2002). Also, the nicotinic antagonist mecamylamine causes drowsiness and decreases the performance in attention-demanding tasks (Pickworth *et al*, 1997). Last, acetylcholine release increases in the rat prefrontal cortex during the performance of a visual attentional task (Passetti *et al*, 2000).

In order to evaluate the specific contribution of nAChRs to the regulation of sleep and wakefulness, mice lacking the β2-subunit (β2-/-) (Picciotto *et al*, 1995, 1998) were investigated. The β2-/- mice are devoid of high affinity nicotinic binding sites in the brain (Picciotto *et al*, 1995; Zoli *et al*, 1998) and exhibit a loss or reduction of nicotine-elicited currents in neurons from various brain regions (Klink *et al*, 2001). Yet, they survive and reproduce without difficulties. Their respiration was monitored by plethysmography (Cohen *et al*, 2002, 2005) and their cerebral electrical activity recorded as polygraphic EEG or EMG (Léna *et al*, 2004).

In a first series of experiments, the respiration of adult β2-/- individuals was followed by plethysmography (Cohen *et al*, 2002). During sleep, the episodes of quiet sleep manifest themselves by regular large amplitude respiratory movements, which contrast with the irregular transient movements of wakefulness. Hypoxic challenge caused in β2-/- mice (compared to WT) an increased ventilatory response with a reduction of transient movements, revealing a contribution of β2*nAChR to awakening under these conditions (Cohen *et al*, 2002).

The plethysmography method was adapted to newborn mice and their breathing and arousal responses challenged following exposure to controlled doses of nicotine during pregnancy. Newborn WT pups exposed to nicotine exhibited unstable breathing and impaired arousal, deficits remarkably similar to those detected in pups lacking β2*nAChR. Chronic nicotine exposure thus, like the lack of β2*nAChR, impairs the arousal response to distress of newborn mice (Cohen *et al*, 2005). The newborn mice maternally exposed to nicotine thus offer a plausible model of Sudden Infant Death Syndrome (SIDS), which prevalence is increased several folds in smoking mothers.

The organisation of sleep was monitored by polygraphic recordings (EEG, EMG) in adult β2-/- vs WT (Léna *et al*, 2004). First nicotine (1-2 mg/kg ip.) produced an increase in the time spent in the awake state in WT mice and correlatively caused a rebound increase in sleep during the second hour after the injection. These two effects of nicotine were not observed in β2-/- mice. In contrast, under normal conditions, the β2-/- mice displayed the same amounts of waking, non-rapid eye movement (NREM) sleep and REM sleep as their WT counterparts. The endogenous physiological activation of β2*nAChR thus does not contribute critically to overall sleep regulation, but plays a discrete role in shaping sleep episodes. In particular, the β2-/- mice exhibited longer REMS episodes and a reduced fragmentation of NREMS by events referred to as "microarousals" characterized notably by a transient increase in EEG power and an EMG activation. The lower frequency of microarousals during NREMS observed in β2-/- mice resembles the deficit in putative arousals reported above after a hypoxic challenge (Cohen *et al,* 2002). These results indicate that nAChRs are involved in the phasic expression of arousal promoting mechanisms (Halàsz, 1998; Terzano and Parrino, 2000) observed in NREMS. They thus control transitions of brain circuits from the sleeping conscious state.

6. Nicotinic receptors and nocturnal frontal lobe epilepsies

Autosomal dominant nocturnal frontal lobe epilepsy (ADNFLE) was the first congenital epilepsy found in humans to segregate as a single gene disorder (Schaeffer *et al,* 1994) and to be caused by point mutations within the genes coding for the α4- and β2-subunits of the neuronal nAChR (Steinlein *et al,* 1995; De Fusco *et al,* 2000; Phillips *et al,* 2001). ADNFLE is characterized by nocturnal motor seizures, which tend to cluster, and often occur several times at night. Seizures are mainly occurring during non-REM sleep, either shortly after falling asleep, or in the early morning hours. The main age of onset is within the first or second decade of age and the frequency of nocturnal attacks tends to decrease with advanced age and even to disappear. The first ADNFLE mutation identified α4 Ser248 Phe (Steinlein *et al,* 1995) hits precisely the particular amino acid which was first identified by chlorpromazine in the *Torpedo* ion channel MII segment (Giraudat *et al,* 1986; Hucho *et al,* 1986). Several other ADNFLE mutations have been identified in α4- and β2-genes (Steinlein, 2004). Even though their physiological phenotype may vary from one mutation to the other, there seems to be a general agreement on their "gain of function" character (Révah *et al,* 1991; Bertrand *et al,* 1992; Devillers-Thiéry *et al,* 1992; Itier and Bertrand, 2002), a phenotype that is simply but not exclusively, accounted for by the MWC model (Changeux and Edelstein, 2005, see Discussion).

Interestingly, the seizures tend to occur in a stage of light sleep where intense cyclic micro-arousals activity takes place (Terzano and Parrino, 2000) and in some ADNFLE cases, a direct temporal correlation was found between microarousals and epileptic manifestations (Zucconi and Ferini-Strambi, 2000).

Mouse models of several of the ADNFLE mutations have been created for α4 (but also for another α7-) subunits (Labarca *et al,* 2001; Broide *et al,* 2002; Gil *et al,* 2002) by *knock-in* technology. Mice homozygous for the α7-L270 T mutation die shortly after birth, while heterozygous animals are viable but more sensitive to the convulsant effects of nicotine (Broide *et al,* 2002; Gil *et al,* 2002). The α4 homologs of known human ADNFLE mutations have been produced and display enhanced susceptibility to epileptic seizures (Tapper *et al,* 2004; Rodrigues-Pinguet *et al,* 2005)..

In conclusion, alteration of the allosteric transitions of the nAChR may cause the production of epileptic seizures. Discrete molecular properties of a receptor molecule thus "gate", in a bottom-up manner, the transitions between distinct brain states known to affect the subject's states of consciousness.

7. Nicotinic receptors and the content of consciousness

Already in the early 90's, a rich body of pharmacological and behavioural observations underlined the importance of cholinergic pathways and their nicotinic component in working memory and attentional processes that rely on the contribution of the prefrontal cortex (ref. Levin, 1992; Nordberg and Winblad, 1986). To evaluate the role of nAChRs in these cognitive - we might even say today "conscious"- tasks, the effect of nAChR blockers was tested in the rat on delayed-response tasks (Granon *et al*, 1994, 1995). The performance on two distinct tasks was compared. First, a delayed-non-matching-to-sample task (NMTS) was used as control. The tasks depend on the tendency of the animal to spontaneously alternate, on a second run, to the branch *it did not visit* on the first time. This spontaneous "instinctive" behaviour makes sense on an evolutionnary basis since the rat will not visit a place where it already consumed the food reward. As a consequence, the delayed-matching to sample tasks (MTS), which consists in visiting the arm where the food reward was, is an effortful, counter–instinctive, task. As anticipated, lesions of the prefrontal cortex selectively impaired the performance of the MTS task, caused errors and perseverations, while preserving the performance of the NMTS tasks (Granon *et al*, 1994). Interestingly, injection of neuronal bungarotoxin into the prelimbic area of the prefrontal cortex produced a significant performance deficit in the MTS, but not in the NMTS. In contrast, an inhibitor of the muscarinic receptor, scopolamine, impaired performance in both tasks. nAChRs in the prefrontal cortex are thus necessary to the successful performance of the delayed-response tasks requiring effortful processing for response selection (Granon *et al*, 1995). Is this a sufficient demonstration of the Turing test for consciousness?

Mice do not pass MTS vs NMTS tasks as easily as rats possibly because, as we shall see, of their compulsive tendency to explore. The specific contribution of particular species of nAChR to such complex cognitive functions was nevertheless examined in mutant vs WT mice (Picciotto *et al*, 1995, 1998; Granon *et al*, 2003). The β2-/-mice were selected because both high affinity nicotine binding sites and electrophysiological responses to nicotine were almost completely absent from the brain (Picciotto *et al*, 1995). The Morris watermaze evaluate spatial orientation learning. The performance of β2-/- mice in the test did not differ from that of WT mice when tested both on the visible platform task and on the hidden one, indicating an intact spatial memory (Picciotto *et al*, 1995).

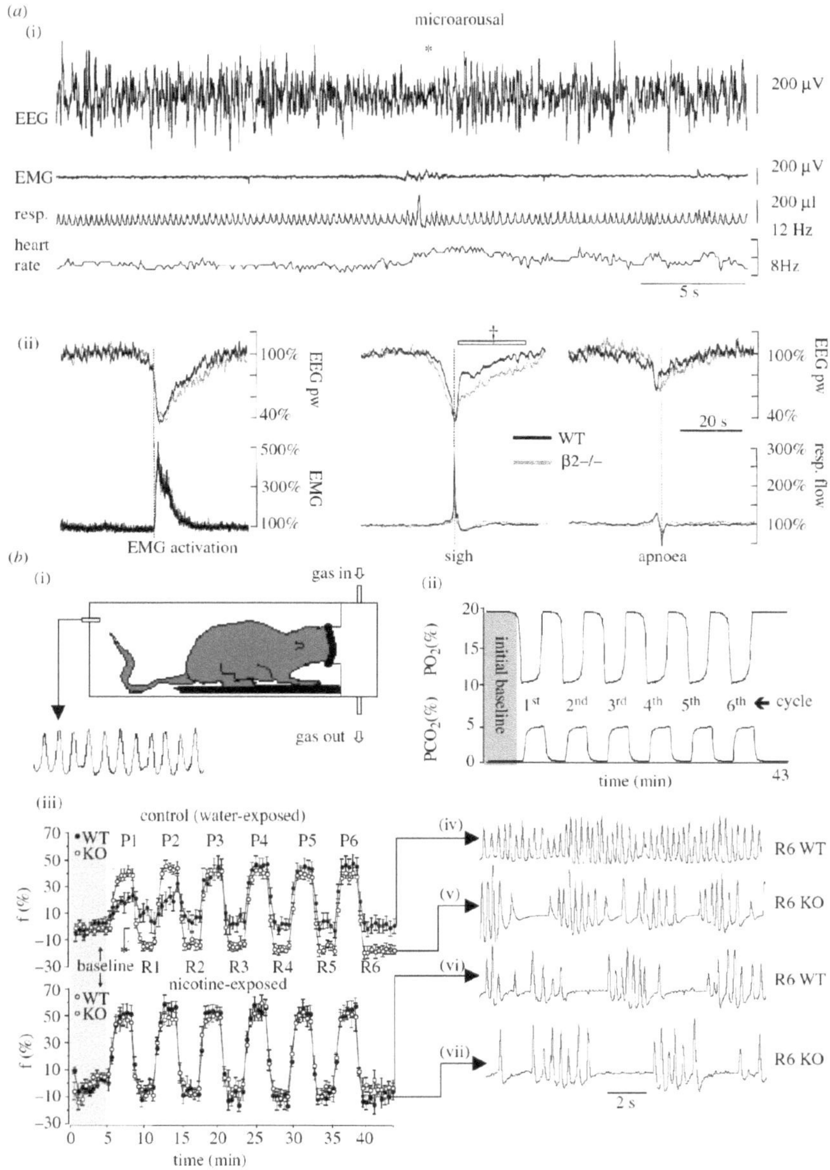

Fig. 6.4 Nicotinic receptors and states of vigilance: consequences of the deletion of the beta 2 subunit on microarousals (from Lena *et al*, 2004) and on the respiratory arousal of newborn mice chronically exposed *in utero* to nicotine (nicotine exposure destabilizes breathing). (From Cohen *et al*, 2005).

Retention of an inhibitory avoidance response was then assessed because of its known sensitivity to nicotine administration. The mouse was placed in a well lit chamber of a shuttle box, and the latency to enter the adjacent dark chamber measured. Upon entry to the dark chamber, a mild inescapable foot shock was delivered and nicotine (vs saline) injected into the mouse. After 24 hours, retention was assessed by measuring the latency to enter the dark chamber. Interestingly, treatment with nicotine after foot shock consistently aided retention and nicotine administration was completely ineffective in β2-/- mice. Control experiments confirmed that the β2-/- mice did not differ from their non mutant siblings by sensory sensitivity, pain threshold or increased emotionality (see also Granon *et al*, 2003 and Granon, unpublished). The effect of nicotine on memory retention may only indirectly concern the issue of consciousness in the mouse, yet, an unexpected though relevant observation was made. Retention latency was significantly longer for mutant mice than for their non-mutant injected siblings (Picciotto *et al*, 1995).

The learning abilities of the β2-/- vs WT mice were further investigated in a maze equiped with four dissociable arms. Animals learned to reach food from any location in a given configuration of the arms. The configuration of the arms was then changed and the starting point as well, without changing the position of the goal arm in the absolute space. Yet, here as well, quite unexpectedly, β2-/- mice learned the task more rapidly than WT mice (Granon *et al*, 2003). Explanation of this paradoxical behaviour noted in both the multiple configuration maze and the avoidance-learning task (Picciotto *et al*, 1995) (which, in addition, becomes amplified with age (Zoli *et al*, 1999; Caldarone *et al*, 2000)), came out from a closer analysis of the organisation behaviour in β2-/- vs WT mice.

Exploratory activity is a spontaneous behaviour that does not involve explicit reinforcement (Renner, 1990; Thinus-Blanc, 1996; Poucet and Hermann, 2001) (even though it mobilizes, as we shall see, endogenous reward processes). It serves to gather and store spatial information which allows allocentric coding of space, itself, necessary for flexible *navigational* processes. Mice were placed in the centre of an empty arena and their spontaneous trajectories recorded with a camera fixed to the ceiling and connected to a video track system. The system was set up to break down, each trajectory into navigation (large movements at fast speeds 14.4 cm/sec), fast exploration (small movements at a speed between 6.8 and 14.4 cm/sec), and slow exploration (small movements at a speed ≤ 6.8 cm/sec). A complementary "automatic" analysis by "symbolic" quantification of trajectories (Faure *et al*, 2003) was carried out in parallel. It included the definition of the position of the animal, its two-dimensional path and its instantaneous velocity. Statistical analysis revealed a modification in the organisation of displacements in the β2-/- mice. The balance between navigation and exploration was shifted in the β2-/- mice in favour of navigation to the detriment of more precise exploration of the environment. The symbolic analysis further revealed a reduction of the trajectories in the centre of the arena for the β2-/- mice, due to high speed crossing without slowing down in the centre. In other words, the β2-/- mice were more rigid and exhibited less behavioural flexibility than the WT (Granon *et al*, 2003, Maskos *et al*, 2005) and the apparent faster learning in the above mentioned situations simply results from the absence of time-consuming exploration of the apparatus used.

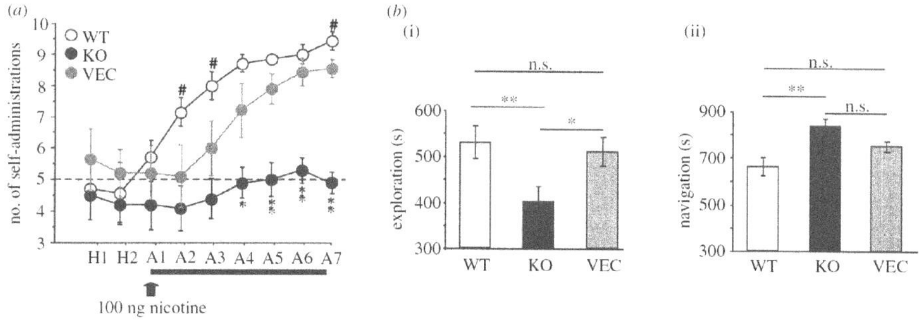

Fig. 6.5 Nicotinic receptors and content of consciousness : consequences of the deletion of the beta 2 subunit and of its stereotaxic re-expression in the Ventral Tegmental Area on self-administration of nicotine and on the navigatory behaviour of the mouse. VEC, re-expressed b2 subunit. (From Maskos *et al*, 2005).

These aspects were further evaluated by placing the mice in more sophisticated environments. First, a single object was presented in the arena. No difference of exploratory behaviour was noted indicating no gross sensory or memory impairment in the recognition process. But, when two objects were placed in the environment, the β2-/- mice exhibited an exploratory behaviour different from WT mice. Symbolic analysis revealed a striking reorganisation of the transitions between sequences of action in the WT, that was absent in the β2-/- mice. WT mice increased the number of trajectories between the two objects while β2-/- mice did not; in other words, they display a spontaneous flexibility of behaviour, an increased *curiosity* (Denton, 2005), that requires the presence of β2*nAChR (Granon *et al*, 2003). Last, the cognitive behaviour of the β2-/- mice was analyzed further in conflict-behaviour paradigms that require strategic choice. Two situations were considered. In a first series of experiments, the multiple arms maze was used and the same learning protocol, except that objects were inserted in the maze as attentional distractors. WT type mice reacted to novelty by increasing exploratory activity, whereas, again, β2-/- mice did not show this adaptative flexibility to a change in the environment.

In a last series of experiments, the sequences of so-called "social" interactions were examined between a test resident mouse and a social intruder. In short, among the diverse types of interactions tested, the β2-/- mice showed significantly higher rates of approach behaviours, and lower rates of escape behaviours. In other words, the β2-/- mice display more "automatic" interactions: faster rates of approach and impaired capacity for interrupting ongoing behaviour (Granon *et al*, 2003).

Lesion experiments (Granon, unpublished) reveal that mice with prefrontal lesions do not show significant alteration in spatial learning but display evident deficit in conflict-resolution situations in the maze paradigm with an unexpected object. The prefrontal lesion creates a behavioural phenotype which displays several features in common with the loss of β2*nAChR, which affects behavioural flexibility and ability to hierarchize competing motivations.

Along these lines, related through different experimental paradigms referred to as delay- vs trace-conditioning (Clark *et al*, 2002) were applied to the mouse by another group, yet without the above pharmacological and genetic analysis (Han *et al*, 2003). They will be mentioned here for methodological reasons. In *trace* conditioning, a *time gap* was introduced between the end of the conditioned stimulus (such as a tone) and the unconditioned stimulus (like a foot shock), while in *delay* fear conditioning, the conditioned stimulus was immediately followed by the unconditioned stimulus. A visual distraction of flash of light was introduced at randomly selected intervals. Interestingly, the distraction interfered selectively with trace- but not delay- conditioning. Trace conditioning, which is thought as requiring conscious awareness, was expectedly found associated with increased neuronal activity in the anterior cingulate cortex and was selectively impaired by lesion of this structure in the medial prefrontal territories at variance with delay-conditioning (Han *et al,*, 2003). Interestingly, similar experiments have been carried out in humans in whom an explicit requirement for attentive awareness was demonstrated for the trace-, but not delay-conditioning task (Carter *et al*, 2003).

Moreover, the phenotypes observed with the β2-/- mice and characterized by alterations of behavioural flexibility and adaptative behaviours coupled with unimpaired memory and anxiety may reproduce the cognitive impairments observed human diseases (Granon *et al*, 2003) for instance, in attention deficit hyperactivity disorders (ADHD) (Granon and Changeux 2006 in press) and, even, in autism (Frith, 2001; Andres, 2002; Booth *et al*, 2003). Both groups of patients show a reduced activation of prefrontal and cingulate cortices (Frith 2001; Giedd *et al*, 2001) and autistic children further show a substantial decrease of α4β2-containing nAChR in their prefrontal cortex (Perry *et al*, 2001). The β2-/- mice thus might represent useful animal model for the study of ADHD and autism (Granon *et al*, 2003; Granon and Changeux, 2006; but see Ch 8).

In conclusion, mice under these experimental conditions do far more than to simply react to sensory information. They engage in complex extended behaviours geared towards far removed goals. They use processes that override or differentially select routines (Thinus-Blanc, 1996) and, according to Denton (2005), orchestrate locomotor behaviours according to "conscious intentions" a behaviour consonant with, though far more elaborate, than Koch's Turing test for consciousness. Further studies are of course needed to further implement, I would even say, justify, these conclusions (see Section 8).

8. The joint recovery of exploratory behaviour and reward function by targeted re-expression of nAChR subunit

To further understand the specific role of nAChR subunits in mediating the effects of nicotine and endogenous acetylcholine on higher cognitive functions in the mouse, a method of selective re-expression of a nAChR-subunit by stereotaxic injection of a lentiviral vector in a defined territory of the brain of a KO-mouse was developed (Maskos *et al*, 2005). Lentivirus based expression systems initially developed for gene therapy purposes (Naldini *et al,* 1996), provide several advantages over other virus-based *in vivo* transgene expression strategies. In particular, they are capable of stable integration into the genome of non-dividing cells, such as neurons, facilitating the potential for long-term, stable transgene expression. The lentiviral expression vector was derived from the pHR' expression vectors of Naldini *et al* (1996), yet with several modifications: it contained a bi-cistronic cassette simultaneously expressing the β2-subunit and the enhanced Green Fluorescent Protein (eGFP) facilitating efficient detection of transduced cells (Maskos *et al,* 2002). The vector was stereotaxically injected into the ventral tegmental area (VTA) of mice carrying β2-subunit deletions. The ligand-binding nAChR protein was found quantitatively re-expressed in dopamine-containing (DA) neurons of the VTA using epibatidine as a ligand. Its functionality was assessed by recording in vivo the effect of nicotine on the electrophysiological activity of DA neurons in the VTA. The well-established rapid 1.5-fold increase in firing frequency upon intravenous injection of nicotine, which disappeared in the β2-/-mice, was recovered. However, instead of lasting on average nearly 10 min as in wild type, this effect did not persist for more than 2 min. These data show that the re-expression of the β2-subunit exclusively in the VTA *suffices* to recover an effect of nicotine on DA neurons. Furthermore, using *in vivo* intracerebral microdialysis in awake, freely moving mice, nicotine-elicited dopamine release from the nucleus accumbeus (NuAcc) was found completely restored by the re-expressed β2-subunit in the β2-/- mice.

To assess whether the nicotine-elicited responses observed in the VTA and NuAcc following reexpression of the β2-subunit were sufficient to support nicotine reinforcement, an intra-VTA nicotine self-administration paradigm in the mouse, as described for morphine self-administration (David and Cazala, 1994), was used. Wild type mice exhibited a clear nicotine-seeking and self-administration behaviour, which disappeared in the β2-/- mice. Interestingly, the lentivirus injected mice recovered intra-VTA nicotine self-administration behaviour and improved their performance over learning sessions, yet they displayed a delay in the acquisition of self-administration and their discrimination performance was slightly lower than that observed for wild type mice. Therefore, re-expression of β2*-nAChR in the VTA is not only *necessary* but also *sufficient* to re-establish sensitivity to nicotine reward in drug-naive mice.

Having established that VTA re-expressed β2*-nAChR respond to nicotine *in vivo*, the spontaneous exploratory behaviour of the injected mice was examined. As mentioned, β2-/- mice exhibit modified spatio-temporal organization of displacements, with increased *navigatory* and decreased *exploratory* behaviour (Granon *et al,* 2003). Interestingly, the lentivirus injected mice showed a selective restoration of exploration, bringing this measure up to the level of wild type without a significant modification of navigation. In addition, they exhibited transitions between fast and slow movements in the central portion of the arena that were similar to wild type. The targeted expression of β2*-nAChR in the VTA thus generated, at least, a partial *dissociation* between exploratory and navigatory behaviour.

These results demonstrate that nAChRs in neurons originating in the VTA and/or their axonal projections suffice for the differential restoration of the cognitive executive function as studied in our paradigm. This behaviour therefore appears to be mediated by endogenous acetylcholine acting on β2*-nAChR expressed on VTA neurons and their axonal projections, in particular to the NuAcc. It suggests the implication of dopamine as the endogenous neuro-modulatory substance for this executive function.

In conclusion, this study shows that re-expression of functional β2*-nAChRs specifically in the VTA and its axonal projections suffices to simultaneously restore the self-administration of nicotine and the ability to perform an executive behaviour under endogenous cholinergic modulation, thus bringing a first experimental demonstration of the proposed theoretical mechanisms implying a causal link between cognitive learning, motivation and reward processes (Dehaene and Changeux, 1998, 2001; see also Damasio 1995). Moreover, it underlines a rather unclassical, yet predicted (Dehaene *et al*, 1998), but still hypothetical connection between reward neurons and consciousness.

9. Is the mouse a useful animal model to investigate consciousness?

The issue of mouse consciousness is not simply philosophical (Block, 1990). The obvious limitation of the mouse model in the scientific study of consciousness is its reduction to a minimal level and thus of the impossibility to investigate some of its most characteristic human features like self-recognition and reflective consciousness. On the other hand, as we saw, it gives access to a large spectrum of methods, in particular to recombinant DNA technologies (gene deletion, gene exchange and targeted gene expression), which provide the means to explore, at the molecular, cellular, neuronal networks and behavioural levels what I referred to as the *fundamentals* of minimal consciousness. In this lecture, data on the contribution of neuronal nAChRs to these fundamentals have been reviewed. The main conclusions may be summarized as follows:

9.1 States of consciousness: sleep & arousal, epilepsy, general anaesthesia

Multiple *states of consciousness* (such as wakefulness, sleep, coma, general anaesthesia, epileptic seizures…) and the regulation of their reversible transitions occur in the mouse as in all mammals, including humans; the circadian sleep-waking cycle is controlled in these species by brain stem reticular formation and intralaminar nuclei of the thalamus (Bogen, 1995; Jones E, 1998; Jones B, 1998) with complex patterned releases of neuromodulatory substances. The use of mice deleted for the β2-subunit gene gave us the opportunity to demonstrate the positive contribution of β2*nAChR to the phasic expression of arousal promoting mechanisms by endogenous ACh (Cohen *et al*, 2002; Léna *et al*, 2004). Similar phenotypes as those noticed in β2-/- mice are observed after chronic intra utero exposure of the foetus to nicotine. These mice thus offer in addition a plausible animal model of Sudden Infant Death Syndrome which prevalence is known to increase in smoking pregnant women (Cohen *et al*, 2005).

Furthermore, as discussed, knock-in mice where point mutations were introduced in the α4-subnit gene by homologous recombination may constitute relevant though still primitive animal models for human familial frontal lobe epilepsies (Labarca *et al*, 2001; Steinlein, 2004). Interestingly, these mutations cause rather paradoxical *"gain of function"*, pleiotropic, phenotypes characterized by higher affinity of ACh and nicotine, loss of desensitization and conversion of the response to some nicotinic antagonists into an activating response (Révah *et al*, 1991; Bertrand *et al*, 1992; Devillers-Thiéry *et al*, 1992). These complex phenotypes are simply accounted for by the Monod-Wyman-Changeux mechanism for signal transduction. Homologous gain of function mutations were also identified as responsible for congenital myasthenic syndromes (Engel *et al*, 2004) and as well viewed as either differentially changing the intrinsic properties of the ion channel in the different conformation of the nAChR or altering the kinetic parameters of the transition from the resting to the active open channel conformation (Edelstein *et al*, 1997; Changeux and Edelstein, 2005a,b). These observations unambiguously demonstrate that the allosteric properties of the nicotinic receptor protein may gate, when altered, the transition from a state of slow wave sleep to an epileptic state of consciousness. In a more general manner, bottom up causal determination may exist between the molecular dynamics of a critical receptor protein and the transitions between states of consciousness.

Last, nAChRs are inhibited by some *volatile anaesthetics* and ketamine (Tassonyi *et al*, 2002). On muscle nAChR, the latter act mostly as open channel blockers but on neuronal nAChRs may also bind to allosteric sites located outside the ion channel where they inhibit activation and/or enhance desensitization. The direct role of brain nAChR ($\alpha4\beta2^*$) as targets of general anaesthetics to cause losses of consciousness has been challenged (see Tassonyi *et al*, 2002). On the other hand, their potential contribution to the analgesic component of general anaesthesia has received support from both pharmacological and *in vivo* animal model experiments (Cordero-Erausquin *et al*, 2000; Cordero-Erausquin and Changeux, 2001), even if $GABA_A$ receptors are thought to be a primary target for general anaesthetics. The mouse and its mutants may thus become convenient animal models to investigate the pharmacology of general anaesthesia. In many instance, both the chemistry and molecular biology of the various states of consciousness are currently under investigations.

9.2 Content of consciousness: behavioural flexibility and the rudiments of "intentionality" in the mouse?

Classically, in the waking conscious state, a "graded *global integration* of multiple cognitive functions is assumed to take place that yields an unified representation of the world, of the body and of the self" (modified from Hobson, 1999, see also Changeux, 1983, 2002; Edelman, 1989, 2004). This notion, which deals now with the *content* of *consciousness*, has been approached, still in a rather rudimentary form, by the qualitative and quantitative analysis of a spontaneous behaviour displayed by mice and rats and referred to as exploratory behaviour (Granon *et al*, 2003; Maskos *et al*, 2005). These behaviours were up to now investigated in the context of animal spatial cognition (Renner, 1990; Thinus-Blanc, 1996; Poucet and Hermann, 2001) but without explicit reference to consciousness. I will take the risk to discuss here the notion that they might be plausibly relevant to the study of what I referred to as *minimal* or "level 1" (Barresi and Moore, 1996) *consciousness*. Exploratory behaviour expresses, in a way or another, a reaction to novelty. It may also, to some extent, implement the spontaneous (Dehaene and Changeux, 2005) or basal activity of the brain (Gusnard and Raichle, 2001). A mouse or a rat placed in a new environment displays intense investigatory activity:

contact with objects, sniffing, stopping, rearing up in hind-paws, and so on. In humans, young children for example display a need to seize and touch any unfamiliar object and extensively visit their environment as soon as they can crawl and walk. Exploratory activity, in general, decreases overtime – it habituates – but it also dishabituates when some spatial relationships between the elements of the situation are modified. It is currently assumed that the organism detects spatial novelty by comparing the actual perceived arrangement and a representation or stored "internal model" of the initial situation (Thinus-Blanc, 1996). Intensive studies of the neurobiological bases of exploratory behaviour are consistent with a multistage processing of spatial inputs: from the visual cortices for the perception of local views, the associative parietal cortex for the extraction of spatial invariants, the prefrontal cortex for holding the information to plan and execute a trajectory and the hippocampus as a comparator (or detector) of mismatches between stored representations and currently perceived situations (Poucet, 1993). In a more general manner, exploratory activity, as its habituation and deshabituation in reaction to spatial changes, are viewed as relying on *multisensory* perception of the environment matched with movement-related information or, in other words, as an "active information seeking structure" (Neisser, 1976), in constant up-dating of its representations, a manifestation of *behavioural flexibility* (Thinus-Blanc, 1996; Granon *et al*, 2003). In its very nature, exploratory behaviour would then be organized along an *egocentric* reference frame (e.g. go ahead, turn left) and sequentially ordered (go ahead, then turn left) bearing of the goal determine on the basis of "abstract" *exocentric* representation of the physical world (see also Berthoz, 2003). In a broader philosophical context, exploratory behaviour might be tentatively considered as a first sign of "intentionality" in a simple animal species (for a discussion of classical literature see Boakes, 1984).

Indeed, according to the Viennese philosopher Brentano, intentionality would be the signature of consciousness as being the tractable feature of recognizing a difference between one's own thoughts and the sensory information coming from the outside. In other words, intentional representations would differ from other purposive thoughts where the aim is clearly in view. Koch (2004), as mentioned earlier, has suggested as a Turing test for conscious behaviour in animals the ability to maintain information through a delay period. In light of what was mentioned about exploratory behaviour, one may further say that, with the course of exploration, as well as in the case of trace conditioning (Clark *et al*, 2002; Han *et al*, 2005), the organism has to keep the information "on line" through delays, but also to exploit delays to elaborate, select and evaluate plans, organize a program of actions, that would ultimately let it respond in a natural environment to inner motivations as, for instance, Denton (2005) primordial emotions. Accordingly, the so-called minimal Turing test should be re-formulated as "the maintenance of information through delay *plus* selection of a *relevant plan*" (Changeux, 2005).

Denton (2005) has underlined that, in the wild, simple species like mice, and sometimes, under special circumstances, human beings themselves might be under the "totalitarian occupancy" of primordial emotions that signals that their very existence is threatened. These primordial emotions basically organize the primitive functions necessary for survival: hunger, thirst, compulsion to sleep or to find a sex partner, which, in a natural environment, requires constant efforts of attentive awareness. At some stage, the drive caused by some physical impairment or imperious sensation – for instance thirst –

may create some organized effort to achieve a "particular purpose", an intentional behaviour – for instance satisfy thirst. But, taking into account stored past memories, such as the map of the environment where for example a source of water would be located, or confronted with the unpredicted sight of a predator, the organism may have to display behavioural *flexibility*, define *action priorities* which match internal motivations and actual environment conditions to reach some kind of "mental synthesis" and take the decision for instance to flee (or to keep going). For the evolutionary biologist as for the philosopher, the behaviour of the "simple-minded" mouse may thus become an outstanding topic to approach the neurobiological bases as well as the origins of minimal consciousness. At least the question is raised!

More specifically, the observations that, in the mouse, the exploratory behaviour may be dissociated from the more automatic navigatory behaviour by the deletion of the ß2-subunit gene of the nAChR (Granon *et al*, 2003; Maskos *et al*, 2005) illustrates: (1) the specialisation of neural circuits engaged in such executive "conscious" functions; (2) the "gating" of these functions (which mobilize in particular the prefrontal cortex) by nAChRs activated through endogenously released ACh; last, (3) the joint recovery of exploratory behaviour and reward functions by targeted re-expression of the ß2-subunit in the VTA dopaminergic nucleus. This important result points to an issue, insufficiently recognized in altogether, behavioural, psychological and philosophical studies on cognition in general and consciousness in particular, that a fundamental connection may exist between the activity of reward processes and the selection of plans, the maintenance of a goal, or in a philosophical and even more speculative manner, the content of consciousness in an intentional behaviour.

On the clinical side, these conclusions may also lead to a plausible interpretation of drug (nicotine) addiction. It may be viewed as an escape from the "voluntary" control of drug taking behaviour for instance as a consequence of the disconnection of a reciprocal-loop linking prefrontal cortex, dopaminergic neurons and striatum, thus uncovering the compulsive non-conscious aspect of drug addiction (for discussion see Gutkin *et al*, 2006). Moreover, as noted earlier, the ß2 -/-mice which are compulsively navigating without pausing for exploration may offer an animal model for human ADHD behaviour which hyperactivity symptoms are known to improve with nicotine treatment (Shytle *et al*, 2002; Granon and Changeux, 2006).

9.3 Social relationships in the mouse?

Behavioural scientists together with philosophers have primarily considered intentionality in the framework of social relationships. Social organisms, including humans, represent intentional relations of themselves and other agents, yet, at different levels. They, unambiguously distinguish their own intentional relations (or first person information) from the qualitatively different information available about other agent's intentional relations (or third person information).

In this respect, one should remember that the analysis of the exploratory behaviour leads to the distinction between allocentric and exocentric motor behaviour (Thinus-Blanc, 1996) pointing to the still highly speculative occurrence of a "self" in the mouse, yet oriented towards the outer physical world. Furthermore, the so-called "social relationship" we examined in wild type and mutant mice were reduced to rather crude sequences of interactions between a resident mouse and a social intruder of the same sex. Higher rates of approach behaviour and lower rates of escape behaviour were noticed in ß2-/-mice compared to wild type, as if again the ß2-/-mice were impaired in their behavioural flexibility for interrupting ongoing behaviours (Granon *et al*, 2003). No evidence was found, at this stage, with the mouse, for comparability between the actions of self and others; no sign of imitating goal directed activity, nor of "understanding" the view point of others (Barresi and Moore, 1996). In other words, the presently available evidence does not support the occurrence of authentic social relationships in the mouse. The human infant at birth seems already at a stage more advanced than the adult mouse in this respect. Indeed, as mentioned earlier, he or she may already distinguish between his own and other's movements, in particular by touch and he/she displays rudiments of imitations (Metzoff and Gopnik, 1993, but Barresi and Moore, 1996; Lagercrantz, 2005). Thus, the mouse cannot be a good animal model to investigate intentional relations and social understanding which highest level is reached exclusively in humans with the theory of mind. On the other hand, it may serve as baseline to define the neural circuits mobilized by these intentional relationships in higher mammals and humans. It may thus hardly become a reliable model of autism (at variance with what was suggested before). If autistic children show a strong need for stereotyped routines in daily life which evokes the dominant navigatory behaviour of the ß2-/-mice, their most characteristic deficits concern the understanding of other minds (Baron-Cohen, 1991; Leslie and Frith, 1990). They do not show social pretend play, which in normal children develop during the first two years of life (Sigman and Ungerer, 1984), neither do they express joint attention and adequate imitation. They fail standard theory of mind tasks. According to Barresi and Moore (1996) they would never attain an understanding of intentional relations at the level 2, "recursive consciousness", level. In other words, following this reasoning the wild type mouse would already constitute some kind of animal model analogue of the autistic child.

9.4 The neuronal workspace hypothesis, blind-sight and the laboratory mouse?

As discussed by Baars (2001) and Dehaene and Changeux (2003), in humans the standard observational index of consciousness is an accurate, verifiable report by pressing a button or any other voluntary response mediated or not by language. Reporting responses have also been used with some animal species. It is, for instance, the "commentary key" report method developed by Weiskrantz (1991) and Cowey and Stoerig (1995) to demonstrate blind-sight in macaque by forced-choice responses. Yet, a still unanswered question is whether or not reportability would be demonstrated in simpler species like the mouse. Since it does not require any kind of social relationship, it looks plausible but a reliable assay still needs to be invented. On the other hand, if the neuronal workspace hypothesis was initially designed to account for access to consciousness and reportability, it obviously deals with important features of minimal consciousness.

It accounts for instance for the active maintenance of "abstract rules" through top down amplification and the flexible control of tasks that require a novel interconnection of existing processors as it typically occurs in the Stroop task (Dehaene and Changeux, 2004). It deals, *a fortiori*, with the active maintenance of information during a delay period and, thus, with the "Turing test" for conscious behaviour of Koch (2004) (see Han *et al*, 2003). Even though the relevant simulations have not been carried out, the neuronal workspace architecture would adequately fit exploratory behaviour and offer an appropriate mechanism for the ultimate stage of spatial processing as introduced by Poucet's model (1993) where unified location-independent representation with one unique reference direction are being built.

The model also offers simple explanation for the observed "disconnection" between exploratory and navigatory behaviour. It may, for instance, result from a deficit of the VTA-Nacc pathway activation, within the more elaborate circuit linking this system to the "prefrontal workspace". Such an interpretation may hold as well for autism as long as it, primarily, affects primary consciousness and secondarily social understanding. Alternatively, as suggested (Changeux and Edelstein, 2005), it may affect the development of the prefrontal workspace itself. The formation of the long-range axonal projection of the supragranular pyramidal neurons might be highly vulnerable to the process of epigenetic selective stabilisation of synapses (Changeux and Danchin, 1976) as found with the visual pathways (Rossi *et al*, 2001; Grubb *et al*, 2003; Mrsic-Flogel *et al*, 2005). This interpretation may fit for the brain phenotype of the adult β2 -/- mouse but also, still in a speculative manner, for that of mentally retarded children, including autistic children (but see before), which postnatal cortical development would be altered by mutations affecting synapse formation and/or stabilisation (X-fragile; Mandel and Biancalana, 2004), autism and Retts syndrome (Neul and Zoghbi, 2004)).

Functional brain imaging data from the mouse and its mutants are still at the stage of being developed (Suarez and Granon, unpublished). In humans and monkeys, the available of MRI and PET data are consistent with the observed link between cortical areas which are found active in conscious effortful tasks and areas particularly rich in workspace neurons with long distance connections (i.e. prefrontal, parietal and superotemporal, cingulate entries) (Goldman-Rakic, 1988; Cohen *et al*, 1997; Pardo *et al*, 1990; Pars *et al*, 1998). These areas show the most intense and consistent spontaneous activity in the awake state (Gusnard and Raichle, 2001; Mazoyer *et al*, 2001) and the greatest drop in metabolism during various types of transitions away from the awake state, whether during anaesthesia, sleep, coma or vegetative state (Shulman *et al*, 2003; Fiset *et al*, 1999; Laurey *et al*, 2000; Maquet *et al*, 1998; Bulkin *et al*, 2002; Paus, 2000). These areas also show the greatest contrast between conscious vs non-conscious processing of visual stimuli (Dehaene *et al*, 2001; Dehaene and Naccache, 2001; Marois *et al*, 2004).

These studies should thus offer the opportunity to further dissect, already with the mouse, the above-mentioned multiplicity of states and of levels in consciousness. This might be of considerable help for the cognitive neuroscientist and the philosopher to clarify situations or experimental paradigms in which the word consciousness is used with rather different meanings (for instance Block, 2005; Dehaene and Changeux, 2004; Fiset *et al*, 1999; Lamme, 2003; Paus, 2000; Zéki, 2003). The use of neurocomputational models might, in this respect, be of some help (Dehaene and Changeux, 2004, 2005). Indeed, one may view the various graded *states* of consciousness (from deep anaesthesia and coma to full awareness) as directly related to the spontaneous activity of recurrent thalamo-cortical loops and reticular thalamic nuclei (see Llinas and Paré, 1991; Steriade *et al*, 1993) under brain stem control. Simulations help to characterize several states of activity (Dehaene and Changeux, 2005): (1) one of *quiet vigilance* with moderate spontaneous activity; I may tentatively suggest - at my own risks - that it suffices for the access of Zéki's levels of "micro- and macro-consciousness" for colour perception or visual motion; (2) another state that one may refer to as full conscious access occurs at higher levels of the neuromodulation parameter, where spontaneous "ignited" states appear originating from higher association cortices and where sensory inputs become self amplified above a threshold; again, it may account for access to Zéki's "unified consciousness". On the other hand, the model predicts that the *content* of consciousness builds up from the processing of both "exogenous" sensory signal and "endogenous" spontaneous activity as long as they give rise to above threshold ignition; in other words, considerable processing may remain below access to consciousness. This is expected to be the case for Zéki's "micro- and macro-levels" for processing of visual perception as well as for all processes of exploratory behaviour, which in Poucet's model do not reach the prefrontal cortex. These different states or levels in the wild type mouse, which would therefore display electrophysiological signals of ignition in its neuronal workspace under conditions of exploration while these ignition signals would disappear in the β2 -/- mice. In other words, the β2 -/- mice would be in a state of "quiet vigilance" below threshold. Could it be called using Koch metaphor "a mouse zombie"? Would it be a simple-minded model of *blind-sight* in an animal species able to carry on minimal vision without having full access to consciousness? In any case, these studies may have as a principal consequence, to re-examine some of the key paradigm of the present scienfic studies on consciousness, in particular on blind-sight, in a phylogenetic and developmental perpective. It is clear from the early Friedrich Goltz experiments on dogs (see Clarke and O'Malley, 1968) that the removal of the cerebral cortex has vastly different consequences on perception in lower vertebrate species than in humans and monkeys. Moreover, recent evidence on face detection in the human newborn reveal a rather unanticipated subcortical processing (Johnson, 2005). Would Goltz statement that "the decerebrated dog is essentially nothing but a child of the moment" bear some truth ? Would the mouse and its mutants be of some help in these investigations? Empirical and theoretical studies on these issues are clearly needed.

On anatomical grounds, as noted earlier, a strong correlation exists between the development of consciousness and the expansion (or maturation) of the neocortex and more specifically of the prefrontal cortex (comparatively to the bulk of the neocortex), both in the course of evolution (see Changeux, 2004; Elston, 2003) and during development (Zelazo, 1995). An interesting correlation moreover has been established between the increased complexity of pyramidal neurons from layer III (largely long axon neuron possibly contributes to the workspace) and their distribution from occipital to temporal and prefrontal areas and, in prefrontal areas, from lower species (rabbit, marmoset) to humans (Elston, 2003). In mouse and rat, the prefrontal cortex occupies a relatively small area of the cerebral cortex. Yet, its role in exploratory behaviour and conflict resolution problems is documented even in these simple species (see Granon *et al*, 1994; Dias and Aggleton, 2000; Robbins, 2000; Granon unpublished).

To conclude, the mouse and its genetically modified variants may become a convenient, though simple, model to investigate the molecular biology of behaviours associated with minimal consciousness but also to test neurocomputational models such as the neuronal workspace hypotheses for states and access to consciousness. It may become a convenient model system, yet with the above mentioned important limitations, to establish a causal relationship – not simply "neural correlates" – between neuronal architectures, their activity processes and the access to minimal consciousness. Would the mouse ever become the *Drosophila* of minimal consciousness?

Acknowledgments

I thank M Besson, S Dehaene, P Faure, S Granon, H Lagercrantz, B Molles for helpful comments. Support was provided by Centre National de la Recherche Scientifique, Collège de France, Ministère de la Recherche, Association pour la Recherche contre le Cancer, European Communities.

References

Andres C (2002) Molecular genetics and animal models in autistic disorder. *Brain Res Bull* **57**:109-19.

Baars BJ (1998) Metaphors of consciousness and attention in the brain. *Trends Neurosci* **21**:58-62.

Baars BJ (2001) The brain basis of a "consciousness monitor": scientific and medical significance. *Consc Cogn*, **10**:159-64

Balfour DJ (2002) Neuroplasticity within the mesoaccumbens dopamine system and its role in tobacco dependence. *Curr Drug Targets CNS Neurol Disord* **1**:413-21.

Balkin TJ, Braun AR, Wesensten NJ, Jeffries K, Varga M, Baldwin P, Belenky G, Herscovitch P (2002) The process of awakening: a PET study of regional brain activity patterns mediating the re-establishment of alertness and consciousness. *Brain* **125**:2308-19.

Baron-Cohen S (1991) The development of a theory of mind in autism: deviance and delay. *Psychiatr Clin North Am* **14**:33-51.

Barresi J, Moore C (1996) Intentional relations and social understanding. *Behav Brain Sci*, **19**:107-122.

Barto AG, Sutton RS, Anderson CW (1983) Neuronlike adaptive elements that can solve difficult learning control problems. *IEEE Transactions on Systems, Man, and Cybernetics*, SMC-**13**:834-846.

Bartocci M, Bergqvist L, Lagercrantz H, Anand KJS (2006) Pain activates cortical areas in the preterm newborn brain. *Pain* **122**:109-117.

Berthoz A (2003) *La Décision*. Paris: O. Jacob.

Bertrand D, Devillers-Thiéry A, Revah F, Galzi J-L, Hussy N, Mulle C, Bertrand S, Ballivet M, Changeux J-P (1992) Unconventional pharmacology of a neuronal nicotinic receptor mutated in the channel domain. *Proc Natl Acad Sci USA*, **89**:1261-1265.

Bianchi L (1921) *La Mécanique du Cerveau et la Fonction des Lobes Frontaux*. Paris:

Louis Arnette.

Block N (1990) Consciousness and accessibility. *Behav Brain Sci* **13**:596-598

Block N (2005) Two neural correlates of consciousness. *Trends Cogn Sci* **9**:46-52.

Boakes R (1984) *From Darwin to Behaviourism: Psychology and the Minds of Animals.* Cambridge UK: Cambridge University Press.

Bogen JE (1995) On the neurophysiology of consciousness: I. An overview. *Consc Cogn* **4**:52-62 and **4**:137-58.

Booth R, Charlton R, Hughes C, Happe F (2003) Disentangling weak coherence and executive dysfunction: planning drawing in autism and attention-deficit/hyperactivity disorder. *Philos Trans R Soc Lond B Biol Sci* **358**:387-92.

Brejc K, van Dijk WJ, Klaassen RV, Schuurmans M, van Der Oost J, Smit AB, Sixma TK (2001) Crystal structure of an ACh-binding protein reveals the ligand-binding domain of nicotinic receptors. *Nature* **411**:269-76.

Broide RS, Salas R, Ji D, Paylor R, Patrick JW, Dani JA, De Biasi M (2002) Increased sensitivity to nicotine-induced seizures in mice expressing the L250T alpha 7 nicotinic acetylcholine receptor mutation. *Mol Pharmacol* **61**:695-705.

Caldarone BJ, Duman CH, Picciotto MR (2000) Fear conditioning and latent inhibition in mice lacking the high affinity subclass of nicotinic acetylcholine receptors in the brain. *Neuropharmacol* **39**:2779-2784.

Carter RM, Hofstotter C, Tsuchiya N, Koch C (2003) Working memory and fear conditioning. *Proc Natl Acad Sci USA* **100**:1399-1404.

Celie PHN, van Rossum-Fikkert SE, van Dijk WJ, Brejc K, Smit AB, Sixma TK (2004) Nicotine and carbamylcholine binding to nicotinic receptors studied by AChBP structures. *Neuron* **41**:907–914.

Chalmers DJ (1998) The problems of consciousness. *Adv Neurol* **77**:7-16

Champtiaux N, Changeux J-P (2004) Knockout and knockin mice to investigate the role of nicotinic receptors in the central nervous system. *Prog Brain Res* **145**:235-251.

Changeux J-P (1981) The acetylcholine receptor: an "allosteric" membrane protein. *The Harvey Lectures* **75**:85-254.

Changeux J-P (1983) *L'Homme Neuronal.* Paris: Fayard.

Changeux J-P (1990) Functional architecture and dynamics of the nicotinic acetylcholine receptor: an allosteric ligand-gated ion channel. *Fidia Research Foundation Neuroscience Award Lectures* **4**:21-168.

Changeux J-P (2002) *L'Homme de Vérité.* Paris: Odile Jacob. *The Physiology of Truth* (2005), Cambridge MA: Harvard University Press.

Changeux J- P (2004) Clarifying consciousness. *Nature* **428**:603-604.

Changeux J-P (2005) Genes, brains and culture: from monkey to human. In: *From Monkey Brain to Human Brain: A Fyssen Foundation Symposium.* The MIT Press, pp 73-94.

Changeux J-P (2005) Préface. In: D Denton. *Les Émotions Primordiales et L'Éveil de la Conscience.* Paris: Flammarion, p 7-12.

Changeux J-P, Danchin A (1976) Selective stabilization of developing synapses as a mechanism for the specificication of neuronal networks. *Nature* **264**:705-712.

Changeux J-P, Edelstein SJ (1998) Allosteric receptors after 30 years. *Neuron* **21**:959-

980.

Changeux J-P, Edelstein SJ (2005) Allosteric mechanisms of signal transduction. *Science* **308**:1424-1428.

Cheng X, Lu B, Grant B, Law RJ, McCammon JA (2006) Channel opening motion of α7 nicotinic acetylcholine receptor as suggested by normal mode analysis. *J Mol Biol* **355**:310-324.

Clark RE, Manns JR, Squire LR (2002) Classical conditioning, awareness, and brain systems.*Trends Cogn Sci* **6**:524-531.

Clarke PB (1993) Nicotine dependence--mechanisms and therapeutic strategies. *Biochem Soc Symp* **59**:83-95.

Clarke E, O'Malley CD (1968) *The Human Brain and Spinal Cord.* Berkeley: UCLA press.

Cohen G, Han ZY, Grailhe R, Gallego J, Gaultier C, Changeux J-P, Lagercrantz H (2002) Beta 2 nicotinic acetylcholine receptor subunit modulates protective responses to stress: A receptor basis for sleep-disordered breathing after nicotine exposure. *Proc Natl Acad Sci USA* **99**:13272-13277.

Cohen G, Roux JC, Grailhe R, Malcolm G, Changeux J-P, Lagercrantz H (2005) Perinatal exposure to nicotine causes deficits associated with a loss of nicotinic receptor function. *Proc Natl Acad Sci USA* **102**:3817-3821.

Cohen G, Malcolm G, Henderson-Smart D (1997) A comparison of the ventilatory response of sleeping newborn lambs to step and progressive hypoxaemia. *J Physiol* **503**:203-213.

Cordero-Erausquin M, Changeux J-P (2001) Tonic nicotinic modulation of serotoninergic transmission in the spinal cord. *Proc Natl Acad Sci USA* **98**:2803-2807.

Cordero-Erausquin M, Marubio LM, Klink R, Changeux J-P (2000) Nicotinic receptor function : new perspectives from knockout mice. *Trends Pharmacol Sci* **21**:211-217.

Cowey A, Stoerig P (1995) Blindsight in monkeys. *Nature* **373**:247-249.

Crick F, Koch C (2003) A framework for consciousness. *Nat Neurosci* **6**:119-126.

Damasio AR (1995) *Descartes' Error: Emotion, Reason, and the Human Brain.* Paris: Quill Ed.

David V, Cazala P (1994) A comparative study of self-administration of morphine into the amygdala and the ventral tegmental area in mice. *Behav Brain Res* **65**:205-211.

De Fusco M, Becchetti A, Patrignani A, Annesi G, Gambardella A, Quattrone A, Ballabio A, Wanke E, Casari G (2000) The nicotinic receptor beta 2 subunit is mutant in nocturnal frontal lobe epilepsy. *Nat Genet* **26**:275-276.

Dehaene S, Changeux J-P (1989) A simple model of prefrontal cortex function in delayed-response tasks. *J Cogn Neurosci* **1**:244-261.

Dehaene S, Changeux J-P (1991) The Wisconsin card sorting test: theoretical analysis and simulation of a reasoning task in a model neuronal network. *Cereb Cortex* **1**:62-79.

Dehaene S, Changeux J-P (1995) Neuronal models of prefrontal cortical functions. *Ann NY Acad Sci* **769**:305-319.

Dehaene S, Changeux J-P (1997) A hierarchical neuronal network for planning behavior. *Proc Natl Acad Sci USA* **94**:13293-13298.

Dehaene S, Changeux J-P (2000) Reward-dependent learning in neuronal networks for planning and decision making. *Prog Brain Res* **126**:217-29.

Dehaene S, Changeux JP (2004) Neural mechanisms for access to consciousness. In: MS Gazzaniga (Ed) *The Cognitive Neurosciences III, X Consciousness*. MIT Press, pp 1145-1158.

Dehaene S, Changeux J-P (2005) Ongoing Spontaneous Activity Controls Access to Consciousness: A Neuronal Model for Inattentional Blindness. *PLoS Biol* **3**:e141.

Dehaene S, Naccache L (2001) Towards a cognitive neuroscience of consciousness: Basic evidence and a workspace framework. *Cognition* **79**:1-37.

Dehaene S, Kerszberg M, Changeux J-P (1998) A neuronal model of a global workspace in effortful cognitive tasks. *Proc Natl Acad Sci USA* **95**:14529-14534.

Dehaene S, Sergent C, Changeux J-P (2003) A neuronal network model linking subjective reports and objective physiological data during conscious perception. *Proc Natl Acad Sci USA* **100**:8520-8525.

Dehaene S, Naccache L, Cohen L, Bihan DL, Mangin JF, Poline JB, Riviere D (2001) Cerebral mechanisms of word masking and unconscious repetition priming. *Nat Neurosci* **4**:752-758.

Denton D (2005) *Les Emotions Primordiales et L'Éveil de la Conscience*. Paris: Flammarion.

Devillers-Thiéry A, Galzi J-L, Bertrand S, Changeux J-P, Bertrand D (1992) Stratified organization of the nicotinic acetylcholine receptor channel. *Neuroreport* **3**:1001-1004.

Dias R, Aggleton JP (2000) Effects of selective excitotoxic prefrontal lesions on acquisition of nonmatching- and matching-to-place in the T-maze in the rat: differential involvement of the prelimbic-infralimbic and anterior cingulate cortices in providing behavioural flexibility. *Eur J Neurosci* **12**:4457-4466.

Di Chiara G (2000) Role of dopamine in the behavioural actions of nicotine related to addiction. *Eur J Pharmacol* **393**:295-314.

Edelman G (1989) *The Remembered Present*. New York: Basic Books.

Edelman G (2004) Biochemistry and the sciences of recognition. *J Biol Chem* **279**:7361-7369.

Edelstein SJ, Schaad O, Changeux J-P (1997) Myasthenic nicotinic receptor mutant interpreted in terms of the allosteric model. *C R Acad Sci III* **320**:953-961.

Ellis JR, Ellis KA, Bartholomeusz CF, Harrison BJ, Wesnes KA, Erskine FF, Vitetta L, Nathan PJ (2005) Muscarinic and nicotinic receptors synergistically modulate working memory and attention in humans. *Int J Neuropsychopharmacol* **9**:1-15.

Elston GN (2003) Cortex, cognition and the cell: new insights into the pyramidal neuron and prefrontal function. *Cereb Cortex* **13**:1124-1138.

Engel AG, Sine SM (2005) Current understanding of congenital myasthenic syndromes. *Curr Opin Pharmacol* **5**:308-321.

Faure P, Neumeiste H, Faber DS, Korn H (2003) Symbolic analysis of swimming trajectories reveals scale invariance and provides a model for fish locomotion. *Fractals* **11**:233-243.

Fiset P, Paus T, Daloze T, Plourde G, Meuret P, Bonhomme V, Hajj-Ali N, Backman SB, Evans AC (1999) Brain mechanisms of propofol-induced loss of consciousness in humans: a positron emission tomographic study. *J Neurosci* **19**:5506-5513.

Franks NP, Lieb WR (1994) Molecular and cellular mechanisms of general anaesthesia. *Nature* **367**:607-614.

Friston KJ, Tononi G, Reeke GN Jr, Sporns O, Edelman GM (1994) Value-dependent selection in the brain: simulation in a synthetic neural model. *Neuroscience* **59**:229-243.

Frith U (2001) Mind blindness and the brain in autism. *Neuron* **32**:969-979.

Gärdenfors P (2003) *How Homo Became Sapiens: On the Evolution of Thinking*. Oxford: Oxford University Press.

Giedd JN, Blumenthal J, Molloy E, Castellanos FX (2001) Brain imaging of attention deficit/hyperactivity disorder. *Ann N Y Acad Sci* **931**:33-49.

Gil Z, Sack RA, Kedmi M, Harmelin A, Orr-Urtreger A (2002) Increased sensitivity to nicotine-induced seizures in mice heterozygous for the L250T mutation in the alpha7 nicotinic acetylcholine receptor. *Neuroreport* **13**:191-196.

Giraudat J, Dennis M, Heidmann T, Chang JY, Changeux J-P (1986) Structure of the high-affinity binding site for noncompetitive blockers of the acetylcholine receptor: serine-262 of the delta subunit is labeled by [3H]chlorpromazine. *Proc Natl Acad Sci USA* **83**:2719-2723.

Goldman-Rakic PS (1988) Topography of cognition: parallel distributed networks in primate association cortex. *Annu Rev Neurosci* **11**:137-156.

Granon S, Changeux J-P (2006) Attention-deficit/hyperactivity-disorder: a plausible mouse model? *Acta pediatrica* **95**:645-649.

Granon S, Faure P, Changeux J-P (2003) Executive and social behaviors under nicotinic receptor regulation. *Proc Natl Acad Sci USA* **100**:9596-9601.

Granon S, Poucet B, Thinus-Blanc C, Changeux J-P, Vidal C (1995) Nicotinic and muscarinic receptors in the rat prefrontal cortex: differential roles in working memory, response selection and effortful processing. *Psychopharmacol* **119**:139-144.

Granon S, Vidal C, Thinus-Blanc C, Changeux J-P, Poucet B (1994) Working memory, response selection, and effortful processing in rats with medial prefrontal lesions. *Behav Neurosci* **108**:883-891.

Grubb MS, Rossi FM, Changeux J-P, Thompson ID (2003) Abnormal functional organization in the dorsal lateral geniculate nucleus of mice lacking the beta2 subunit of the nicotinic acetylcholine receptor. *Neuron* **18**:8008-8018.

Gusnard DA, Raichle ME (2001) Searching for a baseline: functional imaging and the resting human brain. *Nat Rev Neurosci* **2**:685-694.

Gutkin B, Dehaene S, Changeux J-P (2006) A neuro-computational framework for nicotine addiction. *Proc Nat Acad Sci USA* **103**:1106-1111.

Halasz P (1998) Hierarchy of micro-arousals and the microstructure of sleep. *Neurophysiol Clin* **28**:461-475.

Han CJ, O'Tuathaigh CM, van Trigt L, Quinn JJ, Fanselow MS, Mongeau R, Koch C, Anderson DJ (2003) Trace but not delay fear conditioning requires attention and the anterior cingulate cortex. *Proc Natl Acad Sci USA* **100**:13087-13092

Han ZY, Zoli M, Cardona A, Bourgeois J-P, Changeux J-P, Le Novere N (2003) Localization of [3H]nicotine, [3H]cytisine, [3H]epibatidine, and [125I]alpha-bungarotoxin binding sites in the brain of Macaca mulatta. *J Comp Neurol* **461**:49-60.

Hobson JA (1999) *Dreaming as Delirium: How the Brain Goes out of Its Mind.* Cambridge MA – London: A Bradford Book – The MIT Press.

Hucho F, Oberthur W, Lottspeich F (1986) The ion channel of the nicotinic acetylcholine receptor is formed by the homologous helices M II of the receptor subunits. *FEBS Lett* **205**:137-42.

Huxley TH (1874) On the Hypothesis that Animals are Automata, and its History. In: *Collected Essays of T. H. Huxley*, Vol. 1. New York: D. Appleton and Co, p 199.

Itier V, Bertrand D (2002) Mutations of the neuronal nicotinic acetylcholine receptors and their association with ADNFLE. *Neurophysiol Clin* **32**:99-107.

Jasper HH (1998) Sensory in formation and conscious experience. *Adv Neurol* **77**:33-48.

Johnson MH (2005) Subcortical face processing. *Nature Rev Neurosc* **6**:766-774.

Jones BE (1998). The neural basis of consciousness across the sleep-waking cycle. *Adv Neurol* **77**:75-94.

Jones EG (1998) Viewpoint: the core and matrix of thalamic organization. *Neuroscience* **85**:331-45.

Jones MW (2002) A comparative review of rodent prefrontal cortex and working memory. *Curr Mol Med* **2**:639-47

Jones BE (2004) Paradoxical REM sleep promoting and permitting neuronal networks. *Arch Ital Biol* **142**:379-96.

Klink R, de Kerchove d'Exaerde A, Zoli M, Changeux J-P (2001) Molecular and physiological diversity of nicotinic acetylcholine receptors in the midbrain dopaminergic nuclei. *J Neurosci* **21**:1452-63.

Kobayashi Y, Isa T (2002) Sensory-motor gating and cognitive control by the brainstem cholinergic system. *Neural Netw* **15**:731-41.

Koch C (2004) *The Quest for Consciousness: A Neurobiological Approach.* Eaglewood: Roberts and Company Publishers.

Labarca C, Schwarz J, Deshpande P, Schwarz S, Nowak MW, Fonck C, Nashmi R, Kofuji P, Dang H, Shi W, Fidan M, Khakh BS, Chen Z, Bowers BJ, Boulter J, Wehner JM, Lester HA (2001) Point mutant mice with hypersensitive alpha 4 nicotinic receptors show dopaminergic deficits and increased anxiety. *Proc Natl Acad Sci USA* **98**:2786-91.

Lagercrantz H, Hanson M, Evrard P, Rodeck C (Eds) (2002) *The Newborn Brain: Neuroscience and Clinical Applications.* Cambridge UK: Cambridge University Press.

Lagercrantz H (2005) *I Barnets Hjärna.* Stockholm: Albert Bonniers förlag.

Lamme VA (2003) Why visual attention and awareness are different. *Trends Cogn Sci* **7**:12-18.

Laureys S, Faymonville ME, Degueldre C, Fiore GD, Damas P, Lambermont B, Janssens N, Aerts J, Franck G, Luxen A, Moonen G, Lamy M, Maquet P (2000) Auditory processing in the vegetative state. *Brain* **123**:1589-601.

Lawrence NS, Ross TJ, Stein EA (2002) Cognitive mechanisms of nicotine on visual attention. *Neuron* **36**:539-48.

Lee SL, Ralston HP, Drey ED, Partridge JC, Rosen MA (2005) Fetal pain : a systematic multidisciplinary review of the evidence. *JAMA* **294**:947-953

Lena C, de Kerchove D'Exaerde A, Cordero-Erausquin M, Le Novere N, del Mar Arroyo-Jimenez M, Changeux J-P (1999) Diversity and distribution of nicotinic acetylcholine receptors in the locus ceruleus neurons. *Proc Natl Acad Sci USA* **96**:12126-31.

Lena C, Popa D, Grailhe R, Escourrou P, Changeux J-.P Adrien J (2004) Beta2-containing nicotinic receptors contribute to the organization of sleep and regulate putative micro-arousals in mice. *J Neurosci* **24**:5711-8.

Le Novere N, Grutter T, Changeux J-P (2002) Models of the extracellular domain of the nicotinic receptors and of agonist- and Ca^{2+}-binding sites. *Proc Natl Acad Sci USA* **99**:3210-5.

Leslie AM, Frith U (1990) Prospects for a cognitive neuropsychology of autism: Hobson's choice. *Psychol Rev* **97**:122–131.

Levin ED (1992) Nicotinic systems and cognitive function. *Psychopharmacology* **108**:417-31.

Levin ED (2002) Nicotinic receptor subtypes and cognitive function. *J Neurobiol* **53**:633-640.

Llinas RR, Pare D (1991) Of dreaming and wakefulness. *Neuroscience* **4**:521-35.

Maggi L, Le Magueresse C, Changeux J-P, Cherubini E (2003). Nicotine activates immature "silent" connections in the developing hippocampus. *Proc Natl Acad Sci USA* **100**:2059-64.

Maggi L, Sola E, Minneci F, Le Magueresse C, Changeux J-P, Cherubini E (2004) Persistent decrease in synaptic efficacy induced by nicotine at Schaffer collateral-CA1 synapses in the immature rat hippocampus. *J Physiol* **559**:863-74.

Maia TV, Cleeremans A (2005) Consciousness: converging insights from connectionist modeling and neuroscience. *Trends Cogn Sci* **9**:397-404.

Mandel JL, Biancalana V (2004) Fragile X mental retardation syndrome: from pathogenesis to diagnostic issues. *Growth Horm IGF Res* **14 Suppl A**:S158-65.

Maquet P, Phillips C (1998) Functional brain imaging of human sleep. *J Sleep Res* **7 Suppl 1**:42-7.

Marois R, Chun MM, Gore JC (2004) A common parieto-frontal network is recruited under both low visibility and high perceptual interference conditions. *J Neurophysiol* **92**:2985-92.

Maskos U, Kissa K, St Cloment C, Brulet P (2002) Retrograde trans-synaptic transfer of green fluorescent protein allows the genetic mapping of neuronal circuits in transgenic mice. *Proc Natl Acad Sci USA* **99**:10120-5.

Maskos U, Molles BE, Pons S, Besson M, Guiard BP, Guilloux J-P, Evrard A, Cazala P, Cormier A, Mameli-Engvall M, Dufour N, Cloez-Tayarani I, Bemelmans AP, Mallet J, Gardier AM., David V, Faure P, Granon S, Changeux J-P (2005) Nicotine reinforcement and cognition restored by targeted expression of nicotinic receptors. *Nature* **436**:103-7.

Mazoyer B, Zago L, Mellet E, Bricogne S, Etard O, Houde O, Crivello F, Joliot M, Petit L, Tzourio-Mazoyer N (2001) Cortical networks for working memory and executive functions sustain the conscious resting state in man. *Brain Res Bull* **54**:287-98.

McGehee DS, Role LW (1996) Presynaptic ionotropic receptors. *Curr Opin Neurobiol* **6**:342-9.

McMahon LL, Yoon KW, Chiappinelli VA (1994) Electrophysiological evidence or presynaptic nicotinic receptors in the avian ventral lateral geniculate nucleus. *J Neurophysiol* **71**:826-9.

Meltzoff AN (1990) Towards a developmental cognitive science. The implications of cross-modal matching and imitation for the development of representation and memory in infancy. *Ann N Y Acad Sci* **608**:1-31

Meltzoff AN, Gopnik A (1993) The role of imitation in understanding persons and developing a theory of mind. In: S Baron-Cohen, H Tager-Flusberg and DJ Cohen (Eds) *Understanding Other Minds: Perspectives from Autism*. Oxford: Oxford University Press, pp 335-366.

Mitchell BD, Macklis JD (2005) Large-scale maintenance of dual projections by callosal and frontal cortical projection neurons in adult mice. *J Comp Neurol* **482**:17-32.

Monod J, Wyman J, Changeux J-P (1965) On the nature of allosteric transitions: a plausible model. *J. Mol. Biol* **12**:88-118.

Montague PR, Dayan P, Person C, Sejnowski TJ. (1995) Bee foraging in uncertain environments using predictive Hebbian learning. *Nature* **377**:725-8.

Montague PR, Dayan P, Sejnowski TJ (1996) A framework for mesencephalic dopamine systems based on predictive Hebbian learning. *J Neurosci* **16**:1936-47.

Mrsic-Flogel TD, Hofer SB, Creutzfeldt C, Cloez-Tayarani I, Changeux J-P, Bonhoeffer T, Hubener M (2005) Altered map of visual space in the superior colliculus of mice lacking early retinal waves. *J Neurosci* **25**:6921-8.

Mulle C, Changeux J-P (1990) A novel type of nicotinic receptor in the rat central nervous system characterized by patch-clamp techniques. *J Neurosci* **10**:169-75.

Mulle C, Vidal C, Benoit P, Changeux J-P (1991) Existence of different subtypes of nicotinic acetylcholine receptors in the rat habenulo-interpeduncular system. *J Neurosci* **11**:2588-97.

Nagel T (1974) What is it like to be a bat? *Philosophical Review* **83**: 435-450.

Naldini L, Blomer U, Gallay P, Ory D, Mulligan R, Gage FH, Verma IM, Trono D (1996) In vivo gene delivery and stable transduction of nondividing cells by a lentiviral vector. *Science* **272**:263-7.

Neisser U (1976) *Cognition and Reality: Principles and Implications of Cognitive Psychology*. New York: W. H. Freeman and Company.

Neul JL, Zoghbi HY (2004) Rett syndrome: a prototypical neurodevelopmental disorder. *Neuroscientist* **10**:118-28.

Nordberg A, Winblad B (1986) Reduced number of [3H]nicotine and [3H]acetylcholine binding sites in the frontal cortex of Alzheimer brains. *Neurosci Lett* **72**:115-9.

Norman DA, Shallice T (1980) Attention to action: Willed and automatic control of behaviour. In: R Davidson, GE Schwartz and D Shapiro (Eds) *Consciousness and self regulation*. New York: Plenun times, pp 1 - 15.

Paas Y, Gibor G, Grailhe R, Savatier-Duclert N, Dufresne V, Sunesen M, de Carvalho LP, Changeux J-.P, Attali B (2005) Pore conformations and gating mechanism of a Cys-loop receptor. *Proc Natl Acad Sci USA* **102**:15877-82

Pardo JV, Pardo PJ, Janer KW, Raichle ME (1990) The anterior cingulate cortex mediates processing selection in the Stroop attentional conflict paradigm. *Proc Natl Acad Sci USA* **87**:256-9.

Passetti F, Dalley JW, O'Connell MT, Everitt BJ, Robbins TW (2000) Increased acetylcholine release in the rat medial prefrontal cortex during performance of a visual attentional task. *Eur J Neurosci* **12**:3051-8.

Paus T, Koski L, Caramanos Z, Westbury C (1998) Regional differences in the effects of task difficulty and motor output on blood flow response in the human anterior cingulate cortex: a review of 107 PET activation studies. *Neuroreport* **9**:R37-47.

Paus T (2000). Functional anatomy of arousal and attention systems in the human brain. *Prog Brain Res* **126**:65-77.

Perry EK, Perry RH (2004) Neurochemistry of consciousness: cholinergic pathologies in the human brain. *Prog Brain Res* **145**:287-99.

Perry E, Walker M, Grace J, Perry R (1999) Acetylcholine in mind: a neurotransmitter correlate of consciousness ? *Trends Neurosci* **22**:273-80.

Perry EK, Lee ML, Martin-Ruiz CM, Court JA, Volsen SG, Merrit J, Folly E, Iversen PE, Bauman ML, Perry RH, Wenk GL (2001) Cholinergic activity in autism: abnormalities in the cerebral cortex and basal forebrain. *Am J Psychiatry* **158**:1058-66.

Phillips HA, Favre I, Kirkpatrick M, Zuberi SM, Goudie D, Heron SE, Scheffer IE, Sutherland GR, Berkovic SF, Bertrand D, Mulley JC (2001) CHRNB2 is the second acetylcholine receptor subunit associated with autosomal dominant nocturnal frontal lobe epilepsy. *Am J Hum Genet* **68**:225-31.

Picciotto MR, Zoli M, Rimondini R, Lena C, Marubio LM, Pich EM, Fuxe K, Changeux J-P (1998) Acetylcholine receptors containing the beta2 subunit are involved in the reinforcing properties of nicotine. *Nature* **391**:173-7.

Picciotto MR, Zoli M, Lena C, Bessis A, Lallemand Y, Le Novere N, Vincent P, Pich EM, Brulet P, Changeux J-P (1995) Abnormal avoidance learning in mice lacking functional high-affinity nicotine receptor in the brain. *Nature* **374**:65-7.

Pickworth WB, Fant RV, Butschky MF, Henningfield JE (1997) Effects of mecamylamine on spontaneous EEG and performance in smokers and non-smokers. *Pharmacol Biochem Behav* **56**:181-7.

Poucet B (1993) Spatial cognitive maps in animals: new hypotheses on their structure and neural mechanisms. *Psychol Rev* **100**:163-82.

Poucet B, Herrmann T (2001) Exploratory patterns of rats on a complex maze provide evidence for topological coding. *Behav Processes* **53**:155-162.

Preyer W (1894) *Mental development in the child.* New York: D Appleton and Co.

Renner MJ (1990) Neglected aspects of exploratory and investigatory behavior. *Psychobiology* **18**:16-22

Revah F, Bertrand D, Galzi J-L, Devillers-Thiéry A, Mulle C, Hussy N, Bertrand S,

Ballivet M, Changeux J-P (1991) Mutations in the channel domain alter desensitization of a neuronal nicotinic receptor. *Nature* **353**:846-849.

Robbins TW. (2000) From arousal to cognition: the integrative position of the prefrontal cortex. *Prog Brain Res* **126**:469-83.

Rochat P (2003) Five levels of self-awareness as they unfold early in life. *Consc Cogn* **12**:717-31

Rodrigues-Pinguet NO, Pinguet TJ, Figl A, Lester HA, Cohen BN (2005) Mutations linked to autosomal dominant nocturnal frontal lobe epilepsy affect allosteric Ca2+ activation of the alpha 4 beta 2 nicotinic acetylcholine receptor. *Mol Pharmacol* **68**:487-501.

Rossi FM, Pizzorusso T, Porciatti V, Marubio LM, Maffei L, Changeux J-P (2001) Requirement of the nicotinic acetylcholine receptor β2 subunit for the anatomical and functional development of the visual system. *Proc Natl Acad Sci USA* **98**:6453-6458.

Sacco KA, Bannon KL, George TP (2004) Nicotinic receptor mechanisms and cognition in normal states and neuropsychiatric disorders. *J Psychopharmacol* **18**:457-74.

Schaeffer P, Prabonnaud V, Roux M, Gully D, Herbert JM (1994) CCK-JMV-180 acts as an antagonist of the CCKA receptor in the human IMR-32 neuroblastoma cell line. *FEBS Lett* **354**:203- 6.

Schultz C, Koppers D, Braak H, Braak E (1997) Cytoskeletal alterations in the aged human neurohypophysis. *Neurosci Lett* **237**:93-6.

Searle JR (2000) Consciousness. *Ann Rev Neurosci* **23**:557-78.

Semba K (1999) The mesopontine cholinergic system: A dual role in REM sleep and wakefulness. In: R Lydic and HA Baghdoyan (Eds.), *The Handbook of Behavioral State Control: Cellular and Molecular Mechanisms,* Boca Raton: CRC Press, pp. 161-180.

Semba K (2004) Phylogenetic and ontogenetic aspects of the basal forebrain cholinergic neurons and their innervation of the cerebral cortex. *Prog Brain Res* **145**:3-43.

Shallice T (1988) *From Neuropsychology to Mental Structure*. Cambridge University Press.

Shulman RG, Hyder F, Rothman DL (2003) Cerebral metabolism and consciousness. *C R Biol* **326**:253-73.

Shytle RD, Silver AA, Sheehan KH, Sheehan DV, Sanberg PR (2002) Neuronal nicotinic receptor inhibition for treating mood disorders: preliminary controlled evidence with mecamylamine. *Depress Anxiety* **16**:89-92.

Sigman M, Ungerer JA (1984) Cognitive and language skills in autistic, mentally retarded, and normal children. *Developmental Psychology* **20**:293-302.

Steinlein OK (2004) Genetic mechanisms that underlie epilepsy. *Nat Rev Neurosci* **5**:400-408

Steinlein OK, Mulley JC, Propping P, Wallace RH, Phillips HA, Sutherland GR, Scheffer IE, Berkovic SF (1995) A missense mutation in the neuronal nicotinic acetylcholine receptor alpha 4 subunit is associated with autosomal dominant nocturnal frontal lobe epilepsy. *Nat Genet* **11**:201-3.

Steriade M (2004) Slow-wave sleep: serotonin, neuronal plasticity, and seizures. *Arch Ital Biol* **142**:359-67.

Steriade M, McCormick DA, Sejnowski TJ (1993) Thalamocortical oscillations in the sleeping and aroused brain. *Science* **262**:679-85.

Taly A, Delarue M, Grutter T, Nilges M, Le Novere N, Corringer P-J, Changeux J-P (2005) Low frequency Normal Mode Analysis suggest a quaternary twist model for the nicotinic receptor gating mechanism. *Biophys J* **88**:3954-65.

Tapper AR, McKinney SL, Nashmi R, Schwarz J, Deshpande P, Labarca C, Whiteaker P, Marks MJ, Collins AC, Lester HA (2004) Nicotine activation of alpha4* receptors: sufficient for reward, tolerance, and sensitization. *Science* **306**:1029-32.

Tassonyi E, Charpantier E, Muller D, Dumont L, Bertrand D (2002) The role of nicotinic acetylcholine receptors in the mechanisms of anesthesia. *Brain Res Bull* **57**:133-50.

Terzano MG, Parrino L (2000) Origin and Significance of the Cyclic Alternating Pattern (CAP). *Sleep Med Rev* **4**:101-123.

Thinus-Blanc C (1996) *Animal Spatial Cognition: Behavioral and Neural Approaches.* Singapore: World Scientific Publishing Co.

Thorndike EL (1898) *Animal Intelligence: An Experimental Study of the Associative Processes in Animals* (Psychological Review, Monograph Supplements, No. 8). New York : Macmillan.

Trivers R (1985) *Social Evolution.* Menlo Park, CA: Benjamin.

Unwin N (2000) The Croonian Lecture 2000. Nicotinic acetylcholine receptor and the structural basis of fast synaptic transmission. *Philos Trans R Soc Lond B Biol Sci* **355**:1813-29.

Unwin N (2005) Refined structure of the nicotinic acetylcholine receptor at 4A resolution. *J Mol Biol* **346**:967-89.

Vidal C, Changeux J-P (1989) Pharmacological profile of nicotinic acetylcholine receptors in the rat prefrontal cortex: an electrophysiological study in a slice preparation. *Neuroscience* **29**:261-270.

Vidal C, Changeux J-P (1993) Nicotinic and muscarinic modulations of excitatory synaptic transmission in the rat prefrontal cortex in vitro. *Neuroscience* **56**:23-32.

Weiskrantz L (1991) Disconnected awareness for detecting, processing, and remembering in neurological patients. *J R Soc Med* **84**:466-70.

Wonnacott S, Göthert M, Chahl LA, Willow M, Nicholson GM (1995) Modulation of neurotransmitter release by some therapeutic and socially used drugs; Göthert M: Ethanol (alcohol). In: DA Powis and SJ Bunn (Eds) *Neurotransmitter Release and its Modulation.* Cambridge: Cambridge University Press, pp 293-327,

Woolf NJ (1997) A possible role for cholinergic neurons of the basal forebrain and pontomesencephalon in consciousness. *Consc Cogn* **6**:574-96.

Yamamoto KI, Domino EF (1965) Nicotine-induced EEG and behavioral arousal. *Int J Neuropharmacol* **4**:359-73.

Yerkes RM (1916) *The Mental Life of Monkeys and Apes.* New York: H. Holt.

Zeki S (2003) The disunity of consciousness. *Trends Cogn Sci* **7**:214-218.

Zelazo PD (1996) Towards a characterization of minimal consciousness. *New Ideas in*

Psychology **14**:63-80
Zelazo PD (2004) The development of conscious control in childhood. *Trends Cognitive Sci* **8**:12-17.
Zhou WX, Sornette D, Hill RA, Dunbar RI (2005) Discrete hierarchical organization of social group sizes. *Proc Biol Sci* **272** :439-44.
Zoli M, Lena C, Picciotto MR, Changeux J-P (1998) Identification of four classes of brain nicotinic receptors using beta2 mutant mice. *J Neurosci* **18**:4461-72.
Zoli M, Picciotto MR, Ferrari R, Cocchi D, Changeux J-P (1999) Increased neurodegeneration during ageing in mice lacking high-affinity nicotine receptors. *Embo J* **18**:1235-44.
Zucconi M, Ferini-Strambi L (2000) NREM parasomnias: arousal disorders and differentiation from nocturnal frontal lobe epilepsy. *Clin Neurophysiol* **Suppl 2**:S129-35.

Chapter 7

The Emergence of Consciousness in the Newborn

Hugo Lagercrantz

1. Introduction

The quest for the emergence of consciousness is fundamental, but has not been addressed very much until recently. Whether the fetus is conscious about pain and should be given analgesia before e.g. abortion has aroused a lot of public interest and controversy. Using new brain imaging and functional techniques it is now possible to explore the neuronal correlates of consciousness (NCC). Although the knowledge of the developing brain has increased considerably, thanks to these new techniques, surprisingly little is known about its main function, i.e. being conscious. The hard problem is to bridge the gap between the neurobiology of the newborn brain and its subjective feelings (Chalmers, 1996). In this review I will discuss aspects of the development of consciousness, particularly the NCC.

2. Neurons - the atoms of consciousness

Neurons are the atoms of perception, memory, thought and action, and thus the atoms of consciousness (see Koch, 2004). They differ from other cells in, for example, the intestines or the skin, in that they are explicit. Although nearly all cells react to the environment, only the neurons make this information explicit and available for the conscious thinking. The neurons of the immature fetal brain are round and have very few connections with other cells, which explains why they are less explicit. D. Purpura (1982) has demonstrated how cortical pyramidal neurons in the primary visual cortex of the human sprout increasingly more from the 26th week. The neurons branch, acquire dendritic spines and connect with each other. This seems to correspond to the detection of visual evoked potentials (VEP) in preterm infants (Fig. 7.1).

Consciousness Transitions: Phylogenetic, Ontogenetic, and Physiological Aspects
Edited by Hans Liljenström and Peter Århem

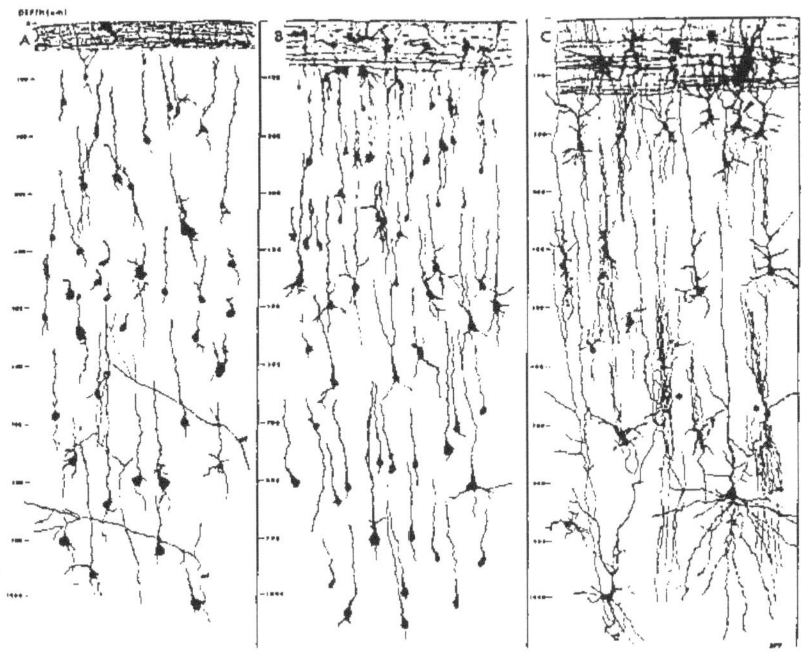

Fig. 7.1 The neurons of the fetal brain are round and have few connections. But already several weeks before birth they sprout, acquire dendritic spines and connect (from Purpura with permission). The samples are taken from week 25, 27, and 32, respectively.

3. The localization of consciousness

The exact anatomical localization of consciousness is not known even for the adult. Thalamus is probably essential as the gateway to neocortex. All sensory modalities except smell are relayed through the thalamus before they reach the cortex (Koch, 2004). If we consider verbal reportability as a hallmark of human consciousness (see Perner and Dienes, 2002), the following areas seem to be involved: the anterior cingulate gyrus of the frontal lobe, the left lateral frontal and posterior cortex in areas involved with processing the meaning of words or sentences, and the right cerebellum (Posner and Rothbart, 1998). Pyramidal cells from layers 2 and 3 in the dorsal lateral prefrontal and inferoparietal cortical structures are probably essential (Dehaene and Changeux, 2004). Long-distance axons link most, if not all the cortical and thalamic regions, forming a neuronal workspace (Baars, 1998; Changeux, 2004). Koch also believes there are specific "consciousness neurons". An alternative view is that there is a "dynamic core" consisting of correlated activities of a large number of neurons in the cortex, thalamus and the limbic system (Edelman and Tononi, 2000).

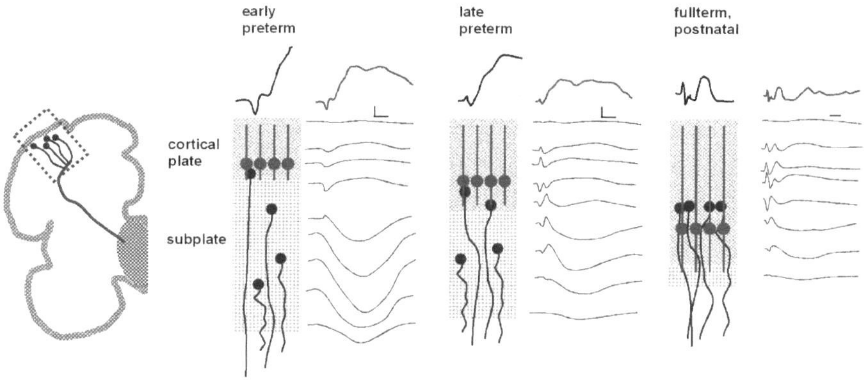

Fig. 7.2 Maturation of the thalamo-cortical connections and the somatosensory evoked potentials (SEP) responses (Reproduced with permission from Sampsala and Lauronen, 2006).

4. Developmental anatomy of consciousness

A primary prerequisite to be conscious is to be aware of sensory impressions, i.e. the neuronal pathways mediating this information must exist and function. Palmar cutaneous sensory receptors appear around the 10th week of gestational age in the human. Spinal reflexes evoked by stimulation of most body areas can be observed from the 14th week. Nociceptive reactions can be recorded from the 19th week (Rees and Rawson, 2002). The fetus reacts to hearing from about the 20th week (see below).

The sensory cells in the nasal cavity and the nasal septum are probably in contact with amniotic fluid from the 22nd week of gestation (Schaal, 2004). The eye lids are closed until the 22nd week and the retina is very immature when the eyes open. Photoreceptors are relatively short and wide at birth. The fovea is also very immature.

To be conscious, the various sensory modalities must get accessible to the cortex. All sensory impressions except olfactory ones are relayed in the thalamus. Thalamic afferents to the cortex appear from about 12-16 weeks gestation. However, these projections only reach the subplate, which is regarded as a "waiting compartment" for afferents to the cortex (Fig. 7.2). At this stage, only very long somatosensory evoked potentials (SEP) from the deep layers can be recorded at the scalp (Vanhatalo and Lauronen, 2006). Only after about 24 weeks there is an ingrowth of thalamocortical axons in the somatosensory, auditory, visual and frontal cortices (Kostovic and Joavnov-Milosevic, 2006). Thalamocortical pathways mediating pain perception do not seem to function before the 29-30 gestational week. At about this time there also seems to be some synchrony of the EEG rhythm of the two hemispheres (Vanhatalo and Kaila, 2006).

The cerebral cortex, particularly the prefrontal area, matures late in the human. The neurons do not become completely myelinated until early adulthood, allowing rapid neuronal activity and mature executive actions (Sowell *et al*, 2004). However, subcortical structures are probably of greater importance for consciousness during early life. The fusiform area for face recognition, the amygdala for emotions etc. seem to function already in the newborn. These areas are of great importance for the social brain and probably also for consciousness (Johnson, 2005). These subcortical structures should not necessarily be regarded as subordinate the cortex, particularly not in the infant.

5. The neurochemistry of consciousness

Excitatory amino acids generate synchronous oscillatory activity, which probably is essential for the maintenance of consciousness. Gamma-aminobutyric acid (GABA) is the dominating excitatory neurotransmitter during fetal life (Letinic *et al*, 2002). Around birth, it becomes the main inhibitory neurotransmitter. This is due to that the immature neurons are depolarized by GABA, while the mature neurons become hyperpolarized. This is caused by the expression of the K+/Cl- co-transporter KCC2, which maintains a low intracellular Cl- concentration. Glutamate and aspartate probably take over the role of GABA as the major excitatory amino acids after birth (see Vanhatalo and Kaila, 2006).

Classical neurotransmitters, like noradrenaline and acetylcholine, may also be involved in the generation of consciousness by stimulating wakefulness and awareness. Noradrenergic neurons originating from the locus coeruleus have been proposed to be involved in arousal. The noradrenaline turn-over was found to be relatively low in the rat fetus, but surged after birth (Lagercrantz, 1996). If we extrapolate to the human newborn baby this increased noradrenaline turnover may explain the arousal of the newborn baby, who is usually awake the first two hours after birth (Fig 7.5). Although increased noradrenaline turnover has not been demonstrated in the human brain, we know that enormously high levels of catecholamines are released after vaginal delivery and that there are strong indications that the central and peripheral catecholamine systems are activated simultaneously (Lagercrantz, 1996).

Acetylcholine also seems to be a more likely candidate as the neurotransmitter of consciousness (Changeux, 2006, Chapter 6 this volume). Cholinergic basal forebrain neurons send their axons to a much wider array of target structures, innervating the thalamus, hippocampus, amygdala and cerebral cortex. The idea of acetylcholine as the transmitter of consciousness is corroborated by the finding of increased activity during the waking state. Furthermore, in diseases like Parkinson, Alzheimer and dementia, associated with depressed consciousness, there is a selective loss of cholinergic neurons. Acetylcholine has been found to promote cortical processing of incoming stimuli. Newborn mice lacking the ß2 nicotinic acetylcholine receptor subunit showed impaired arousal response during hypoxia (Cohen *et al*, 2005).

6. Methods to study consciousness in the fetus and the infant

Functional magnetic resonance imaging (fMRI) has been used to study how the human fetus processes sensory input (Gowland and Fulford, 2004). However, it is a difficult method, since the fetus must be immobilized. Another technique is magnetoencephalography (MEG), which has been used by some research groups (Preissl *et al*, 2004). The principle is to record magnetic signals corresponding to the electrical activity in the brain. It is completely passive and noninvasive with superior temporal resolution; however, it does not provide any anatomical information. This has to be obtained by combining MEG with ultrasound, or other techniques. The method has been used to study auditory and visual evoked response in the fetus (Huotilainen, 2006).

Conventional EEG, amplitude integrated EEG, event-related potentials can be used to assess neonatal consciousness (see Fellman *et al*, 2006). Functional magnetic resonance imaging (fMRI) is considered the leading technique to explore the function of the brain. However, the infant must be immobilized and usually asleep, which makes it difficult for studies of consciousness.

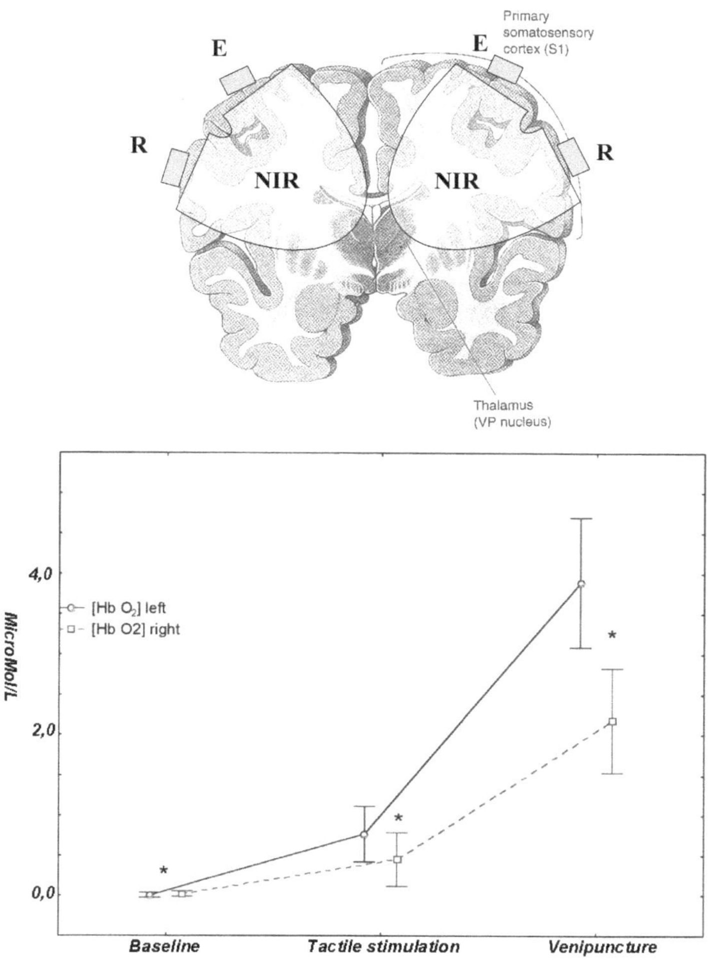

Fig. 7.3 Recording of NIRS over the sensorimotor area of the responses to tactile and noxious (venipuncture) stimulation at the somatosensory area in preterm infants. No response was seen over the occipital area. (From Bartocci *et al*, 2006).

This is not necessary when using near-infrared spectroscopy (NIRS). This method is noninvasive, relatively simple and a useful method to assess how the neonatal brain processes various sensory signals (Meek, 2002). NIRS is based on measuring the hemodynamic responses over the cortical areas. Near-infrared light which is transmitted by optodes placed on the skull is reflected by oxyhemoglobin and deoxyhemoglobin and measured. Changes in hemoglobin oxygenation and blood volume and flow can be computed by algorithms and used as indexes of neural activation. Using this method it is at least possible to study how sensory input is processed in the brain, although it only indirectly indicates whether the infant is aware or conscious of the stimuli. The spatial resolution is 1-2 cm and the temporal sampling resolution 0.01 s which is better than fMRI. One limitation is that structures situated deeper than 2-3 cm under the skull cannot be studied with NIRS.

With the NIRS method, positive responses to visual, auditory and olfactory stimulation have been documented (Meek *et al*, 1998; Bartocci *et al*, 2000). This method has also been used to study how the infant perceives human speech (Dehaene-Lamberts *et al*, 2002).

7. Components of consciousness

A catalogue of conscious experiences or components of consciousness can be listed for the fetus and the neonate:

7.1 Sensory experiences and pain

Tactile and painful stimuli (e.g. venipuncture) elicit specific hemodynamic responses in the somatosensory cortex (Fig. 7.3), implying conscious sensory perception already in the preterm neonates (Bartoccci *et al*, 2006; Slater *et al*, 2006). The mean gestational age was 32 weeks (range 28-36) in the study by Bartocci *et al*. A more pronounced response was seen in the youngest infants, consistent with the finding that the pain threshold is lower in preterm infants. On the other hand, there was a positive correlation between pain response and postnatal age, consistent with a postnatal decay of fetal inhibition. The latency between venipuncture and cortical activity was comparable to that of adults.

Higher responses to noxious stimulation was seen in awake infants (Slater *et al*, 2006), confirming the idea of being conscious about the pain. The lateralization of pain processing, the latency and duration of these responses and their gradations across gestational age and postnatal age, and the neuroanatomical location of these responses suggest that preterm infants may be consciously processing the acute pain from venipuncture.

7.2 Smell and taste

Early infant behaviour seems to be influenced by olfactory cues, many originating from the intrauterine environment (Schaal *et al*, 2004; Varendi *et al*, 2002). Infants seem to be more attractive to the smell of amniotic fluid than to other odours. Exposure to am-

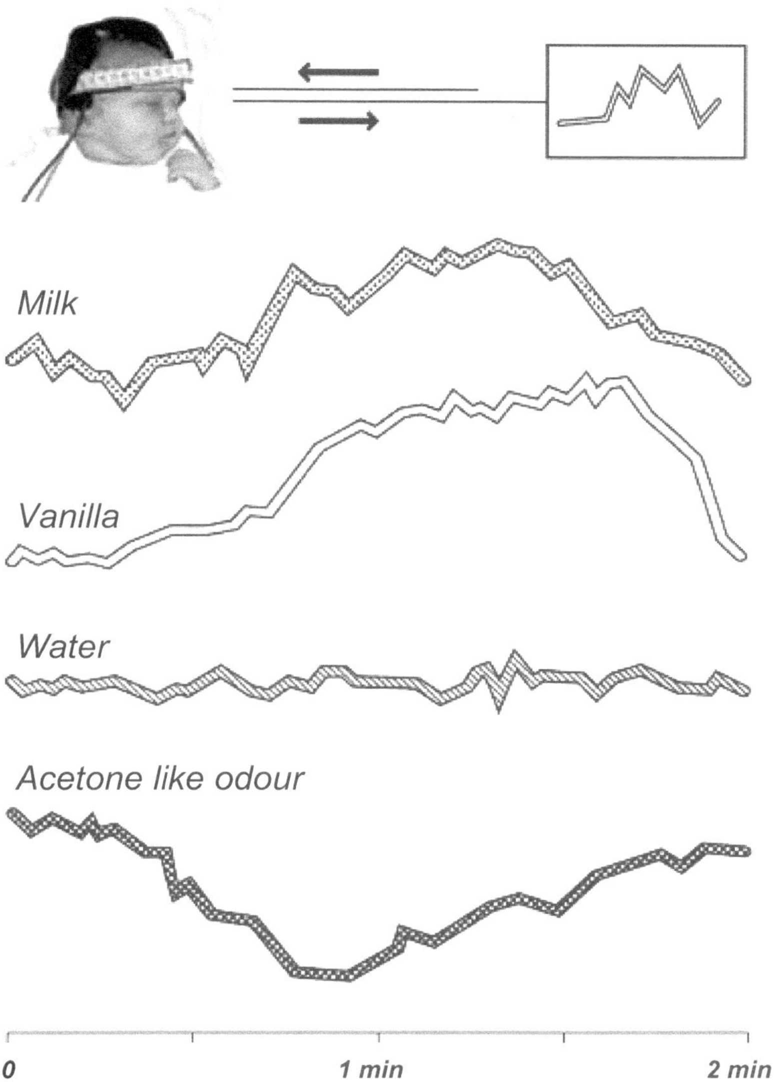

Fig. 7.4 Recording of the responses to smell by Near-Infrared Spectroscopy (NIRS) from the olfactory cortex. This shows that various smells are processed at a cortical level.

niotic fluid and other maternal odours seems to have a soothing effect in newborns. When babies were exposed to clothes with their own mother's odours they stopped crying. Infants also seemed to prefer tastes that they were exposed to during fetal life through their mother's diet (Schaal *et al*, 2004). Odours were found to be processed in the orbito-frontal olfactory area, by the use of NIRS (Bartocci *et al*, 2000, 2001). (See Fig. 7.4).

7.3 Hearing and seeing

The fetus does not react to sounds in general until the 20th week, when tachycardia can be elicited by noise (Counter, 2002). External sound is reduced to about half of its strength when it reaches the fetal cochlea. However, it is plausible that the maternal voice is transmitted also by direct conduction. This may explain why the newborn infants seem to be able to discriminate between his mother and unfamiliar women's voices (Gray and Philbin, 2004). The full-term infant can orient visually to auditory signals by turning the head and the eyes towards the sound. If an infant is shown for example an object at the same time as being presented with a sound, it will turn the eyes towards the sound, suggesting that hearing is more mature at birth. On the other hand, the preterm infant seems to have difficulties to orient towards an auditory stimulus.

Fetal brain activity has been studied during visual stimulation. Bright light was shone at the maternal abdomen for short periods (8 s) repetitively and brain activity was monitored with fMRI. Activity could be recorded in the frontal eye fields but not in the primary visual cortex in the occipital region (Preissl *et al*, 2004).

Visual acuity in the full-term newborn infants is only 1/40 visual acuity but they can recognize faces and imitate (Johnson, 2005). The ability to recognize different colours develop even later.

7.4 Wakefulness

The human fetus is mainly asleep, although it can be observed with ultrasound that it sometimes opens its eyes (see Mellor *et al*, 2005). However, it may be awake without being conscious, like in some patients in vegetative states. The fetus is mainly asleep in REM-sleep, which is characterized by rapid eye movements. Non-REM sleep increases successively during maturation.

It can be argued that one is conscious also during dream or REM-sleep, at least after birth. However, if we assume that wakefulness is required for being fully conscious, there are indications that the fetus is never awake and conscious. H. Rigatto *et al* (1986) recorded electrocorticograms, eye and breathing movements, in parallell with behaviour in fetal sheep, which was monitored by video camera observations. No wakeful periods were discovered when analyzing videotapes recordings of more than 5000 hours during 8 years. Furthermore, a number of inhibitory substances have been found in the placenta, which suppress the fetus and promote sleep. The low partial oxygen level of the fetus (Mount Everest in utero) may also contribute to suppress the fetal brain. This may, for example, increase the endogenous levels of adenosine (Irestedt *et al*, 1989), which inhibits neuronal activity.

The normal newborn infant seems to be awake the first two hours after birth (Fig. 7.5). The eyes are wide-open and the pupils are usually big. This waking state was probably missed a few decades ago. Because of this, the babies received Crédé prophylaxis. After a couple of hours they usually fall asleep. The awakening of the newborn baby may be due to activation of the locus coeruleus (see above). Extremely preterm infants (< 25 w) usually have closed eyes, indicating that they are asleep (McMillen *et al*, 1987). By gentle stimulation, they can be encouraged to open their eyes briefly. After 26 weeks of gestation, longer periods of wakefulness can be observed (Fig. 7.6). It is now possible to briefly fix the sight of its mother. According to the EEG pattern, wakefulness can be seen first after 30 weeks. However, it is somewhat dubious to identify wakefulness only by EEG, since patients who have lost their autonomy may exhibit an EEG typical for wakefulness (see D´Allest *et al*, 2002).

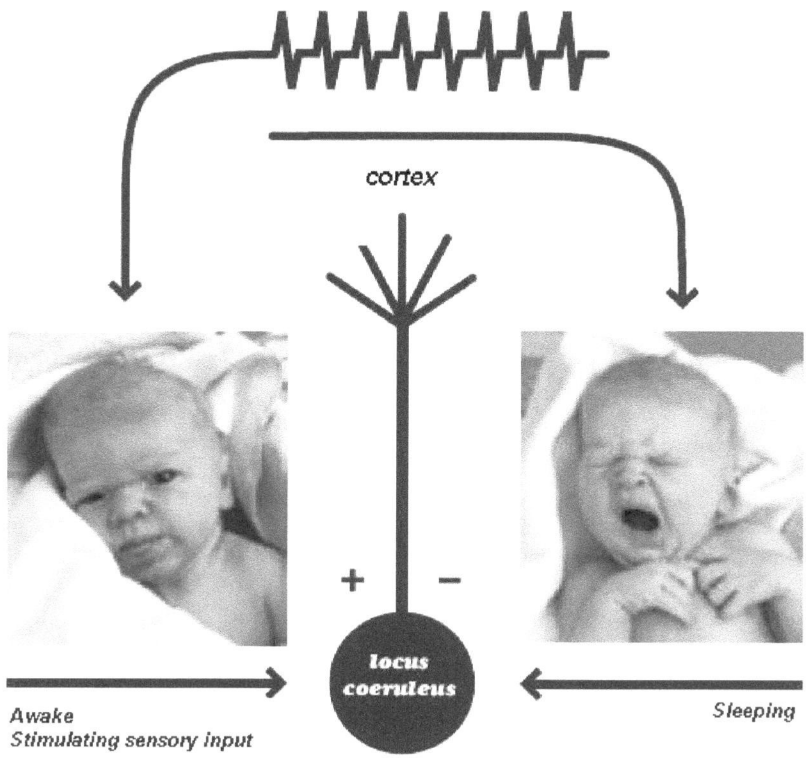

Fig. 7.5 The newborn is awakened and aroused at birth possibly due to activation of the locus coeruleus.

7.5 Self-awareness

According to classical studies self-awareness does not appear until the end of the second year, when the child recognizes itself in a mirror and says "I". However, already a newborn baby seems to have some degree of self-awareness. It reacts differently to tactile stimulation by the mother, as compared to self-stimulation, which it does not respond to (Rochat 2003).

7.6 Showing emotions

Emotions can be defined as external events perceived by the individual results in bodily responses, particularly from the autonomic nervous system. By this definition the newborns certainly show emotions. They seem to be able to activate the same facial muscles as adult and express primary emotions such as joy, disgust, surprise and distress (Tronick, 1989). It reacts with negative emotions to pain (Slater *et al*, 2006), cold temperature, bad smells (Bartocci *et al*, 2001), and tastes. It can also react with positive emotions to sensory stimulation, such as when breastfeeding and seeing their mother. Newborns also react with positive emotions when exposed to harmonic music, but with apnea when listening to disharmonic music or noise (Zentner and Kagan, 1996; Arnon *et al*, 2006).

It was earlier believed that infants younger than six months do not recognize emotions. However, even younger infants seem to react more positively to happy faces than to sad faces (Johnson, 2005).

7.7 Imitation

Imitation is necessary for a higher level of consciousness. Newborn infants can imitate grimaces of an adult, as established in a seminal study by Meltzoff and Moore (1977). This finding was a paradigm shift from the previous view that infants perceived the environment only as a blooming buzzing confusion (James, 1890). It is probably not just a reflex response, because there is some latency.

7.8 *Face processing*

The subcortical face-detection system seems to function at an early stage. Newborns react to figures showing faces with eyes and a mouth presented in a high contrast. We do not know whether face processing is also functioning in extremely preterm infants. There are separate routes for face detection and face identification. Blind-sighted patients seem to detect faces at a subcortical level. Thus, it is likely that the infant perception of faces is subcortical (Johnson, 2005).

Infants like adults seem to have a preference for attractive faces. They are sensitive to the presence of eyes in the face, and dislike scrambled faces (Mehler *et al*, 2002).

7.9 Memory

The most primitive form of memory, i.e. habituation, appears around the 22nd to 23rd gestational week in the human fetus. If the fetus is exposed to a repetitive stimulus, like the vibration of an electric tooth brush, it reacts by movements. However, after multiple stimuli, it seems to remember the stimulus and does not react any longer. There are also reports demonstrating that infants can remember rhythms or musical jingles, which they have been exposed to during fetal life (Hepper, 1996). However, a real representative memory probably does not emerge until the second to fourth month, when babies can experience sensations and emotions, and start to think about objects and events.

Working memory appears around seven months, according to most textbooks in developmental psychology. However, even younger babies seem to remember for example a soother which suddenly disappears. There seems to be some kind of mental representation of faces and things at about two months. Thus, the previously held view that "out of sight is out of mind", may not be true any longer.

7.10 Language

The newborn brain responds specifically to normal speech after a few hours of experience with speech signals outside the womb. This was tested by using NIRS. Newborns were exposed to normal forward speech, as compared with reversed backward speech. A significantly greater activity was found in the left hemisphere when the babies listened to normal speech (Dehaene-Lamberts *et al*, 2002). Babies seem to be born with a language instinct and a built-in number sense (Marcus, 2004).

7.11 Self-regulation

Infants less than three months are difficult to soothe if they are distressed by e.g. pain. Holding and rocking the infant may help. After three months the infant can be quiet by distracting them, bringing attention to other stimuli (Posner and Rothbart, 1998).

8. Integration of the components

It is plausible that the sensory signals are not integrated in the newborn brain to the same extent as later in life. Maybe there is an evolutionary parallel. For example, reptiles are not able to integrate the sight and the smell of a prey (Sjölander, 1999). The snake is governed by sight to strike a mouse, but to start to swallow the head the snake must use the smell of the dead mouse. The snake has no concept of a mouse, no object constancy, according to Piaget. A mouse disappearing behind an obstacle has simply disappeared, it does not exist for the reptile. Reptiles are not able to combine sensing, vision and hearing as the mammal.

Fullterm human infants seem to be able to connect what they see with what they hear (Morrongillo *et al*, 1998). One well-known example of the ability to combine sensory sensations is that newborn infants feeling pain can be calmed by sucking sucrose (Zeifman *et al*, 1996). To achieve an optimal effect, a four week old baby must also see his caregiver. This was not necessary in the two week old babies, indicating that the integration of sensory inputs from different modalities have to mature.

9. Cyber-babies

A virtual baby has been constructed by R. Cotterill in Copenhagen. It cries when it is hungry i.e. when the blood sugar level is low. By pressing a button on the computer, milk is given and the cyber-baby stops to cry. It may remain calm for about 30 min, when it starts to cry again, when the diaper becomes wet. By pressing another button, the diaper can be changed and the baby is happy again. To simulate feelings of hunger and a wet diaper is not so difficult. But we do not believe that the cyber-baby has subjective feelings as the real baby. This is the hard problem (Chalmers, 1996).

The metaphore of the newborn brain as a computer is completely wrong. The newborn brain cannot be compared with some hardware, which is programmed to become a mature adult. The newborn brain is rather a jungle with redundant neurons (Edelman and Tononi, 2000). Maturation is due to programmed cell death and successive selection and organization of the best neuronal pathways with successive construction of new neuronal circuits (Changeux, 2004).

10. When does minimal consciousness emerge?

Even the preterm infant born from 14-12 weeks before term, i.e. the 26th to 28th gestational week, seems to have reached a minimal level of consciousness (Zelazo, 2004). It reacts to tactile, auditory and visual stimulation, as well as to pain at a cortical level, as discussed above. It is awake short periods, shows avoidance reactions to harmful stimuli and seems to establish a minimal eye contact with its mother (Fig. 7.6).

However, the preterm infant born before 26 weeks (< 800 g) does not fulfill these criteria of minimal consciousness, probably due to that the thalamic connections have not yet penetrated the cortical plate. However, the somatosensory cortex does not seem to be essential to perceive pain even in adults. It is possible that the extremely preterm infant can be conscious of pain at a subcortical level.

Preterm infants with a gestational age of 26 weeks and above are usually resuscitated at birth and receive full intensive care, if necessary, in most rich countries. However, there is a lot of controversy concerning the handling of infants born before 26 weeks. Most neonatologists agree that fetuses below 23 weeks should not be resuscitated, since they rarely survive. However, more than 50 % of the infants born at 23 weeks may now survive, although at a high price. Nearly 50 % develop neurological handicap and many of the other infants who look intact at about two years may suffer from cognitive deficiencies (Marlow *et al*, 2005).

In some countries, like the Netherlands, infants below 26 weeks are not taken care of at all. In France, treatment of these extremely preterm infants is terminated, although this practice is not officially recognized. According to the US Supreme Court the fetus is protected from abortion after 22 weeks, why preterm infants down to 22 weeks, or even below, may receive full intensive care in some states. In Canada all

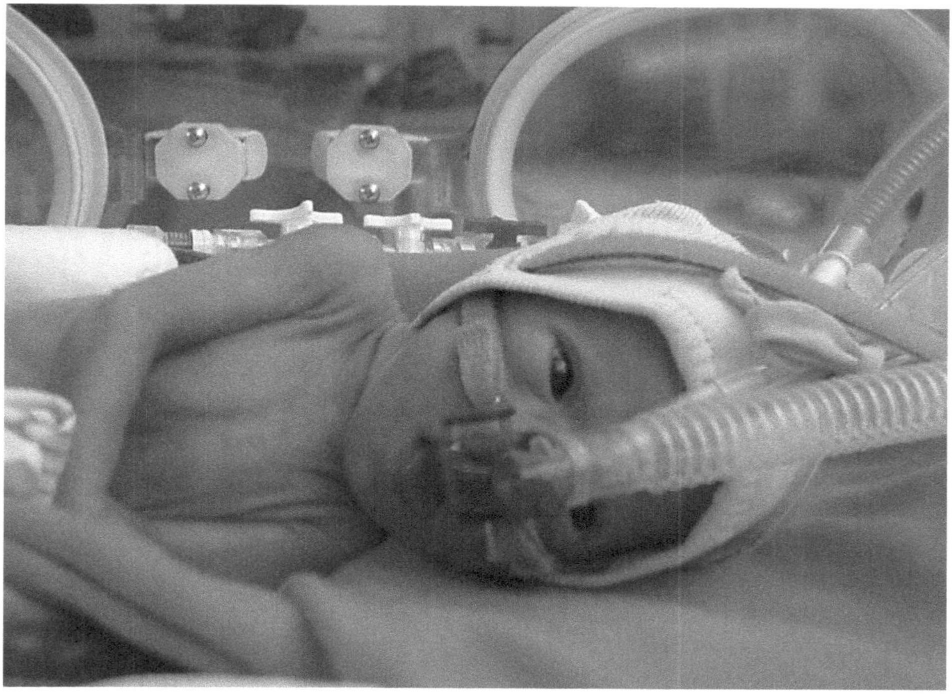

Fig. 7.6 A preterm infant at 26 weeks is awake for short periods and seem to fix the gaze for short moments towards his mother. It also seems to be able to move its arms for protection (self-awareness). Photo by Ann-Sofi Gustafsson. (The parents have given permission to use this photo for publication).

infants from 25 weeks are resuscitated, while infants born in the 23-24th week may receive full treatment if the parents wish. In Sweden, most infants who are born viable after 22 weeks receive tertiary care.

If there is consensus that infants are conscious, although at a minimal level, from the 26th gestational week, they should be treated as any human patients, receiving full intensive care. Maybe it is necessary with a margin, so let's say that the preterm infant has reached a minimal level of consciousness at 25 weeks, i.e. it has a mind and some contact with the mother can be established. Infants born 23rd -24th gestational week seem to lack even a minimal level of consciousness.

They "are designed for intra-uterine existence but surviving without it by virtue of high technology" (Hull *et al*, 1988). The decision that these infants should be resuscitated at birth and receive full intensive care must be taken by the parents, who may have tried *in vitro* fertilization etc. to become pregnant. However, it must also be possible to be less active when caring for these exteriorized fetuses without consciousness or a mind.

Acknowledgement

My own research reviewed in this article has been supported by the Torsten and Ragnar Söderberg foundation. A similar article will be published in Fetal and Neonatal Neurology and Neurosurgery by M. Levene *et al.*

References

Arnon S, Shapsa A, Forman L, Regev R, Bauer S, Litmanovitz I, Dolfin T (2006) Live music is beneficial to preterm infants in the neonatal intensive care unit environment. *Birth* **33**:131-6

Baars BJ (1998) Metaphors of consciousness and attention in the brain. *Trends Neurosci* **21**:58-62.

Bartocci M, Bergqvist LL, Lagercrantz H, Anand KJS (2006) Pain activates cortical areas in the preterm newborn brain. *Pain* **122**:109-117.

Bartocci M, Winberg J, Papendieck G, Mustica T, Serra G, Lagercrantz H (2001) Cerebral hemodynamic response to unpleasant odors in the preterm newborn measured by near-infrared spectroscopy. *Pediatr Res* **50**:324-330.

Bartocci M, Winberg J, Ruggiero C, Bergqvist LL, Serra G, Lagercrantz H (2000) Activation of olfactory cortex in newborn infants after odor stimulation: a functional near-infrared spectroscopy study. *Pediatr Res* **48**:18-23.

Bergson H (1920) *Mind-Energy: Lectures and Essays*. New York: Kolt Co.

Chalmers DJ (1996) *The Conscious Mind: In Search of a Fundamental Theory*. New York: Oxford University Press.

Changeux J-P (2004) *The Physiology of Truth.* Cambridge, MA: Harvard University Press.

Changeux J-P (2006) The Ferrier lecture 1998. The molecular biology of consciousness investigated with genetically modified mice. *Phil Trans R Soc Lond B* **1098**:1-21.

Cohen G, Roux J-C, Grailhe R, Malcolm G, Changeux J-P, Lagercrantz H (2005) Perinatal exposure to nicotine causes deficits associated with a loss of nicotinic receptor function. *PNAS* **102**:3817-3821.

Counter SA (2002) Fetal and neonatal development of the auditory system. In: H Lagercrantz, M Hanson, P Evrard, CH Rodeck (Eds) *The Newborn Brain. Neuroscience and Applications*. Cambridge: Cambridge University Press, pp 226-251.

DÁllest AM, André M (2002) Electroencephalography. In: H Lagercrantz, M Hanson, P Evrard, CH Rodeck (Eds) *The Newborn Brain. Neuroscience and Applications.* Cambridge: Cambridge University Press, pp 339-367.

Dehaene S, Changeux J-P (2004) Neural mechanisms for access of consciousness. In: MS Gazzaniga (Ed) *The Cognitive Neurosciences III.* Cambridge, MA: MIT Press, pp 1145-1154.

Dehaene-Lamberts G, Dehaene S, Hertz-Pannier L (2002) Functional neuroimaging of speech perception in infants. *Science* **298**:2013-2015.

Derbyshire S W G (2006) Can fetuses feel pain. *BMJ* **332**:909-912.

Edelman GM, Tononi GA (2000). *A Universe of Consciousness. How Matter Becomes Imagination.* New York: Basic Book.

Fellman V, Huotilainen M (2006). Cortical auditory event-related potentials in newborn infants. *Semin Fetal Neonatal Med* **11**:452-458.

Gowland P, Fulford J (2004) Initial experience of performing fetal fMRI. *Exp Neuro* **190**:S22-S27.

Gray L, Philbin MK (2004) Effects of the neonatal intensive care unit on auditory attention and distraction. *Clin Perinatol* **31**:243-260.

Hepper PG (1996) Fetal memory: Does it exist? What does it do? *Acta Paediatr Suppl* **416**:16-20

Hull MG, Eddowes HA, Fahy U, Abuzeid MI, Mills MS, Cahill DJ, Fleming CF, Wardle PG, Ford WC, McDermott A (1992) Expectations of assisted conception for infertility. *BMJ* **304**:1465-1469.

Huotilainen M (2006) Magnetoencephalography of the newborn brain. *Semin Fetal Neonatal Med* **11**:437-443

Irestedt L, Dahlin I, Hertzberg T, Sollevi A, Lagercrantz H (1989) Adenosine concentration in umbilical cord blood of newborn infants after vaginal delivery and cesarean section. *Pediatr Res* **26**:106-108.

James W (1890) *The Principles of Psychology* (2 vols) Dover Publications 1950.

Johnson M H (2005) Subcortical face processing. *Nature Neurosci Rev* **6**:766-774.

Koch C (2004) *The Quest for Consciousness: A Neurobiological Approach*. Englewood, CO: Roberts & Company Publishers, 429 p.

Kostovic I, Rakic P (1984) Development of prestriate visual projections in the monkey and human fetal cerebrum revealed by transient cholinesterase staining. *Neurosci* **4**:25-42.

Kuhl PK (2004) Early language acquisition: cracking the speech code. *Nat Rev Neurosci* **5**:831-43

Lagercrantz H (1996) Stress, arousal, and gene activation at birth. *New Physiol Sc* **11**:214-218.

Lagercrantz H, Slotkin T (1986) The stress of being born. *Sci Am* **254**:100-107.

Letinic K, Zoneu R, Rakic P (2002) Origin of GABAergic neurons in the human neocortex. *Nature* **417**:645-649.

Lewis M (2003) The emergence of consciousness and its role in human development. *Ann N Y Acad Sci* **1001**:104-133.

Marcus GF (2004) Something old, something new. *J Child Lang* **31**:934-5

Marlow N, Wolke D, Bracewell MA, Samara M; EPICure Study Group (2005) Neurologic and developmental disability at six years of age after extremely preterm birth. *N Engl J Med* **352**:9-19

McMillen IC, Kok JS, Adamson TM (1987) Development of circadian sleep-wake rhythms in preterm and full-term infants. *Pediatr Res* **29**:381-384.

Meek JH, Firbank M, Elwell CE, Atkinson J, Braddick O, Wyatt JS (1998) Regional hemodynamic responses to visual stimulation in awake infants. *Pediatr Res* **43**:840-843.

Meek J (2002) Optical imaging of infants. *Dev Sci* **5**:271-380.

Mehler J, Dupoux W (2002) *Naitre Humain*. Paris :Odile Jacob.

Mellor D, Diesch TJ, Gunn AJ, Bennet L (2005) The importance of 'awareness' for understanding fetal pain. *Brain Research Reviews* **49**:455-471.

Meltzoff AN, Moore MK (1977) Imitation of facial and manual gestures by human neonates. *Science* **198**:75-78.

Morrongillo BA, Fenwick KD, Chana G (1998) Crossmodel learning in newborn infants. *Infant Behav Dev* **21**:543-553.

Perner J, Dienes Z (2003) Developmental aspects of consciousness. *Conscious Cogn* **12**:63-82

Piaget J (1954) *The construction of reality in the child.* Basic Books N Y.

Posner MI, Rothbart MK (1998) Attention, self-regulation and consciousness. *Phil Trans R Soc Lond B* :1915-1927.

Preissl H, Lowery CL, Eswaran H (2004) Fetal magnetoencephalography: current progress and trends. *Exp Neurol* **190**:S28-S36.

Purpura D P (1982) Normal and abnormal development of cerebral cortex in man. *Neurosci Res Program Bull* 4:569-577.

Rees S, Rawson J (2002) Development of the somatosensory system. In: H Lagercrantz, M Hanson, P Evrard, CH Rodeck (Eds) *The Newborn Brain. Neuroscience and Applications.* Cambridge: Cambridge University Press, pp 177-203.

Rigatto H, Moore M, Cates D (1986) Fetal breathing and behavior measured through a double-wall Plexiglas window in sheep. *J Appl Physiol* **66**:106-114.

Rochat P (2003) Five levels of self-awareness as they unfold early in life. *Conscious Cogn* **12**:717-731.

Schaal B, Hummel T, Soussignan R (2004) Olfaction in the fetal and premature infant: Functional status and clinical implications. *Clin Perinatol* **31**:261-285.

Sjölander S (1999) How animals handle reality. In: A Rigler, AV Stein, M Pesche (Eds) *Does Representation Need Reality*. New York: Plenum Press.

Slater R, Cantarella A, Gallella S, Worley A, Boyd S, Meek J, Fitzgerald M (2006) Cortical pain responses in human infants. *J Neurosci* **26**:3662-3666.

Sowell ER, Thomson PM, Leonard CM, Welcome SE, Kan E, Toga AW (2004) Longitunal mapping of cortical thickness and brain growth in normal children. *J Neurosci* **24**:8223-8231.

Tronick EZ (1989) Emotions and emotional communication in infants *Am Psychol* **44**:112-19

Vanhatalo S, Kaila K (2006) Development of neonatal EEG activity: From phenomenology to physiology. *Semin Fetal Neonatal Med* **11**:471-478.

Vanhatalo S, Lauronen L (2006) Neonatal SEP – Back to bedside with basic science. *Semin Fetal Neonatal Med* **11**:464-470.

van Heteren CF, Boekkooi PF, Jongsman HW, Nijhuis JG (2000) Fetal learning and memory. *Lancet* **356**:1169-1170.

Varendi H, Porter RH, Winberg J (2002) The effect of labor on olfactory exposure learning within the first postnatal hour. *Behav Neurosci* **116**:206-211.

Warnock M (1992) Green College lecture. Ethical challenges in embryo manipulation. *BMJ* **304**:1045-9

Zeifman D, Delaney S, Blass EM (1996) Sweet taste, looking and calm in 2- and 4-week-old infants: The eyes have it. *Developmental Psych* **32**:1090-1099.

Zelazo P D (2004) The development of conscious control in childhood. *Trends Cogn Sci* **8**:12-17.

Zentner MR, Kagan J (1996) Perception of music by infants. *Nature* **383**:29

Chapter 8

An Inside-Out Paradigm for Consciousness and Intelligence

Rodney M.J. Cotterill

What we hope ever to do with ease, we must first learn to do with diligence.
Samuel Johnson (1709-1784)

1. Introduction

This communication is going to be unusual in two respects. First, it is going to contain a number of colloquialisms. Some readers may take this to indicate lack of substance, and I would naturally disagree. I can choose no better an authority than Francis Crick, who once wrote that the average scientific paper is rendered virtually illegible because of its excessive use of specialised jargon and its employment of long convoluted sentences. It has been my experience that scientists tend to be suspicious of any paper that can be immediately understood. This is nonsense; if more authors strove to make their texts easy to read, the scientific literature would be far less cumbersome than it has regrettably become. Secondly, it is not my purpose here to present a debate. I hold some very strong views about consciousness and intelligence, and more than fifteen years of reading the ideas of others have not shaken my belief in the essential correctness of the theory I first tentatively put forward in 1989. This paper presents that theory in its starkest and up-to-date form. If it does not capture every grain of truth about consciousness and intelligence, I believe that it comes very close to doing so nevertheless.

Analyses of consciousness and intelligence are dominated by the stimulus-response paradigm: that consciousness intervenes between detection of sensory input and *possible* preparation of a response. The reader scans these words — a visual stimulus — contemplates their significance, and responds or remains passive. The conscious contemplation seems distinct from the possible response. Recent stimulus-response theories (Eccles, 1992; Edelman, 2003; Crick and Koch, 2003) link consciousness exclusively (or *primarily* — see Crick and Koch, 2003) to signal processing in the cerebral cortex's sensory-receiving regions — the occipital, temporal and parietal lobes, interact-

Consciousness Transitions: Phylogenetic, Ontogenetic, and Physiological Aspects
Edited by Hans Liljenström and Peter Århem

ing with the thalamus, certain nerve cells perhaps playing a vital role (Crick and Koch, 2003) — a *possible* consequence being dispatch of signals to the frontal lobe's muscle-activating regions (Fig. 8.1a). In basic reflexes (Fig. 8.1a) like the knee-jerk or instinctive looking toward a sudden sound or light flash — automatic and unconsciousness — the paradigm *must* be correct. But we will argue that this cannot be validly extended to consciousness, which thus requires a new paradigm.

Treatments of intelligence typically invoke conscious cognition (Anderson, 1992), tacitly supporting the stimulus-response paradigm. It is interesting to include autism in any discussion of intelligence because although there are autistic people with (apparently) normal intelligence, the majority of people having the syndrome are retarded to a certain degree. Lack of a *theory of mind* (see Barkley, 2001) for an updated version) in many autistic people could explain their impaired interaction with others, difficulties with learning and paucity of imaginative play. But the theory-of-mind deficit — brilliantly conceived (Leslie, 1987) and ingeniously discovered (Baron-Cohen *et al.*, 1985) in autism — cannot account for their faulty muscular coordination, bizarre ritualistic behaviour and stubborn resistance to change (Kanner, 1943). As we will see, such aspects of the syndrome are more relevant to the issues of consciousness and intelligence than one might imagine.

2. The new paradigm

We suggest consciousness embellishes a behavioural mode diametrically opposed to the stimulus-response scenario: the relevant stimulus is the animal's own muscular movement, the relevant response being the sensory feedback from the surroundings (Figure 1b). Fundamental for all creatures since life began (Cotterill, 2001), this predated the reflex. But that non-reflex mechanism doesn't automatically produce consciousness. Which evolutionary developments were vital for consciousness's appearance? Why is consciousness advantageous? And how could we believe that the stimulus-response paradigm is reliable? The vital developments were working memory (Baddeley, 1992) and an attention mechanism simultaneously controlling all the sensory pathways; consciousness is an asset because it lets an animal learn and exploit new behavioural patterns within its lifetime (Cotterill, 1994); and an *apparently* passive variant of consciousness — thought — exists because the underlying mechanism involves signalling around closed loops within the brain.

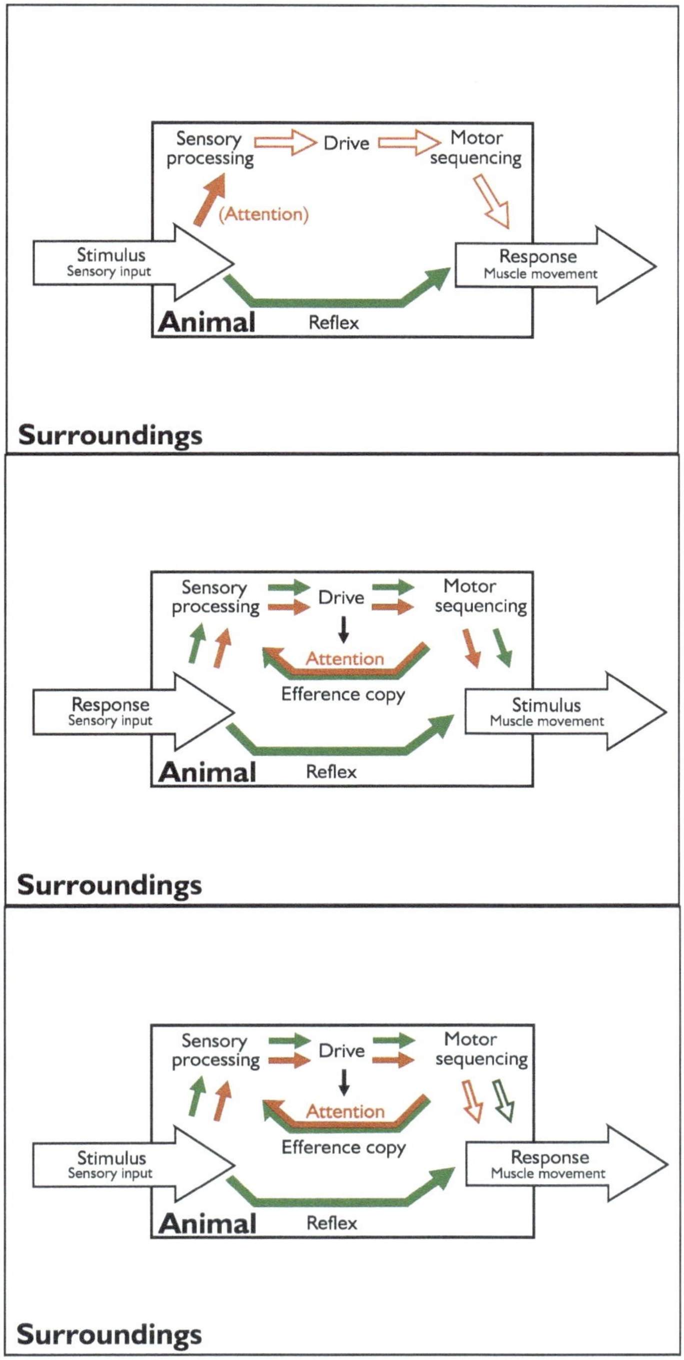
Sensory processing
Drive
Motor sequencing
(Attention)
Stimulus
Sensory input
Response
Muscle movement
Animal
Reflex
Surroundings
Sensory processing
Drive
Motor sequencing
Attention
Efference copy
Response
Sensory input
Stimulus
Muscle movement
Animal
Reflex
Surroundings
Sensory processing
Drive
Motor sequencing
Attention
Efference copy
Stimulus
Sensory input
Response
Muscle movement
Animal
Reflex
Surroundings

Fig. 8.1 (a) The traditional stimulus-response paradigm for consciousness. Consciousness arises solely from processing of sensory input (full red arrow), a *possible* consequence (open red arrows) being activation of a drive mechanism, which selects and activates an appropriate sequence of muscular movements — the response. The drive is provoked by the surroundings, as in the simple reflex (full green arrow). **(b)** The new paradigm for conscious exploration. Muscular movements are the stimulus, as the animal probes its surroundings, and the response is the resulting sensory feedback from those surroundings. The drive comes from within, and controls attention. When an animal consciously probes its surroundings, an internal simulation guides its muscular movements (full red arrows). This simulation precedes the actual movements by a fraction of a second, and it mimics those movements and the anticipated feedback from the surroundings. It thus invokes the appropriate drive-activated schema (upper full green arrows), and any mismatch between the anticipated and actual sensory feedback is available for correcting the movement, and revising the schema. The nervous system is thereby able to update its knowledge of the surroundings. **(c)** The new paradigm for conscious perception. Consciousness does not prevail unless nerve signals have reached the movement-sequencing region of the brain, and activated the appropriate schema (full red arrows), a possible consequence being actual muscular movement (open red arrow). Even apparently passive conscious perception has to activate the movement-sequencing region, which in humans is the premotor cortex (together with the supplementary motor cortex and frontal eye field). The additional time — compared with the shorter time course suggested by Fig. 8.1a — required for the incoming sensory signals to penetrate through to the movement-sequencing region, set up signals returning via the efference-copy route, and thereby permit the signals to self-organise and home in on an appropriate schema, may be the origin of the delay-to-consciousness (Hoover and Strick, 1999). Those self-organising signals circulating around the full red arrow route will display a certain resistance to instantaneous change, this inertial effect underlying the sense of effort, even in the absence of overt muscular movement.

Nerve activity's sole *external* product is muscular movement — an animal's only means of acquiring life's essentials. These are unevenly distributed in the surroundings, so foraging movements — consuming energy and other biochemical resources — must not be haphazard. Information is needed, this also being obtained through muscular movement, which need only be covert (virtual) if the animal is sufficiently sophisticated and also sufficiently experienced. The overt (actual) need for movement is obvious to a blind person, tactile feedback signals announcing that a reached-for object has been located. Even in vision, which seems muscularly passive, the eyes abruptly change direction about four times a second, searching for visual clues (Yarbus, 1967). The driving force for (possible) muscular movement originates externally in the stimulus-response paradigm (Fig. 8.1a), and *always* in the reflex. It stems from *within* the animal in the new paradigm (Fig. 8.1b), the drive ultimately serving homeostasis — the body's aim to maintain the internal status quo; hence *inside-out.*

Putting it colloquially, every muscular movement asks questions, though not necessarily consciously. Each step taken by an experienced walker asks *Is the floor still there*? And the system just as unconsciously records the answer *Yes, the floor's still there*! Only if the feet (or eyes) encounter something unexpected does the attempted step become the focus of consciousness. But novice walkers and experienced walkers negotiating unusual terrain must be conscious of every step; diligence is then paramount.

The coupling of an unconsciously posed question with an unconsciously recorded answer is implicit in the term *schema* (plural: *schemata*), an updated definition (Cotterill, 2001) being *concurrently activated motor planning and anticipated feedback patterns corresponding to a cognitive element*, the words *planning* and *anticipated* permitting inclusion of thought in the absence of overt movement. So each successful step when walking invokes the schema *Is the floor still there*? *Yes, the floor's still there*! A schema is thus a question equipped with its own answer, and provides a basis for all cognition; we recognise things by fitting them to the relevant schema, and our knowledge is the totality of our remembered schemata. A schema can be used automatically and unconsciously, as a *context-specific reflex*. Reflexes, innate or learned, are *ballistic* whereas conscious behaviour is a special type of *guided* (or servo) mode.

3. The consciousness mechanism

Schema acquisition and use must be fundamental in behaviour. How are new ones learned? Primarily through consciousness, we believe, though subtle augmenting mechanisms may exist (Craig, 2003). Internal drive chooses the sequence of movements best suited to achieving the system's immediate goal, the link between goal and sequence being stored in memory through prior experience (or inheritance). But the surroundings may have changed since that experience, so there must be a check that the feedback is as expected; so again putting it colloquially, *an activated schema is like a hypothesis seeking confirmation or revision*. The system does this by alerting (Cotterill, 1995) the sensory receiving areas to the imminent arrival of feedback, the apprising mechanism involving dispatch to those areas of signals that are duplicates — *efference copies* — of the ones sent to the muscles. Other signals arising from any mismatch between the actual and anticipated feedback (Schmahmann and Sherman, 1998; Allen, *et al,* 2004; Schmahmann and Sherman, 1998) are available for correcting the ongoing muscular movements, in the standard servo manner (Cotterill, 1996). Such corrections do not require consciousness, however, a typical unconscious example being the mechanism that holds the image of an object at a fixed position on the retina (von Holst, 1957), despite head movement.

The vital extra ingredients required for consciousness are working memory — because an answer is useless if the question has been forgotten — and a mechanism to control attention — because a question may be useless if misdirected, attention thus being more related to the question than the answer. When conscious, we can (a) think without moving, (b) move while thinking about that movement, and (c) move while thinking about something else. In mode (c), thinking requires more of our attention than the movement; *thoughts are never unattended.* In mode (b), attending to our drive-provoked movement, schema activation will have alerted the sensory receptors, through the above-described efference-copy mechanism (Cotterill, 1995). But detection of simultaneous sensory input that is *not* a consequence of our movement could lead to acquisition of a false schema, without utility for the system. (Tapping an infant's forehead every time its hand touched the side of its cot could make its nervous system conclude that the forehead was located there.)

The system distinguishes between concurrent provoked and unprovoked inputs — between touching and being touched, for example — by subjecting the just-activated schema to attention stemming from the drive, and briefly holding a record of the former in working memory. Unprovoked sensory input will then be perceived as such because it will not have been associated with efference copy — it will not satisfy the sensory expectations of the already-activated schema. The resulting mismatch may then shift the attention. So our consciously performed overt actions are immediately preceded by covert (imagined) versions, the latter guiding the former (Fig. 8.1b).

But how do we recognise sensory input we have not provoked? It has to generate signals penetrating to the brain's movement-sequencing region, this returning efference-copy signals to the sensory receptors (Fig. 8.1c), the looping signals permitting the system to lock on to an appropriate schema — inevitably consuming time, and producing the observed delay-to-consciousness (Libet *et al,* 1979). The octave illusion endorses this, perception of whether a two-tone sequence is ascending or descending correlating with the frequency range in the subject's own voice (Deutsch, 1992). But how do we perceive input we are *unable* to provoke, like sound above our vocal range? It too generates looping signals, and when the system learns to recognise them, the corresponding schema permits us to say *high pitch*, for example. Perceptual limitation is thus in the ranges of the sensory receptors, not in the movement-sequencing region — and this leads to a prediction.

We predict the existence of neurons in the premotor cortex — *premotor metalink neurons* — involved in perception despite not necessarily dictating movement sequences mimicking what is perceived — as when certain neurons mediate perception of sound lying outside one's vocal range, subsequently directing intonation of those words *high pitch.* Metalink neurons belonging to a (minority) simple default subset *would* mimic what is perceived, and these are probably identical with the *mirror neurons* already reported (Rizzolatti *et al,* 1996). The existence of premotor metalink neurons that are *not* mirror neurons possibly explains the recently discussed inability of the latter to account for all types of perception (Jacob and Jeannerod, 2005).

When drive-provoked movement produces the desired feedback from the surroundings, the drive is presumably discontinued. Mismatch between the anticipated and actual feedback, conversely, must maintain the drive. Deposition in working memory of the record of a movement-embodied question — actual or merely imagined — must precede arrival of the corresponding answer — actual or merely anticipated — so if working memory labels time of arrival, the system will link cause to effect, serving as time's arrow, and underlying recognition that one is the agent of one's actions. But the sense of self must stem from the system's gradually acquired ability to distinguish between self-provoked and externally arising sensory input.

4. Anatomical and physiological basis

One learns little by merely touching unseen objects — much more when exploring them through sequenced movements. Preconscious creatures commonly use sequenced movements (Cotterill, 2001), their nervous systems comprising a *sequence-orchestrating component* (counterpart of the mammalian premotor, supplementary motor and frontal eye areas) in addition to the *muscle-activating component* (counterpart of the mammalian primary motor area). The former controls the latter, when the system uses all its options. The sequence-orchestrating component is thus better suited as the source of efference-copy paths, and in animals possessing consciousness these feedback routes provide links to all the sensory receptor areas, and are subject to a global attention mechanism acting close to the sequence orchestrator.

In mammals, the (composite) thalamic intralaminar nucleus — candidate for attention's focal point — receives signals from the premotor, supplementary motor and frontal eye areas (Cotterill, 2001), and projects to all sensory input areas (Groenewegen and Berendse, 1994) and the basal ganglia (Hoover and Strick, 1999). It is indispensable to consciousness, its output being controlled by the nucleus reticularis thalami, which serves attention (Yingling and Skinner, 1977). That latter nucleus is controlled by the brainstem reticular formation (Singer, 1977), which governs the sleep-wake cycle. The basal ganglia presumably mediate drive (Schultz, 1999); like the intralaminar nuclei, they have widespread connections to the cerebral cortex (Schultz, 1999).

The sequence orchestrator selects muscular movement patterns not threatening the skeletal musculature's overall integrity (Cotterill, 1995). It guides the system through muscular hyperspace — hyperspace because we must consider *all* the muscles simultaneously — tracing a trajectory that produces the serial nature of consciousness. Such trajectory tracing and the *rapid* detection of sensory feedback signals, makes the animal thereby aware that it is the agent of its own actions. Just as the more familiar form of navigation would be impossible if no record were kept of the path recently followed, so is viable navigation through muscular hyperspace dependent upon brief (working) memory of what the muscles have been recently doing, in the service of self-paced probing of the surroundings.

So this mechanism acts globally but is centrally (and thus *dictatorially* (Cotterill, 1995)), not democratically (Eccles, 1965)) controlled, possibly (Cotterill, 1995) underlying the unity of conscious experience. It precludes a binding problem by forcing the senses to cooperate, permitting cross-modal perception; the animal can recognise feedback received exclusively through one type of sensory receptor on the basis of what has previously been received only through a different type (as in recognising solely by touch, when blindfolded, something previously only seen). In conscious animals, schemata must be stored in a distributed and superimposed manner. But what can one say about the dynamics of their selection?

This must occur within the time-span of a cognitive event. How brief may this be? We noted the roughly 250 ms between movements of the eyes (Yarbus, 1967), but what about other muscles, as they mediate probing of the surroundings? Top male sprinters take 45-50 steps when covering 100 m in about 10 s, their movements being guided because they can avoid a suddenly appearing obstacle; this too indicates around

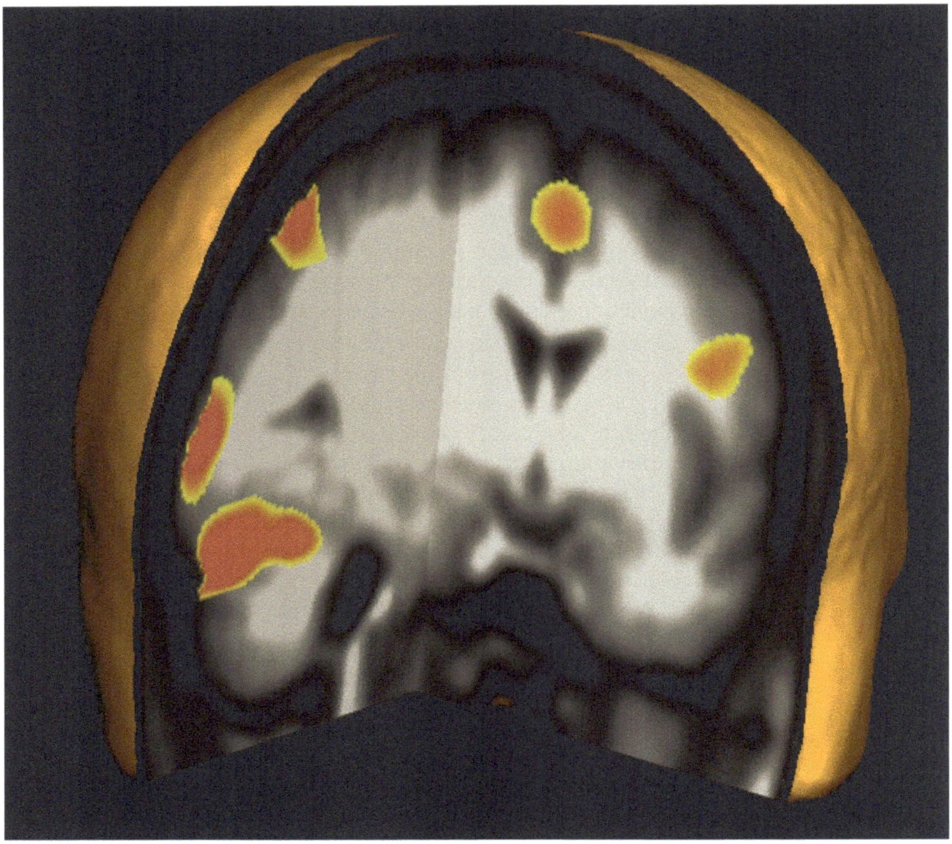

Fig. 8.2 Brain regions activated during perception. This fMRI image shows the brain regions activated when a subject passively observes an abstract visual pattern (Siegel *et al,* 1998), and reveals that the movement-sequencing region of the brain — the premotor cortex — is indeed activated under these circumstances, in agreement with Fig. 8.1c. It shows a cut-away view of the brain from the right rear of the head, the red and yellow colours indicating the areas maximally activated. Progressing from right to left, the areas shown being activated are the right premotor cortex; the right and left anterior cingulates (which lie medially, and which are known to be implicated in attention); the left parietal cortex (which has been linked to visual processes serving action); and finally, various early visual areas.

200 ms for a cognitive event. And cellular responses in a monkey's frontal eye field fail to differentiate between targets and distractors until around 200 ms (Thompson, 1996); selection in the sequence-orchestrating area is clearly the rate-limiting process. Even seemingly passive conscious perception involves activation of the premotor, supplementary or frontal eye areas (Cotterill, 1997), as recently demonstrated (Christensen *et al* (2004) by functional magnetic resonance imaging (fMRI) (Fig. 8.2). But why the observed delay-to-consciousness (Libet *et al,* 1979) is about *twice* the 200 ms remains unclear.

Schema consolidation could be dangerous. When the surroundings are being explored, schemata provisionally stored relate only to the current context, the accompanying evaluation guaranteeing against *immediate* mutual conflict. But there could be clashes with the numerous schemata already in long-term memory, some relevant to vital muscular movements like those serving breathing. (I will hazard a guess that this could be the cause of sudden cot death in human infants, and am currently preparing a manuscript which discusses that possibility.) The system could engineer compatibility with these well-entrenched schemata while the skeletal musculature is immobilised during dreaming. There is evidence (Stickgold *et al,* 2001) of integration of memories during human sleep, including the dreaming phase, particularly for *procedural memory* – memory for oft-used unconscious muscular sequences. The porpoise would risk drowning if conflict with its breathing mechanism arose during schema consolidation involving the entire brain, so the cerebral hemispheres take turns in sleeping, one side always being awake.

Presumably only those animals sufficiently high on the evolutionary scale possess consciousness, but where is the dividing line? Perhaps the watershed species are the *monotremes*, that mammalian sub-class comprising the duck-billed platypus and the two genera of *spiny anteater*. Very primitive, and only distantly related to other mammals, they lay eggs and display many other reptilian characteristics, though having hair and secreting milk. Zoo keepers often say that spiny anteaters never learn from experience. Their brains lack properly developed thalamic intralaminar nuclei, and the associated nucleus reticularis thalami (Jones, 1985) – the key component of the attention apparatus, without which consciousness should be precluded. Lack of frontal lobe activity during the spiny anteater's sleep (Siegel *et al,* 1998) indicates absence of dreaming, though certain brainstem activity does suggest rapid eye movements (Siegel *et al,* 1998), which perhaps evolved earlier, possibly serving to massage the corneas' vulnerable blood vessels. Two reliable indicators of consciousness may thus be presence of *dreaming* and presence of *play*. When playing, animals explore their surroundings and interact with their peers, challenging both to react in novel ways, thereby providing consciousness with its *raison d'être*. In so doing, they acquire new schemata that are candidates for permanent storage in their behavioural repertoires — hence the need for dreaming. All true mammals play, and all dream; they may all be conscious, therefore, despite their varied levels of intelligence.

5. Intelligence and autism

Intelligence may be the ability to acquire and manipulate schemata, one factor being working memory span. fMRI reveals a (linear) correlation [38] between intelligence and activation of the prefrontal cortex, which is involved in working memory (Fuster, 1985); Klingberg *et al,* 2002) (though working memory capacity additionally involves an executive attention component (Conway *et al,* 2003)).

Although autism and low intelligence often appear together, they are distinct (Bartak and Rutter, 1976). Autism may have a broader aetiology (and thus occur less frequently). The guided mode epitomising consciousness may be undermined by its complexity, compared with the ballistic mechanism. Several brain components must cooperate for optimal behaviour, and the possibility of autism may stem from its sophistication (Just *et al*, 2004; Minshew and Goldstein, 2001; Courchesne *et al*, 1994).

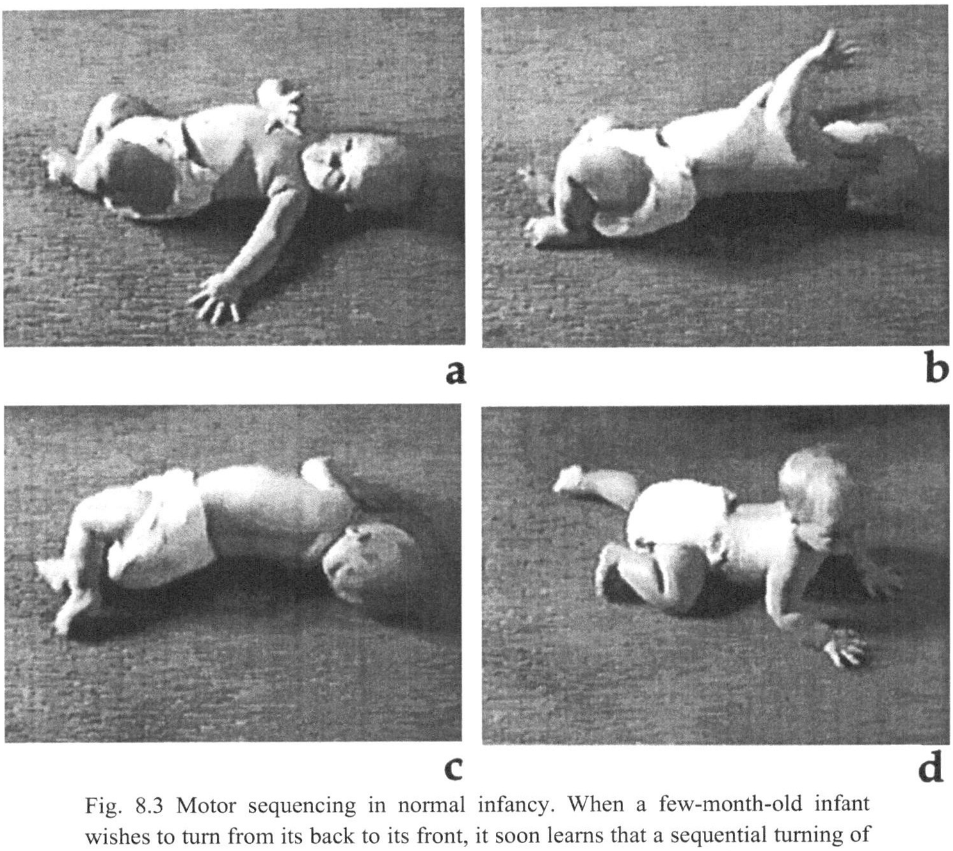

Fig. 8.3 Motor sequencing in normal infancy. When a few-month-old infant wishes to turn from its back to its front, it soon learns that a sequential turning of the head, then the shoulders, and finally the hips, produces the supine-prone transition with the minimum of effort. (Picture sequence reproduced with the kind permission of Philip Teitelbaum (Teitelbaum *et al,* 1998)).

Working memory, attention, drive and the efference-copy route can all malfunction. Difficulty controlling attention has been observed in autistic people (Courchesne *et al*, 1994), and diminution of (drive-providing) basal ganglia function impairs learning (Brainard and Doupe, 2000). Our proposed link between movement and cognition is endorsed by the under-developed cerebellum in many autistic people (Ito, 1993; Bauman, 1991).

There is a genetic factor in autism (Folstein and Rutter, 1977), and multiplicity of vulnerable sites in the schema-manipulating mechanism is consistent with autism stemming from faults at multiple genetic loci (Pickles *et al*, 1995), as verified by genetic screening (International Molecular Genetic Study of Autism Consortium, 2001). Finally, a study of movement impairment in infants later diagnosed with autism supports this conjecture, the victims encountering difficulty in learning sequences of movements (Teitelbaum *et al,* 1998) (compare Figs. 8.3 and 8.4). They apparently *do* find it hard to acquire new schemata. This could underlie their apparent preoccupation with repetitive movement, and their resistance to change (Kanner, 1943). It could also be the reason why even high functioning autistic people tend to take language too literally.

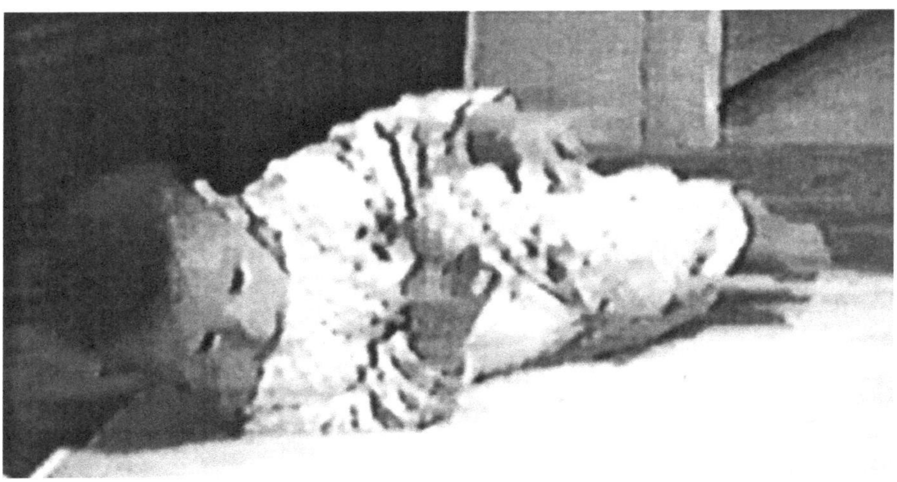

Fig. 8.4 Impaired motor sequencing in autistic infancy. The few-month-old autistic infant encounters great difficulty in smoothly executing the supine-prone transition. Instead, he/she awkwardly arches the back, and ultimately falls onto his/her front. Such an autistic child clearly has difficulty in acquiring the schemata that govern whole-body movement, just as he/she will later encounter problems in learning the schemata involved in language, during the second year of life. (Picture reproduced with the kind permission of Philip Teitelbaum (Teitelbaum *et al,* 1998)).

In sum, all this evidence strengthens our belief that the time is ripe for the proposed change of paradigm, and that the mechanism of schema manipulation holds the key to understanding consciousness and intelligence, and indeed autism to a certain extent.

6. Epilogue: A physics of thought?

There are hints in this article of what could be called a physics of thought. It might sound like an unlikely marriage, given thought's ethereal nature and the sheer down-to-earthedness of physics. But when we check the dictionary definition of physics — the treating of properties of matter and energy, or the action of different forms of energy on matter — we immediately glimpse the possibility of a connection. Thoughts, as we have seen, are mediated by the flow of nerve signals, which undeniably use electrochemical energy. And the movement of muscles — actual or merely imagined — is simply an example of matter in motion. But we can go much further than this. We have already seen that a time period of about 200 ms characterises the movements both of the darting eyes and the running legs. And the same period is encountered when we contemplate speech; the record for articulation of English words lies at between four and five per second, and it is highly relevant that this is also the highest rate at which English can be understood. Finally, we have seen that a period of about 200 ms is required by the signals in the nervous system if a target is to be distinguished from a distractor. This is highly suggestive of a need for the signals to self-organise.

The really important thing in all this is the equality of the period required for that process of self-organisation and the maximum rate at which the muscles can move in a guided — as opposed to a ballistic — manner. There is the suggestion here of what is generally referred to as impedance matching. This is certainly down-to-earth, and a far cry from the ephemeral quality that thought is widely believed to possess. It is as if the nervous system has evolved in such a manner as to match the inertial properties of the skeletal musculature, and not to waste effort on functioning much faster than the muscles can move.

One last point can be made, despite the risk of overstretching these ideas. The existence of a minimum period for both guided muscular movement and the self-organisation of the neural signals that accomplish that guiding shows that we may not imagine time as being strictly continuous in the cognitive domain. It comes in quanta of about 200 ms, and we can thus never have full knowledge of anything occurring on a smaller time scale. The system has clearly compensated for this shortcoming by developing components such as the cerebellum, which mediate interpolation and extrapolation, but the movements they permit have to be of the ballistic type, not the guided type that is the province of consciousness. There is thus what we could call an indeterminacy principle. It is not like the indeterminacy principle that applies to the quantum domain, but it is indeterminacy nonetheless.

These first remarks about the possibility of a physics of thought are clearly very tentative, but they seem to hold the promise of further elucidation of consciousness if adequately pursued.

Acknowledgements

Warm thanks are due to Ann Butler, Erik Fransén, John Geake, Hans Liljenström, Ingemar Lindahl, Paul Manger, Bjørn Nielsen, Claus Nielsen, John Skoyles, Philip Teitelbaum, Peter Århem and the Carlsberg Foundation.

References

Allen G, Müller R, Courchesne E (2004) Cerebellar function in autism: functional magnetic resonance image activation during a simple motor task. *Biol Psychiatry* **56**:269-278.

Anderson M (1992) *Intelligence and Development.* Blackwell.

Baddeley AD (1992) Working memory. *Science* **255**:556-559.

Barkley RA (2001) The executive functions of self-regulation: an evolutionary neuropsychological perspective. *Neuropsych Rev* **11**:1-29.

Baron-Cohen S, Leslie A, Frith U (1985) Does the autistic child have a "theory of mind"? *Cognition* **21**:37-46.

Bartak L, Rutter M (1976) Differences between mentally retarded and normally intelligent autistic children. *J Autism and Childhood Schizophrenia* **6**:109-120.

Bauman ML (1991) Microscopic neuroanatomic abnormalities in autism. *Pediatrics* **87**:791-796.

Belmonte MK, Allen G, Beckel-Mitchener A, Boulanger LM, Carper RA, Webb SJ (2004) Autism and abnormal development of brain connectivity. *J Neurosci* **24**:9228-9231.

Brainard MS, Doupe AJ (2000) Interruption of basal ganglia-forebrain circuits prevents plasticity of learned vocalizations. *Nature* **404**:762-766.

Christensen MS, Madsen KH, Lund TE, Sidaros K, Cotterill RMJ (2004) Ventral premotor cortex involved in conscious perception. *NeuroImage* **22**:614-617.

Conway ARA, Kane MJ, Engle RW (2003) Working memory capacity and its relation to general intelligence. *Trends Cogn Sci* **7**:547-552.

Cotterill RMJ (1994) Autism, intelligence and consciousness. *Biol Skr Dan Vid Selsk* **45**:1-93.

Cotterill RMJ (1996) Prediction and internal feedback in conscious perception. *J Consciousness Studies* **3**:245-266.

Cotterill RMJ (1997) On the mechanism of consciousness. *J Consciousness Studies* **4**:231-247.

Cotterill RMJ (2001) Cooperation of the basal ganglia, cerebellum, sensory cerebrum and hippocampus: possible implications for cognition, consciousness, intelligence and creativity. *Prog Neurobiol* **64**:1-33.

Courchesne E, Lincoln AJ, Townsend JP, James HE, Akshoomoff NA, Saitoh O *et al* (1994) A new finding: impairment in shifting attention in autistic and cerebellar patients. In: SH Broman and J Grafman (Eds) *Atypical Cognitive Deficits in Developmental Disorders: Implications for Brain Function.* Mahwah NJ: Erlbaum, pp 101-137.

Crick F, Koch C (2003) A framework for consciousness. *Nat Neurosci* **6**:119-126.

Deutsch D (1992) Some new pitch paradoxes and their implications. *Philos Trans R Soc Lond Ser B* **336**:391-397.

DeVito JL (1969) Projections from cerebral cortex to intralaminar nuclei in monkey. *J Comp Neurol* **136**:193-201.

Eccles JC (1965) *The Brain and the Unity of Conscious Experience.* Cambridge: Cambridge University Press

Eccles JC (1992) Evolution of consciousness. *Proc Natl Acad Sci USA* **89**:7320-7324.

Edelman GM (2003) Naturalizing consciousness: a theoretical framework. *Proc Natl Acad Sci USA* **100**:5520-5524.

Folstein S, Rutter M (1977) Infantile autism: a genetic study of 21 twin pairs. *J Child Psychology and Psychiatry* **18**:297-321.

Fuster JM (1985) The prefrontal cortex, mediator of cross-temporal contingencies. *Hum Neurobiol* **4**:169-179.

Gao JH, Parsons LM, Bower JM, Xiong J, Li J, Fox PT (1996) Cerebellum implicated in sensory acquisition and discrimination rather than motor control. *Science* **272**:545-547.

Geake JG, Hansen PC (2005) Neural correlates of intelligence as revealed by fMRI of fluid analogies. *NeuroImage* **26**:555-564.

Graybiel AM (2000) The basal ganglia. *Curr Biol* **10**:R509-R511.

Groenewegen HJ, Berendse HW (1994) The specificity of the "non-specific" midline and intralaminar thalamic nuclei. *Trends Neurosci* **17**:52-57.

Hoover JE, Strick PL (1999) The organization of cerebellar and basal ganglia outputs to primary motor cortex as revealed by retrograde intraneuronal transport of herpes simplex virus type I. *J Neurosci* **19**:1446-1463.

International Molecular Genetic Study of Autism Consortium. (2001) A genomewide screen for autism: strong evidence for linkage to chromosomes 2q, 7q and 16p. *Am J Hum Genet* **69**:570-581.

Ito M (1993) Movement and thought: identical control mechanisms by the cerebellum. *Trends Neurosci* **16**:448-450.

Jacob P, Jeannerod M (2005) The motor theory of social cognition: a critique. *Trends Cogn Sci* **9**:21-25.

Jones EG (1985) *The Thalamus.* Plenum

Just MA, Cherkassky VL, Keller TA, Minshew NJ (2004) Cortical activation and synchronization during sentence comprehension in high-functioning autism: evidence of underconnectivity. *Brain* **127**:1811-1821.

Kanner L (1943) Autistic disturbances of affective contact. *Nervous Child* **2**:217-250.

Klingberg T, Forssberg H, Westerberg H (2002) Increased brain activity in frontal and parietal cortex underlies the development of visuospatial working memory capacity during childhood. *J Cogn Neurosci* **14**:1-10.

Leslie AM (1987) Pretence and representation: the origins of "theory of mind". *Psychol Rev* **94**:412-426.

Libet B, Wright EW, Feinstein B, Pearl DK*l* (1979) Subjective referral of the timing for a conscious sensory experience: a functional role for the somatosensory specific projection system in man. *Brain* **102**:193-224.

Minshew NJ, Goldstein G (2001) The pattern of intact and impaired memory functions in autism. *J Child Psychology and Psychiatry* 42:1095-1101.

Pickles A, Bolton P, Macdonald H, Bailey A, Le Couteur A, Sim CH, Rutter M (1995) Latent-class analysis of recurrence risks for complex phenotypes with selection and measurement error: a twin and family history study of autism. *Am J Hum Genet* 57:717-726.

Rizzolatti G, Fadiga L, Gallese V, Fogassi L (1996) Premotor cortex and the recognition of motor actions. *Cogn Brain Res* **3**:131-141.

Schmahmann JD, Sherman JC (1998) The cerebellar cognitive syndrome. *Brain* **121**:561-579.

Schultz W (1999) The primate basal ganglia and the voluntary control of behaviour. *J Consciousness Studies* **6**:31-45.

Siegel JM, Manger PR, Nienhuis R, Fahringer HM, Pettigrew JD (1998) Monotremes and the evolution of rapid eye movement sleep. *Phil Trans R Soc Lond B Biol Sci* **353**:1147-1157.

Singer W (1977) Control of thalamic transmission by corticofugal and ascending reticular pathways in the visual system. *Physiol Rev* **57**:386-420.

Stickgold R, Hobson JA, Fosse R, Fosse M (2001) Sleep, learning and dreams: off-line memory reprocessing. *Science* **294**:1052-1057.

Thompson KG (1996) Perceptual and motor processing stages identified in the activity of macaque frontal eye field neurons during visual search. *J Neurophys* **76**:4040-4055.

Teitelbaum P, Teitelbaum O, Nye J, Fryman J, Maurer RG (1998) Movement analysis in infancy may be useful for early diagnosis of autism. *Proc Natl Acad Sci USA.* **95**:13982-13987.

von Holst E (1957) Aktive Leistungen der Menschlichen Gesichtswahrnemung. *Studium Generale* **10**:231-243.

Yarbus AL (1967) *Eye Movements and Vision.* Plenum

Yingling CD, Skinner JE (1977) Gating of thalamic input to cerebral cortex by nucleus reticularis thalami. In: JE Desmedt (Ed) *Attention, Voluntary Contraction and Event-Related Potentials.* Basel: Karger, pp 30-69.

Chapter 9

Consciousness Without a Cerebral Cortex†

Björn Merker

1. Introduction

The four semi-independent pacemakers of the non-cephalised nervous system of the cubomedusa equip this predatory jellyfish with flexible directional locomotor responsiveness to asymmetric sensory inputs (Satterlie and Nolen, 2001). There is no reason to assume that the environmental guidance thus supplied by its radially arranged nerve net, involves or gives rise to experience of any kind. Our own environmental orientation, on the other hand, commonly takes place in a state of wakefulness we call conscious, which typically involves seeing, hearing, feeling, or other kinds of experience. Somewhere between medusa and human there is a transition to conscious function, and the nature of the capacity it bestows has exercised psychology, neuroscience, and cognitive studies virtually since their inceptions (Adrian *et al*, 1954; Baars, 1988; James, 1890/1983; Mandler, 1975).

There is no compelling reason to think that nervous systems more complex than those of the medusa, and capable of performing more sophisticated functions, should not also perform in a perpetual night of unconsciousness. The fact that not all of them do so suggests that consciousness has some role or function to fill in the neural economy of brains thus endowed (Searle, 1992). In exploring what this might involve, the exclusive concern throughout what follows will be with consciousness in its most basic and general sense, that is, as the state or condition presupposed by any experience whatsoever. Given recent proliferation of terminology surrounding the concept of consciousness (see Morin, 2006 for a useful analysis and integration), the following additional remarks should help place this usage in context.

---

† "Consciousness without a cerebral cortex: A challenge for neuroscience and medicine" by Björn Merker appeared in *Behavioral and Brain Sciences* **30**:63-81 (2007). Reprinted with kind permission of Cambridge University Press.

Consciousness Transitions: Phylogenetic, Ontogenetic, and Physiological Aspects
Edited by Hans Liljenström and Peter Århem

As employed here, the attribution of consciousness is not predicated upon any particular level or degree of complexity of the processes or contents that constitute the conscious state, but only upon whatever arrangement of those processes or contents makes experience itself possible. To the extent that any percept, simple or sophisticated, is experienced, it is conscious, and similarly for any feeling, even if vague, or any impulse to action, however inchoate. This agrees well with the type of dictionary definition that renders consciousness as "the state or activity that is characterized by sensation, emotion, volition, or thought" (*Webster's Third New International Dictionary*, unabridged edition, 1961). In this basic sense, then, consciousness may be regarded most simply as the "medium" of any and all possible experience.

With regard to the way in which this medium might be implemented neurally, the present treatment is committed to an architectonic rather than a quantitative (or "graded") view. That is, as here conceived, a conscious mode of functioning is dependent upon quite specific neural arrangements creating interfaces of particular kinds between specific domains of neural function, rather than a result of a general increase in informational capacity or complexity achieved by expansion of a structural substrate which below a certain size does not support consciousness. Thus, what disqualifies the medusa nerve net in this regard is *not* its simplicity, but its lack of specific structural arrangements required to support conscious function. Given an arrangement capable of supporting consciousness, its contents may differ widely in complexity or sophistication. The range of possibilities in this regard is felicitously captured by the "scale of sentience" of Indian tradition (Bagchi, 1975), as follows:

> "This."
> "This is so."
> "I am affected by this which is so."
> "So this is I who am affected by this which is so."

Each "stage" in this scale, from mere experienced sensation to self-consciousness, falls within the compass of consciousness as here defined, and presupposes it. Accordingly, to see, to hear, to feel, or otherwise to experience something is to be conscious, irrespective of whether in addition one is aware that one is seeing, hearing, and so forth, as cogently argued by Dretske (1993; see also Merker, 1997; Searle, 1992). Such additional awareness, in reflective consciousness or self-consciousness, is one of many contents of consciousness available to creatures with sophisticated cognitive capacities. However, as noted by Morin (2006), even in their case, it is present only intermittently, in a kind of time-sharing with more immediate, unreflective experience. To dwell in the latter is not to fall unconscious, but to be unselfconsciously conscious. Reflective awareness is thus more akin to a luxury of consciousness on the part of certain big-brained species, and not its defining property.

The exploration of the constitution of the conscious state to be pursued here will yield a conception of its functional role revolving around integration for action. As such, its functional utility will turn out to be independent of the level of sophistication at which the contents it integrates are defined. This opens the possibility that the evolution of its essential mechanisms did not have to await advanced stages of cortical development, but took place independently of it. As we shall see, certain fundamental features of vertebrate brain organization suggest that key mechanisms of consciousness are implemented in the midbrain and basal diencephalon, while the telencephalon serves as a medium for the increasingly sophisticated elaboration of conscious contents.

With some notable exceptions (e.g., Bogen, 1995; Brown, 1989; Panksepp, 1982; Parvizi and Damasio, 2001; Scheibel and Scheibel, 1977; Sewards and Sewards, 2000; Thompson, 1993; Watt, 2000), brainstem mechanisms have not figured prominently in the upsurge of interest in the nature and organization of consciousness that was ushered in with cognitivism in psychology and neuroscience (Baars, 1988; Mandler, 1975; Miller, 1986). Few cognitivists or neuroscientists would today object to the assertion that "cortex is the organ of consciousness."[1] This is, in a sense, a return to an older view of the supremacy of the cerebral cortex from which a fundamental discovery of the late 1940s had stimulated a partial retreat. In keeping with the sense that the cerebral cortex is the organ of higher functions, it had been widely assumed that the regulation of its two primary states – sleep and wakefulness – was a cortical function, as well (see, e.g., the critical discussion of this stance in Gamper, 1926). Then, in the late 1940s, Moruzzi and Magoun (1949) discovered that local stimulation of circumscribed cell groups in the pons and midbrain of experimental animals exerts a global activating influence on the cerebral cortex as well as on behavioural state, and that experimental lesions in these brainstem sites are capable of rendering animals somnolent and even comatose (Magoun, 1954; cf. Parvizi and Damasio, 2003). This came as a shock to the corticocentric perspective, and stimulated an avalanche of research on brainstem regulation of sleep and wakefulness and its relationship to the conscious state (summarized in symposium volumes edited by Adrian *et al*, 1954; Jasper *et al*, 1958; Eccles, 1966).

These efforts proved to be so successful that the once daring proposal that the brainstem regulates cortical state is unproblematic today. The same cannot be said of an allied, largely neglected, but even more radical proposal that emerged from the same pioneering wave of consciousness studies. Some of the principals in these developments – notably the neurosurgeon Wilder Penfield and his colleague Herbert Jasper – went on to re-examine the routine assumption that another "higher function," closely allied to that of sleep and wakefulness, namely consciousness, is an exclusively cortical affair (Penfield and Jasper, 1954). On the basis of a set of clinical and physiological observations centred on the epilepsies, these authors proposed that *the highest integrative functions of the brain are not completed at the cortical level, but in an upper brainstem system of central convergence supplying the key mechanism of consciousness* (Penfield, 1952). As their proposal is the natural point of departure for the present one, which elaborates and updates it in the light of subsequent developments, a brief review of its history follows.

2. Clinical beginnings

Penfield and Jasper left the anatomical definition of the upper brainstem system they invoked somewhat vague, but it was suggested to include the midbrain reticular formation and its extension into what was then known as the "nonspecific" thalamus (a nuclear grouping encompassing the midline, intralaminar, and reticular thalamic nuclei). They regarded this anatomically subcortical system to be functionally supra-cortical in the sense of occupying a superordinate position relative to the cerebral cortex in functional or control terms (Penfield and Jasper, 1954). They called it the "centrencephalic system," and assigned it a crucial role in the organization of conscious and volitional functions. Fig. 9.1 is based on a figure illustrating A. Fessard's lucid account of the conceptual setting for these ideas, included in the first of the symposium volumes cited earlier (Fessard, 1954).

The Penfield and Jasper proposal emerged from extensive experience derived from an innovation in neurosurgical practice: they routinely removed sizeable sectors of cortex in conscious patients for the control of intractable epilepsy (Penfield and Jasper, 1954). By performing the surgery under local anaesthesia only, the authors ensured that their patients remained conscious, cooperative, and capable of self-report throughout the operation. This allowed the neurosurgeons to electrically stimulate the exposed cortex while communicating with the patient, in order to locate functionally critical areas to be spared when removing epileptogenic tissue. They then proceeded to remove cortical tissue while continuing to communicate with the patient. They were impressed by the fact that the removal of sizeable sectors of cortex such as those diagrammed in the composite of Fig. 9.2 never interrupted the patient's continuity of consciousness even while the tissue was being surgically removed.

Penfield and Jasper note that a cortical removal even as radical as hemispherectomy does not deprive a patient of consciousness, but rather of certain forms of information, discriminative capacities, or abilities, but not of consciousness itself (Penfield and Jasper, 1954; cf. Devlin *et al,* 2003). That does not mean that no cortical insult is capable of compromising consciousness. In adult humans massive bilateral cortical damage will typically issue in a so-called persistent vegetative state (Jennett, 2002). This by itself does not, however, allow us to make an equation between cortical function and consciousness, because such damage inevitably disrupts numerous brainstem mechanisms normally in receipt of cortical input, as discussed further in subsequent sections (see Shewmon, 2004 for the conceptual and empirical complexities of the vegetative state). What impressed Penfield and Jasper was the extent to which the cerebral cortex could be subjected to acute insult without producing so much as an interruption in the continuity of consciousness. Their opinion in this regard bears some weight, in that their magnum opus of 1954 – *Epilepsy and the Functional Anatomy of the Human Brain* – summarizes and evaluates experience with 750 such operations.

When the exposed cortex was stimulated electrically to assess functional localization, stimulation parameters were adjusted so as to avoid triggering epileptic seizures in the patient. From time to time seizures were nevertheless triggered inadvertently. Over the large number of operations performed, every variety of seizure was thus produced by cortical stimulation, except one: Penfield and Jasper never saw the complete electrographic pattern that accompanies absence epilepsy induced by electrical stimulation of any part of the cerebral cortex (Penfield and Jasper, 1954). This pattern of 3 per second trains of "spike and wave" discharges evolves synchronously in the two hemispheres, down to a coincidence in the two hemispheres of the very first abnormal spike detectable in the electroencephalogram (Gibbs *et al,* 1936, 1937; Penfield and Jasper, 1954).

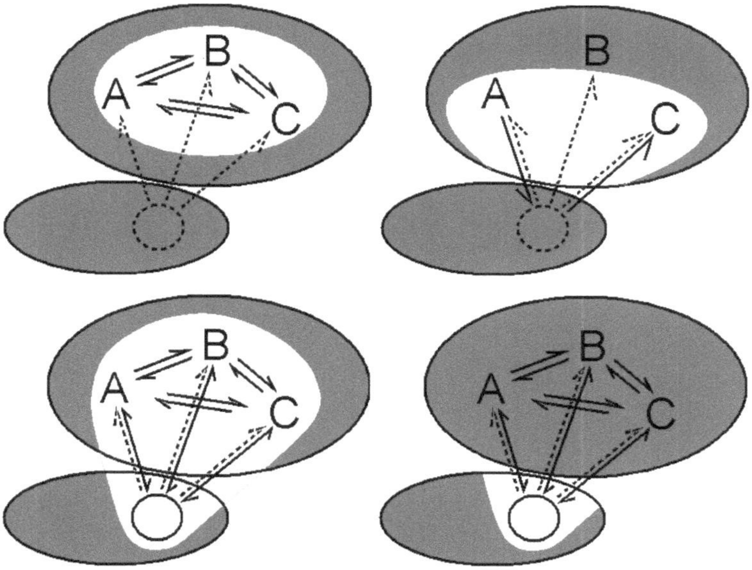

Fig. 9.1 Four principal alternatives regarding interactions between cortex and brainstem in the constitution of the conscious state. Cortex (large oval) and brainstem (small oval) in highly schematic side (saggittal) view. Small circle: "centrencephalic system." In each alternative, normal waking cortical function is assumed to require "enabling" activation originating in the brain stem, marked by three dashed arrows radiating from brainstem to cortex. **Upper left**: the "corticocentric" alternative, in which integration through cortico-cortical connections alone is sufficient to constitute the conscious state. **Upper right**: Cortical integration via a subcortical relay, such as might occur via the dorsal thalamus. Only one such relay is depicted for the sake of clarity. The scheme is still corticocentric, since integration is cortical, albeit dependent upon extracortical relays for its implementation. **Lower left**: Centrencephalic hypothesis, based on diagram IV in Fessard (1954). Here an essential functional component of consciousness is supplied by brainstem mechanisms interacting with the cortex. **Lower right**: Primary consciousness implemented in the brainstem alone, as in cases of cortical removal or damage discussed in Sections 4.4 and 5 of the text.

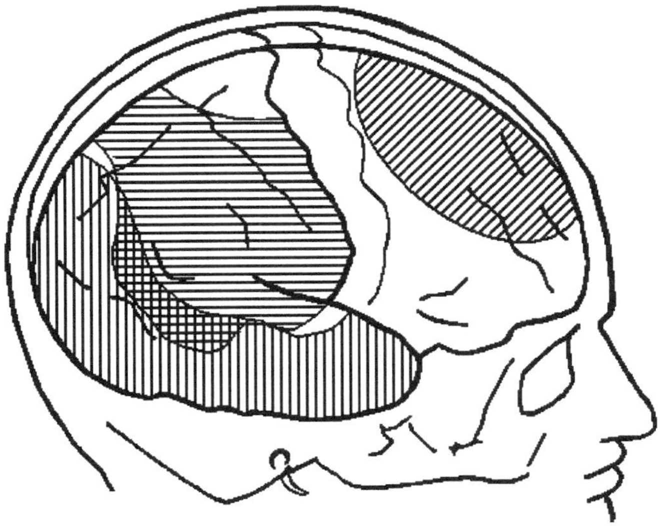

Fig. 9.2 Large cortical excisions performed under local anesthesia by W. Penfield for the control of intractable epilepsy in three patients, entered on a single diagram. The patients remained conscious and communicative throughout the operation. All removals extended to the midline. The two posterior cases were right-sided, whereas the frontal removal was left-sided, and has been mirror-imaged. In no case was the removal of cortical tissue accompanied by a loss of consciousness, *even as it took place*. (Redrawn after figures VI-2, XIII-2, and XVIII-7 of Penfield and Jasper, 1954.)

Seizures of this type bear directly on our topic because of their conspicuous association with disturbances of consciousness (Penfield and Jasper, 1954). In fact, they are often *initiated* by a lapse of consciousness, and in pure form they "consist almost solely of a lapse of consciousness". Without a preceding "aura" or other warning, and in the midst of normal activities, the patient assumes a vacant expression ("blank stare") and becomes unresponsive. Ongoing activities may continue in the form of automatisms (as complex as automatic speech, implying organized cortical activity), or they may arrest for the duration of the often-brief seizure episode. At the end of such a seizure, which may last no more than a few seconds, the patient, who typically remains upright throughout, sometimes actively moving, resumes conscious activities where they were interrupted, has amnesia for what transpired during the episode, and may have no knowledge that the episode took place except indirectly, by means of evidence for the lapse of time available to the discursive, post-seizure, intellect.

Penfield and Jasper recognized in these seizures "a unique opportunity to study the neuronal substratum of consciousness" (Penfield and Jasper, 1954; cf. Blumenfeld and Taylor, 2003). The coincident bilateral onset and cessation of these seizures suggested to the authors an origin in a centrally placed upper brainstem site of paroxysmal induction (Penfield and Jasper, 1954). Though in their experience the pattern was not triggered by cortical stimulation, it could be evoked experimentally in the cat by stimulation of the midline thalamus (Jasper and Droogleever-Fortuyn, 1947). Modern methods have added both detail and qualifications to the Penfield and Jasper account (see review by Meeren *et al,* 2005), yet upper brainstem involvement in absence epilepsy has stood the test of time, and is still being actively pursued both clinically and through research employing animal models (Blumenfeld and Taylor, 2003; Danober *et al,* 1998; Derensart *et al,* 2001; McCormick and Contreras, 2001; Stefan and Snead, 1997; Strafstrom, 2006). We shall return to this matter in Section 4.5.3.

Penfield and Jasper stressed that the postulated centrencephalic system is *symmetrically related to both cerebral hemispheres* (in the sense of radial rather than bilateral symmetry (see Penfield and Jasper, 1954). They denied that this system "functions by itself alone, independent of the cortex" and suggested instead that it "functions normally only by means of employment of various cortical areas" (Penfield and Jasper, 1954). They conceived of it as a convergently innervated upper brainstem system serving to coordinate and integrate the functional economy of the forebrain as a whole, intimately involved in conscious and volitional functions, as well as in the laying down of memories across the lifespan (Penfield and Jasper, 1954).

3. Bringing the centrencephalic proposal up to date

A valuable review of the centrencephalic proposal, in light of developments up till the end of the 1980s, is provided by Thompson (1993, published posthumously). He calls attention to the relevance of the clinical literature on so called "subcortical dementia" to the centrencephalic theory, and further suggests that animal evidence for a subcortical "general learning system" may supply some of the anatomical detail left unspecified by Penfield and Jasper. This "general learning system" is defined by neural structures which, when damaged, produce deficits in *each* member of a set of highly diverse learning tests for rats. As identified through a long-term research program conducted by Thompson and colleagues, it consists of the basal ganglia, including the *substantia nigra* and ventral tegmental area, ventrolateral thalamus, superior colliculus, median raphé, and pontine reticular formation. The functional significance of key members of this constellation (which has access to sensory information independently of the cortex) is considered in some detail in Section 4 of the target article, for which the following preliminary considerations will set the stage.

The central claim of the Penfield and Jasper hypothesis is a claim regarding systems-level organization of neural functions. The idea that a system can be "anatomically subcortical but functionally supra-cortical" is a statement about brain macrosystems and how they relate and interact with one another. It is most easily approached from the side of the "final common path" of all brain output as far as actual behaviour is concerned, namely brainstem and spinal motoneuron pools. Not only are these clusters of final output cells invariably innervated by multiple sources of afference (Graf *et al,*, 2002; Kuypers and Martin, 1982; Nudo and Masterton, 1988; Ugolini, 1995), but individual motoneurons receive synaptic input from diverse sources utilizing different transmitters (Holstege, 1991; Wentzel *et al,* 1995). These sources include spinal and brainstem pattern generators (Grillner, 2003), various territories of the brain stem reticular formation (Jordan, 1998), and a multitude of both direct and indirect brainstem and forebrain afferents, among which the indirect ones often are relayed via the reticular formation (Zahm, 2006).

Thus, the fact that the motor cortex maintains direct connections with brainstem and spinal motoneurons by no means implies that it ever is in sole command of behaviour. At every level of its descending innervation of motoneuron pools it is only one of many inputs determining final outcomes. Moreover, the motor cortex accounts for just a fraction of descending cortical output, and is responsible for only select forms of distal behaviour (Lang and Schieber, 2003; Lawrence and Kuypers, 1968; Kuypers, 1982, 1987). In such a setting, the idea that the output of a subcortical structure might override a cortical one, and in this sense could exercise supra-cortical control over behaviour, is hardly controversial. When an act of deliberate effort (say driven by prefrontal executive systems) is successful in overriding or inhibiting a given behavioural tendency, the cortex is in command of behaviour, temporarily exercising determining control over its course. The fact that such effort does not always succeed (say in the face of sufficient magnitudes of fear, hunger, or pain) means that the frontal executive can be overridden by more primitive mechanisms. When a subcortical source prevails in such competitive interactions, an anatomically subcortical system has exercised supra-cortical functional control over behaviour.

It is necessary, in other words, to distinguish "higher" in the sense of cognitive sophistication from "higher" in control terms. In this light, the Penfield and Jasper proposal amounts to a claim that certain upper brainstem systems in receipt of convergent cortical projections occupy a superordinate position in the latter sense. As I detail further in subsequent sections, the diverse hemispheric as well as brainstem input to these structures equips them for the kind of superordinate decision making crucial for the global sequencing and control of behaviour (Prescott *et al,* 1999). It is also within processes dedicated to "integration for action" that we can find a well-defined functional role for a particular mode of neural organization that qualifies as conscious, in good agreement with the Penfield and Jasper proposal. To set the stage for a treatment of that more demanding topic in Sections 4 and 5, two lines of evidence regarding brainstem function that bear on their proposal are briefly reviewed.

3.1 The Sprague effect

Complete removal of the posterior visual areas of one hemisphere in the cat (parietal areas included) renders the animal profoundly and permanently unresponsive to visual stimuli in the half of space opposite the cortical removal (Sprague, 1966; see also, Sherman, 1974; Wallace *et al,* 1989). The animal appears blind in a manner resembling the cortical blindness that follows radical damage to the geniculostriate system in humans. Yet inflicting *additional* damage on such a severely impaired animal at the midbrain level *restores* the animal's ability to orient to and to localize stimuli in the formerly blind field (Sprague, 1966; cf. Sherman, 1977; Wallace *et al,* 1989). This is accomplished by removing the contralateral superior colliculus or by an intervention as small as a knife-cut that severs fibres running in the central portion of the collicular commissure. That is, adding a small amount of damage in the brainstem to the cortical damage "cures" what appeared to be a behavioural effect of massive cortical damage. The restored visual capacity is limited essentially to the ability to orient to and approach the location of moving visual stimuli in space (Wallace *et al,* 1989). Visual *pattern* discrimination capacity does not recover after the midbrain intervention (Loop and Sherman, 1977), though the midbrain mechanism can be shown to play a role even in such tasks (Sprague, 1991).

The Sprague effect is a consequence of secondary effects generated at the brainstem level by the unilateral cortical removal (Hikosaka and Wurtz, 1989; Hovda and Villablanca, 1990; Jiang *et al,* 2003). The damage not only deprives the ipsilateral superior colliculus of its normal and profuse cortical input (Berson and McIlwain, 1983; Harting *et al,* 1992; Palmer *et al,* 1972; Sprague, 1975), but it unbalances collicular function via indirect projection pathways. Chief of these is the powerful inhibitory projection from the substantia nigra to the colliculus, which crosses the midline in a narrow central portion of the collicular commissure (McHaffie *et al,* 1993; Sprague, 1996; Wallace *et al,* 1990; for additional possibilities, see Durmer and Rosenquist, 2001). The "restorative" interventions partially correct this imbalance, allowing the collicular mechanism to resume at least part of its normal functional contribution to behaviour, with partial restoration of visually guided behaviour as a result.

The point is underscored by the analogous circumstances pertaining to the neglect of one half of space (unilateral neglect) that follows more limited inactivation of the cortex (by reversible cooling) at the junction of occipital, parietal, and temporal lobes in one hemisphere of the cat. This neglect also lifts upon inactivation (by reversible cooling) of the superior colliculus opposite to the cortical inactivation (Lomber and Payne, 1996). Analogous restorative effects of midbrain damage on neglect caused by frontal cortical damage have been observed in a human patient (Weddell, 2004). Though the unawareness featured in cases of unilateral neglect in humans is far from a simple entity (see review by Mesulam, 1999), it bears on our topic by being perhaps the closest approximation to an impairment that includes specific effects on consciousness produced by localized cortical damage (Driver and Vuilleumier, 2001; Rees, 2001; see also Jiang *et al,* 2003).

The Sprague effect demonstrates that hidden in the hemianopia or neglect caused by cortical damage lies a deficit on the part of a brainstem visual mechanism disabled as a secondary effect of the cortical removal. This means that a functional deficit following damage limited to the cortex cannot, as a matter of course, be taken to reflect an exclusively cortical contribution to functional capacity, because the deficit may reflect "remote" effects on brainstem systems, as well. As Sprague originally expressed it:

> The heminanopia that follows unilateral removal of the cortex that mediates visual behaviour cannot be explained simply in classical terms of interruption of the visual radiations that serve cortical function. Explanation of the deficit requires a broader point of view, namely, that visual attention and perception are mediated at both forebrain and midbrain levels, which interact in their control of visually guided behaviour. (Sprague, 1966, p. 1547)

That conclusion agrees well with the Penfield and Jasper perspective reviewed in the foregoing; and it tells us that without cognizance of potential subcortical contributions to a deficit caused by cortical damage, the scope of functions attributed to the cortex will be counterfactually inflated.

3.2 Target selection in the midbrain

Although superficially inconspicuous, the superior colliculus in the roof ("tectum") of the midbrain exhibits considerable structural and functional complexity. Long known to play a role in "visual grasping" or "foveation" (Hess *et al,* 1946; Schiller and Koerner, 1971), further study has revealed unexpected sophistication in its functional organization (Keller *et al,* 2005; Krauzlis *et al,* 2004; May, 2005; Sparks, 1999). It is the only site in the brain in which the spatial senses are topographically superposed in laminar fashion within a common, premotor, framework for multi-effector control of orienting (Merker, 1980). Its functional role appears to centre on convergent integration of diverse sources of information bearing on spatially triggered replacement of one behavioural target by another, and evidence is accumulating for a collicular role in target selection (Basso and Wurtz, 1998, 2002; Carello and Krauzlis, 2004; Cavanaugh and Wurtz, 2004; Fecteau and Munoz, 2006; Glimcher and Sparks, 1992; Horowitz and Newsome, 1999; Krauzlis *et al,* 2004; McPeek and Keller, 2004; Schlag-Rey *et al,* 1992; Wurtz and Mohler, 1974; see also Grobstein, 1988). Such a role has direct implications for the topic of superordinate control functions.

A collicular role in target selection is unlikely to be a passive reflection of decisions taken in other structures. It is not fully accounted for by the powerful input it receives from the substantia nigra (Basso and Wurtz, 2002), and the diversity of collicular afferents precludes any one of them from exercising sole control over collicular function. These afferents include a wide range of brainstem (Edwards, 1980; Edwards *et al,* 1979) and visual as well as nonvisual cortical sources (Collins *et al,* 2005; Harting *et al,* 1992, 1997; Kawamura and Konno, 1979; Sherman *et al,* 1979). Cortical afferents are monosynaptic, originating in layer V pyramidal cells, placing the colliculus as close to the cortex as two cortical layers are to one another. In the cat they include some 17 visual areas (Harting *et al,* 1992), and in primates there are contributions from both the dorsal (parietal cortex) and the ventral (temporal cortex) "streams" of the visual system (Fries, 1984; Steele and Weller, 1993; Webster *et al,* 1993). Any sensory modality used in phasic orienting behaviour appears to receive obligatory representation in the colliculus. Besides the major spatial senses of vision, audition, and somesthesis, they include pain (Wang and Redgrave, 1997) and exotic senses such as infrared (Hartline *et al,* 1978), electroceptive (Bastian, 1982), magnetic (Nemec *et al,* 2001), and echolocation systems (Valentine and Moss, 1997), depending on species.

In the colliculus these diverse convergent inputs are arranged in topographically organized sheets layered one upon the other through the depths of the colliculus (Harting *et al,* 1992; May, 2005). Intrinsic collicular circuitry distributes excitatory as well as inhibitory collicular activity within and across layers and across major collicular subdivisions (Behan and Kime, 1996; Bell *et al,* 2003; Binns, 1999; Doubell *et al,* 2003; Lee *et al,* 1997; Meredith and King, 2004; Meredith and Ramoa, 1998; Mize *et al,* 1994; Özen *et al,* 2000; Zhu and Lo, 2000). There is therefore no dirth of complex intrinsic collicular circuitry – only beginning to be systematically charted – for collicular decision-making based upon its diverse sources of afference.

The collicular role in target selection is accordingly likely to be causal (Carello and Krauzlis, 2004; McPeek and Keller, 2004; see also Findlay and Walker, 1999; Yarrow *et al,* 2004). This would place the colliculus at the functional top rather than bottom of control processes in its domain. The selection of a target for behaviour is the brain's final output in that regard. It is the pivotal event for which all other processes are but a preparation, summing them up in the actual decision to settle on one target for action rather than another (Allport, 1987; Brooks, 1994; Dean and Redgrave, 1984; Isa and Kobayashi, 2004; McFarland and Sibly, 1975; Tyrrell, 1993; see Fecteau and Munoz, 2006 for collicular "priority mapping" in relation to action).

The functional prediction from the loss of such a structure is not the absence of target acquisition, but its impoverishment. Not only is the brain redundantly organized in this regard (Lomber *et al,* 2001; Schall, 1997; Schiller *et al,* 1979; Tehovnik *et al,* 1994), but the loss of a superordinate function in a layered control architecture does not disable the system as a whole (Brooks, 1986, 1989; Prescott *et al,* 1999), just as a well organized army need not cease functioning on the loss of its commander. A macaque with experimental collicular lesions is not incapable of moving its eyes onto targets, but exhibits a reduced variety of eye and orienting movements and is indistractible, a common finding in other species as well (Albano and Wurtz, 1978; Casagrande and Diamond, 1974; Denny-Brown, 1962; Goodale and Murison, 1975; Merker, 1980; Mort *et al,* 1980; Schiller *et al,* 1979; Schiller and Lee, 1994; Schneider, 1967). This may reflect a compromised *scope* and *sophistication* of target selection, and the role of the intact colliculus would accordingly instantiate the Penfield and Jasper conception of a *highest integrative function* which, while *anatomically subcortical*, is *functionally supra-cortical.*

4. Integration for action

As noted in Section 3, in drawing the contrast between "higher" in cognitive terms and "higher" in control terms, competition for control over behaviour ends only at the stage of the "final common path" of motoneurons. It is along that approach, among upper brainstem mechanisms of "integration for action," that we shall identify a prototype organization for conscious function. The issue takes us to the very origin of the vertebrate brain plan, which is not only cephalised, but centralized. Not all animals rely on centralized neural organization to control behaviour, even when possessed of a brain. A number of invertebrate forms, including insects, concentrate considerable neural resources to segmental ganglia. Their brain is in a sense no more than the anterior-most of these ganglia, in receipt of the output of the specialized receptors of the head. It does not necessarily exercise a command function in the sense of central control of behaviour (see Altman and Kien, 1989).

The decentralized neural control of an insect such as the ant allows its body to survive without its brain. Moreover, if given adequate somatic stimulation in this condition, it will perform many of the complex behaviours in its repertoire with apparent competence, though naturally without relation to the distal environment (Snodgrass, 1935). A vertebrate, on the other hand, does not survive for more than seconds after the loss of its brain, because in vertebrates even vital functions are under central brain control. The difference with respect to insects is underscored by the contrasting disposition of motor neurons. In insects, they are concentrated to segmental ganglia but are rare in the brain (Snodgrass, 1935), whereas in vertebrates they populate the brain in sets of distinctively organized motor nuclei. Motor control in vertebrates has "moved up," as it were, to that end of the neuraxis which leads in locomotion and is in receipt of the output of the chief exteroceptors (cf. Grillner *et al,* 1997).

The basic organizational features of the vertebrate brain are highly conserved across taxa despite unequal development of one or another of its senses or subdivisions (Nieuwenhuis *et al,* 1998). All vertebrates, that is, have "in outline" the same brain plan, assembled from primitive beginnings in chordate ancestry (Butler and Hodos, 1996; Holland and Holland, 1999; Northcutt, 1996b). The prominent role of large, image-forming eyes and their central connections in this development came to exert a profound effect on the manner in which the vertebrate brain plan was centralized, with implications for our understanding of the way in which "higher" in cognitive terms relates to "higher" in control terms. That development involves the integrative machinery straddling the so-called synencephalon, or junction between midbrain and diencephalon – to which we now turn.

4.1 The synencephalic bottleneck and how the vertebrate brain came to be centralized around it

There was a time in prevertebrate ancestry when the midbrain and diencephalon alone, or rather the first rostral differentiations of the neural tube that can be homologized with the vertebrate midbrain and diencephalon, constituted the functionally highest and also anatomically most rostral subdivision of the neuraxis (Holland and Holland, 1999;, 2001; Holland *et al,* 1994; Lacalli, 1996, 2001; Wicht, 1996). It housed the neural circuitry connecting a primitive, unpaired "frontal eye" and other rostral sensory equipment (Lacalli, 1996) with premotor cells in cephalochordate filter feeders (represented today by *Amphioxus*, the lancelet). As far as is known, cephalochordate filter feeders lacked a sense of smell, and they were without a telencephalon altogether (Butler, 2000; Holland *et al,* 1994).

Though our brain nomenclature historically groups the diencephalon together with the telencephalon to make up the forebrain, there is nothing fundamental about such a grouping, as the just mentioned phylogenetic circumstances show. Rather, for what follows it will be convenient to retain the primitive grouping of midbrain and diencephalon together under the label mesodiencephalon or "optic brain." In all vertebrates these two segments of the neuraxis, along with the transitional "synencephalon" (pretectum) wedged between them, house the primary terminations of the optic tract (cf. Butler, 2000). The latter covers their external surfaces in the form of a ribbon of fibres running obliquely from the optic chiasm beneath the hypothalamus across the diencephalon and mesencephalon up to the latter's roof ("tectum"). Along the way it innervates structures as different as the hypothalamus, ventral thalamus, dorsal thalamus, pretectum, accessory optic nuclei, and superior colliculus (tectum). The same territory also houses some of the major *integrative structures of broad functional scope common to all vertebrates* (see Fig. 9.3).

The principal poles of this integrative machinery are the hypothalamus forming the floor of the diencephalon, on the one hand, and the superior colliculus forming the roof of the midbrain, on the other. The former is an intricate nuclear aggregate critical for the mutual regulation and integration of a vertebrate's entire repertoire of goal-directed, motivated behaviour covering exploratory, foraging, ingestive, defensive, aggressive, sexual, social, and parental modes of behaviour (Swanson, 2000), to name the principal ones. The other pole, colliculus/tectum, serves the intermodal integration of the spatial senses by which vertebrates relate to their surroundings via coordinated orienting movements of eyes, head, and body, as already summarized in Section 3.2. Between these two is wedged additional integrative machinery in the form of the midbrain reticular formation, ventral thalamus, the periaqueductal gray, the ventral tegmental/substantia nigra pivot of the striatal system, as well as "locomotor centres" and basic mechanisms serving navigation. I will return to some of these in subsequent sections.

This concentration of conserved integrative machinery to the mesodiencephalon, I suggest, reflects the costs and benefits of evolving image-forming eyes in the ancestors of vertebrates (cf. Northcutt, 1996a). Full use of the potential powers of visual guidance meant evolving solutions to an intricate set of sensorimotor problems. The confounding of sensory information by the sensory consequences of movement ("re-afference"; von Holst and Mittelstaedt, 1950) is particularly problematic for image-forming eyes, requiring their stabilization with respect to the world during movement. This is done by vestibular counter-rotation punctuated by quick resets of the eyes, which concentrates blurring-time to the brief reset episodes. Thus, *vision alone among all the senses features independent spatial mobility of the receptor array itself*, and a full-fledged oculomotor system evolved in the immediate ancestors of true vertebrates (Braun, 1996; Fritsch *et al,* 1990; Wicht, 1996). The reflex circuitry connecting vestibular and oculomotor nuclei, centred on the medial longitudinal fasciculus, is also among the most conservative and basic features of the brainstem in all vertebrates (Carpenter, 1991; Windle and Baxter, 1936).

Yet, with eyes free to turn in their orbits, there is no longer a fixed relation between retinal location and spatial direction relative to body or head, nor to the localizing function of any sensory modality which (in whole or in part) bears a fixed relation to the head. Hence the need for intermodal integration, for which the sensory integrating mechanism of colliculus/tectum – present in the roof of the midbrain of even jawless vertebrates – provides the basic, early and conserved solution (Iwahori *et al,* 1999; Zompa and Dubuc, 1996). But once these basic problems of vision were solved, a bonus was within reach: Mobile eyes present a highly efficient means for sampling the environment, provided their control can be linked to motivational mechanisms ensuring their appropriate deployment in accordance with shifting needs.

It appears, in other words, that as the vertebrate brain plan took shape in prevertebrate ancestry under pressure of the evolution of mobile, image-forming eyes, a central association between optic control circuitry and major neural mechanism for the integration of behaviour/action were forged in segments of the neuraxis covered and innervated by the optic tract (cf. Fig. 3). At the time when this optic orienting machinery and associated integrative mechanisms evolved, the forebrain was still dominated by olfaction (Braun, 1996; Northcutt and Wicht, 1997; Wicht and Northcutt, 1992). The sense of smell added no fundamentally new control requirements comparable to those of vision, and olfaction accordingly could be integrated with the mesodiencephalic control system by caudally directed fibre projections. These simply happen to arrive at the "optic brain" from an anterior direction, whereas other sensory afferents reach it from a caudal direction (somatosensory, octavolateral, i.e., vestibular/auditory/lateral line/electrosensory, etc.), or directly "from the side" through the optic tract (cf. Butler, 2000).

Indeed, however much the telencephalon subsequently expanded, even to the point of burying the mesodiencephalon under a mushrooming mammalian neocortex, no other arrangement was ever needed, and that for the most fundamental of reasons. No efferent nerve has its motor nucleus situated above the level of the midbrain. This means that the very narrow cross-section of the brainstem at the junction between midbrain and diencephalon (synencephalon, marked by arrows in the main part of Fig. 9.3 and by a black bar in the inset) carries the total extent of information by which the forebrain is ever able to generate, control, or influence behaviour of any kind. If, therefore, integration is for action, as proposed here for the mesodiencephalic control system, information-theory poses no obstacle to having an expansive neocortex make its contribution in this regard by convergent projections onto the highly conserved and preexisting machinery of the midbrain and basal diencephalon, which therefore could retain its old integrative functions (see Fig. 9.3). Indeed, a bottleneck of this kind is exactly what is needed in order to convert the massively parallel and distributed information capacity of the cerebral hemispheres into a limited-capacity, sequential mode of operation featured in action selection for coherent behaviour (Allport, 1987; Baars, 1993; Cabanac, 1996; Cowan, 2001; Mandler, 1975, 2002; McFarland and Sibly, 1975; Tyrrell, 1993).

That is, one need not know anything more about the vertebrate brain than the fact that its most rostral motoneurons are located *below* the synencephalic bottleneck, to know that the total informational content of the forebrain must undergo massive reduction in the course of its real-time translation into behaviour. In the setting of such obligatory "data reduction" in a stretch of the neuraxis hosting major systems for the global regulation of behaviour, a so far unrecognized optimizing principle lies hidden in the mutual dependency that links the motivational, the sensory, and the action selection requirements of the brain's control tasks. They form a "selection triangle," the principle of which is introduced here for the first time. The efficient neural implementation of this principle may harbour the secret of conscious function itself.

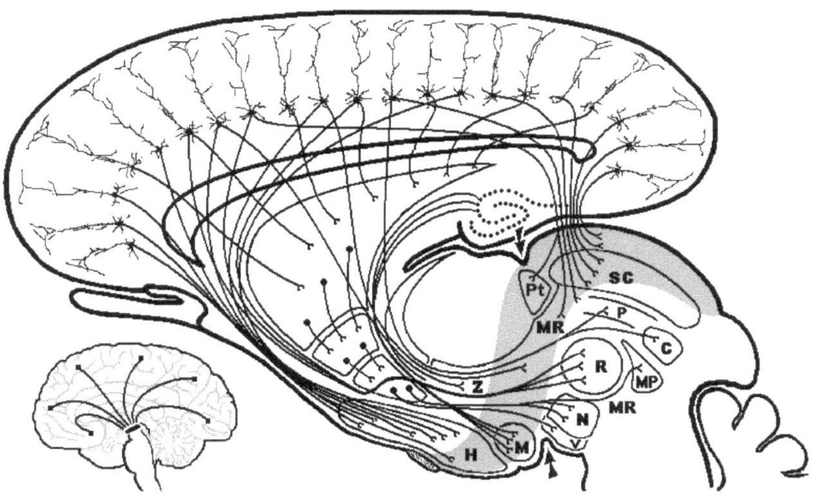

Fig. 9.3 Schematic saggittal diagram depicting cortical convergence (in part via the basal ganglia) onto key structures in the region of the "synencephalic bottleneck" (marked by thick arrows in the main figure and by a black bar in the inset). Abbreviations: **C**, nucleus cuneiformis; **H**, hypothalamus (preoptic area included); **M**, mammillary bodies; **MP**, "mesopontine state control nuclei" (locus coeruleus, pedunculopontine and laterodorsal tegmental nuclei, and dorsal raphé); **MR**, midbrain reticular formation; **N**, substantia nigra; **P**, periaqueductal gray matter; **Pt**, pretectum; **R**, red nucleus; **SC**, superior colliculus; **V**, ventral tegmental area; Z, zona incerta. The dual axon seen issuing from some of the pyramidal cells of cortical layer 5 is an illustrative convenience only. Shaded region marks the surface course of the optic tract.

4.2 The "selection triangle": A proposed key to conscious function

Elementary necessities of animal existence such as food, shelter, or mates are not typically found in the same place at any given time, and they each require different and often incompatible behaviours. An animal's activities accordingly unfold under constraint of multiple goals or motives derived from the evolved and acquired needs it must fill through the sequence of its diverse actions over time (Baerends, 1976; Tinbergen, 1951). The tasks set by these goals compete for an animal's behavioural resources, and because the actions by which they are implemented are always confined to the present (where they typically are executed one at a time), their scheduling (action selection) features perpetual trade-offs in the time and effort that is allocated to them (McFarland and Sibly, 1975). The ethological insight, that animal behaviour rests upon a foundation of diverse goal functions that sometimes entail incompatible tasks or behaviours requiring sequencing/selection, entered the so-called behaviour-based approach to robotics under the name "action selection" (Blumberg, 1994; Brooks, 1986; McFarland and Houston, 1981; Maes, 1990; Prescott *et al,* 1999; Tyrell, 1993; see also Meyer and Wilson, 1991).

The needs reflected in the time budget of an animal's task allocations are, however, only one side of the equation of efficient decision-making. The fulfilment of needs is contingent on available *opportunities*. These are scattered in the world as ever-shifting targets of approach and avoidance among lively and often unpredictable contingencies within which they must be detected, located, and identified, often among multiple competing alternatives, all in real time. Interposed between the needs and their fulfilment through action on the world is the body with its appendages and other resources for getting about in the world and manipulating its objects. In concrete terms an action is a time series of bodily locations and conformations. These are what connect needs with opportunities. In so doing they themselves become a factor in singling out a given opportunity (target) for action (target selection). This is so because determining which one of several available potential targets is the best current choice for action will often depend, not on current needs alone, but additionally on the disposition of the body relative to those targets (in terms of its posture and position, movement trajectory, energy reserves, etc.; cf. Körding and Wolpert, 2006).

In principle, each of the decision domains just invoked – action selection, target selection, and motivational ranking – may be defined in its own terms, without regard to the others. They may even make their contributions to behaviour independently of one another (Altman and Kien, 1989; Brooks, 1986). But from the inherent functional relationship just sketched, that is, the fact that in terms of optimal performance *target selection is not independent of action selection, and neither of these is independent of motivational state* (reflecting changing needs), it follows that savings are achievable by exploiting that triangular dependency. It is not possible to reap the benefits of those savings short of finding some way of interfacing the three state spaces – each multidimensional in its own right – within some common coordinate space (decision framework) allowing their separate momentary states to interact with and constrain one another. This extends to such a tripartite interaction the principle already derived for the efficient management of motivational trade-offs, namely, the convertibility of different motives through a motivational "common currency" and their convergence among themselves at some point in the system (McFarland and Sibly, 1975; see also Cabanac, 1992, and further in the present article).

The principle of a centralized brain system dedicated to this decision domain follows from this, though not the particulars of the three-way interface that must form its centrepiece. Evolving such an interface is far from a trivial problem, all the more so since its decisions must be made in real time. The brain, of course, has no direct access to either the target states of the world or the action states of the body that must be compared and matched in the light of motivational priorities. It is saddled with an inverse problem on both sensory and motor sides of its operations (Gallistel, 1999; Kawato *et al,* 1993). The indirect reflections of relevant parameters to which it does have access, come to it, moreover, in diverse data formats. The differences between the spatial senses among themselves in this regard are mild compared to those between any one of these senses and the various musculoskeletal articulations and configurations they serve to control. How then might the former be compared with the latter? Add to this the already mentioned circumstance that every movement confounds the sensory information needed to guide behaviour, and that the needs to be taken into account differ not only in urgency, but in kind, and the size of the design problem begins to emerge in outline.

To exploit the savings hidden in the functional interdependence between target selection, action selection, and motivation, this confounded complexity must be radically recast, to allow the three domains to interact directly in real time for the determination of "what to do next." It is the principal claim of the present target article that the vertebrate brain incorporates a solution to this decision problem, that it takes the general form of a neural *analogue reality simulation* of the problem space of the tri-partite interaction, and that the way this simulation is structured constitutes a conscious mode of function. It equips its bearers with veridical experience of an external world and their own tangible body manoeuvring within it under the influence of feelings reflecting momentary needs, that is, what we normally call reality.[2] To this end it features an analogue (spatial) mobile "body" (action domain) embedded within a movement-stabilized analogue (spatial) "world" (target domain) via a shared spatial coordinate system, subject to bias from motivational variables, and supplying a premotor output for the control of the full species-specific orienting reflex. The crucial separation of body and world on which this arrangement hinges has recently been worked out in formal terms by David Philipona and colleagues (Philipona *et al,* 2003, 2004).

We have already seen in Sections 3.2 and 4.1 that the roof of the midbrain of vertebrates houses a sophisticated laminar superposition of the spatial senses in a premotor framework for orienting. It appears to contain the essential signals for bringing these senses into registry (Groh and Sparks, 1996; Jay and Sparks, 1987; Krauzlis, 2001; Populin and Yin, 1998; Van Opstal *et al,* 1995; Zella *et al,* 2001) and for stabilizing the world relative to the body. Such stabilization is likely to utilize not only vestibular information (Bisti *et al,* 1972; Horowitz *et al,* 2005), but cerebellar "decorrelation" as well (Dean *et al,* 2002, 2004; cf. Guillaume and Pélisson, 2001; Hirai *et al,* 1982;

May *et al,* 1990; Niemi-Junkola and Westby, 2000). The layered spatial maps in the roof of the midbrain would, in other words, represent the vertebrate brain's first bid for an analogue simulation of a distal "world" (Scheibel and Scheibel, 1977). We also saw that the other pole of the "optic brain," the hypothalamus, houses the basic circuitry for regulating and integrating motivational states related to goal-directed behaviours. Its output is brought to bear on the intermediate and deep layers of the superior colliculus not only by direct projections (Beitz, 1982; Rieck *et al,* 1986), but indirectly, via massive and organized projections from hypothalamic nuclei to different sectors of the periaqueductal gray substance (Goto *et al,* 2005; see refs. in Swanson, 2000).

The periaqueductal gray is a midbrain territory intimately related to the deeper collicular layers. It surrounds the cerebral aqueduct, and plays a critical role in the expression of a variety of emotion-related behaviors such as defensive, aggressive, sexual, vocal, and pain-related ones (Adams, 1979; Behbehani, 1995; Fernandez de Molina and Hunsperger, 1962; Holstege *et al,* 1996; Jurgens, 1994; Kittelberger *et al,* 2006; Lonstein *et al,* 1998; Mouton, 1999; Panksepp, 1982,, 1998a; Watt, 2000). Its longitudinal columns are functionally organized in terms of high-level tasks, goals, strategies, or contexts, such as "inescapable versus escapable pain" (Keay and Bandler, 2002). It achieves particular prominence in mammals, and stimulating it electrically in conscious humans evokes powerful emotional reactions (Heath, 1975; Iacono and Nashold, 1982; Nashold *et al,* 1969). Functionally the periaqueductal gray is continuous and reciprocally interconnected with the immediately overlying deep layers of the superior colliculus (Bittencourt *et al,* 2005; Cadusseau and Roger, 1985; Gordon *et al,* 2002; Grofova *et al,* 1978; Harting *et al,* 1992; Sprague *et al,* 1961; Wiberg, 1992). Here, then, in the intermediate and deep collicular connections with hypothalamus and periaqueductal gray, lies a connective interface between the brain's basic motivational systems and the orienting machinery of the collicular analogue "world."

The third member of the selection triangle enters this system through the prominent projections from the substantia nigra to the intermediate collicular layers (Jiang *et al,* 2003; Mana and Chevalier, 2001; see also sects. 3.1 and 3.2). Here the final distillate of basal ganglia action-related information is interdigitated with the latticework of histochemically defined compartments that organize the input-output relations of the intermediate colliculus (Graybiel, 1978; Harting *et al,* 1997; Illing, 1992; Illing and Graybiel, 1986). It appears, in other words, that the territory extending from the dorsal surface of the midbrain to the aqueduct houses the connectivity needed to implement a three-way interface of the kind outlined in the foregoing discussion, and it is hereby proposed to do so. The elements of this scheme are sketched in Fig. 9.4.

Such a conception fits seamlessly with the proposed role of the superior colliculus in target selection outlined in Section 3.2. As noted there, the selection of a target for action is the final event in the brain's real-time decision-making regarding "what to do next." The significance of gaze control, moreover, goes far beyond the matter of

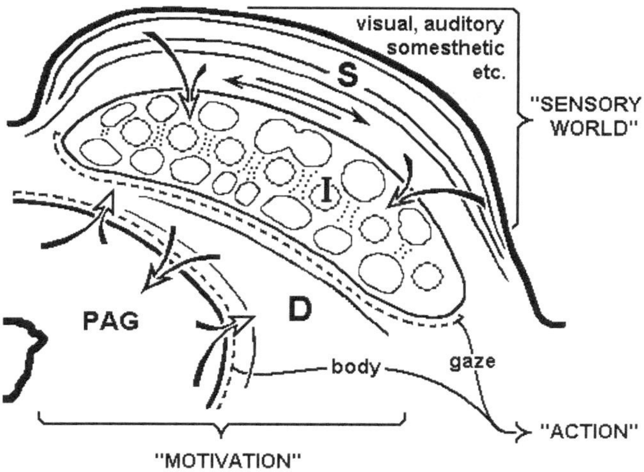

Fig. 9.4 The three principal domains of "world" (target selection), "body" (action selection), and "motivation" (needs) that must interact to optimize decision processes in real time, portrayed in their proposed "primary" implementation in the roof of the midbrain. The extension of its logic into the forebrain, and the cerebral cortex of mammals in particular, can be conceived in terms of this primary system "writ large," as follows (cf. Fig. 6 in particular): A dorsolateral to ventromedial path from the surface of the colliculus to the midbrain aqueduct corresponds to a posterior to frontal to medial path in the cortex. In the reverse direction, and in functional terms, it reads "motivation," "action," and "world." **S**, **I**, and **D**: superficial, intermediate, and deep layers of the superior colliculus, respectively. **PAG**: the periaqueductal gray matter surrounding the midbrain cerebral aqueduct. Bidirectional arrow aligned with the collicular lamina stand for compensatory coordinate transformations. (Drawing based in part on Harting *et al,* 1997).

moving eyes-and-head in space. The gaze plays an organizing role in a wide range of behaviors by "leading" many forms of action, as has been shown in exquisite detail for manual reaching and manipulation (Johansson *et al,* 2001; see also Courjon *et al*, 2004; Jackson *et al,* 2005; Schneider and Deubel, 2002; Stuphorn *et al,* 2000; Werner *et al,*, 1997). Nor is the output of the tecto-periaqueductal system limited to the species-specific orienting reflex: it includes escape behaviour (Dean *et al,* 1989; Merker, 1980; Sprague *et al,* 1961) as well as a number of innate postural schematisms associated with behaviors under periaqueductal control (Holstege *et al,* 1996; Lonstein *et al,* 1998).

In its primitive beginnings, the "world" of the proposed neural reality simulator presumably amounted to no more than a two-dimensional screen-like map of spatial directions on which potential targets might appear as mere loci of motion in an otherwise featureless noise field, defined more by their displacement than by any object features (see Stoerig and Barth, 2001, for a plausible simulation). Advances on this primitive arrangement apparently proceeded by adding to it more sophisticated information from a rostral direction. Thus, the ability of a frog to side-step stationary barriers during prey-catching is dependent upon input to the tectum from the region of the caudal thalamus and pretectum, just anterior to the tectum (Ewert, 1968; Ingle, 1973). With the elaboration of the telencephalon, culminating in the neocortex of mammals, the arrangement was expanded further (see Section 4.5), into a fully articulated, panoramic three-dimensional world composed of shaped solid objects: the world of our familiar phenomenal experience.

4.3 Inhabiting a neural simulation

Whether primitive or advanced, the fundamental simplifying device of the proposed simulation space is to associate the origin of its shared body-world coordinate system for orienting with the head representation of its analogue body. This does *not* mean that the coordinate system itself is head centred (i.e., moves with the head). At brainstem levels it appears, rather, to be oculocentric (Klier *et al,* 2001; Moschovakis, 1996; Moschovakis and Highstein, 1994). It means only that the coordinate system origin is lodged in the head representation of the simulated analogue *visual* body, say in close proximity to its analogue eye region. With such a location, a number of sensory-sensory mismatches and the contamination of sensory information by movement caused by the largely rotary displacements of eyes and head involved in perpetual orienting movements can be remedied – to a first approximation – by spherical coordinate transformations. This economy of control helps explain the fact that at the brainstem level not only eye movements, but also head movements, despite their very different musculo-skeletal demands, utilize a common intermediate control system organized in separate horizontal and vertical, that is, spherical, coordinates (Grobstein, 1989; Masino, 1992; Masino and Grobstein, 1989; Masino and Knudsen, 1990; see also Isa and Sasaki, 2002). In humans, covert orienting of attention, as well as the visuomotor map for reaching (Gawryszewski *et al,* 2005; Vetter *et al,* 1999), appear to be framed in spherical coordinates,[3] perhaps reflecting collicular involvement in both functions (Müller *et al,* 2004; Werner *et al,* 1997).

There is reason to believe that the implicit "ego-centre" origin of this coordinate space is the position we ourselves occupy when we are conscious, and that the analogue body and analogue world of that space is what we experience as and call our tangible, concrete body and the external world (cf. Note 2). This would explain the irreducible asymmetry adhering to the relation between perceiving subject and apprehended objects defining the conscious state. The ego-centre places the conscious subject in an inherently "perspectival," viewpoint-based, relation to the contents of sensory consciousness. It is from there that objects are apprehended; objects do not apprehend the subject (cf. Merker, 1997). By the same token, the one necessary constituent of consciousness that can never be an object of consciousness is that very vantage point itself, namely, the origin of the coordinate system of the simulation space. It cannot be an object of consciousness any more than an eye can see itself (Schopenhauer, 1819, vol. 2; see Baars, 1988, for this and other "contextual" aspects of consciousness).

Should these reasons appear somewhat abstract and rarefied, there is a far more concrete indication to the same effect. Our very body bears a tell-tale sign allowing us to recognize it as the product of a neural simulation. Vision differs topologically from somesthesis and audition by its limited angular subtense, particularly in animals with frontally directed eyes. The other two senses can be mapped *in toto* onto a spherical coordinate system for orienting, whereas vision is only partially mapped in this way. This is not in itself a problem, but becomes one given that vision can be directed, not only to the external world, but to the body itself. This necessitates some kind of junction or transition between the distal visual world and the proximal visual body, and there a problem does arise.

Though, as we have seen, the ego-centre is present in consciousness by implication only, its location can be determined empirically (Cox, 1999; Hering, 1879/1942; Howard and Templeton, 1966; Neelon *et al,* 2004; Roelofs, 1959). It is single, and located behind the bridge of the nose inside our head. From there we *appear* to confront the visible world directly through an empty and single cyclopean aperture in the front of our head (Hering, 1879/1942; Julesz, 1971). Yet that is obviously a mere appearance, since if we were literally and actually located inside our heads we ought to see, not the world, but the anatomical tissues inside the front of our skulls when looking. The cyclopean aperture is a convenient neural fiction through which the distal visual world is "inserted" through a missing part of the proximal visual body, which is "without head" as it were or, more precisely, missing its upper face region (see Harding, 1961). Somesthesis by contrast maintains unbroken continuity across this region. The empty opening through which we gaze out at the world betrays the simulated nature of the body and world that are given to us in consciousness. The essentials of the arrangement are depicted in highly schematic form in Fig. 9.5.

The simulated nature of our body and world is further supported by a number of phenomena that alert us to the synthetic nature of what we typically take to be physical reality itself, that is, phenomena such as inattention blindness, change blindness, and allied effects (O'Reagan *et al,* 2000; Rensink, 2002; Rensink *et al,* 1997; Simons and Chabris, 1999). Such "deletions from consciousness" can be countered by appropriately placed microstimulation of the superior colliculus (Cavanaugh and Wurtz, 2004; see also Müller *et al,* 2005). These various indications all support the conclusion that what we confront in sensory consciousness is indeed a simulated (synthetic) world and body.

As central residents of that simulation, we are subject to ever shifting moods, feelings, urges, emotions, and impulses. These, then, would be those aspects of the brain's motivational dynamics that reach consciousness (cf. Cabanac, 1992; Panksepp, 1982, 1998a). The reason they do so, according to the present proposal, is their rele-

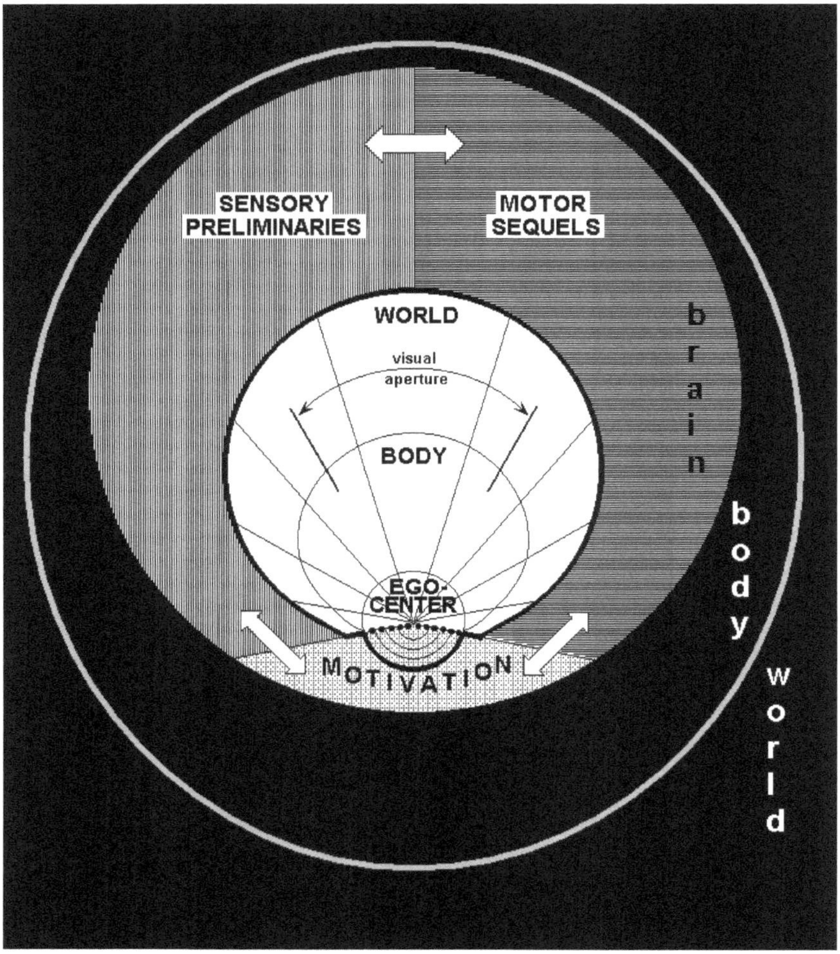

Fig. 9.5 Highly schematic depiction of the nested relation between ego-centre, neural body, and neural world constituting the analogue neural simulation ("reality space") proposed as a solution to the tri-partite selection problem described in the text. Black depicts the physical universe, one part of which is the physical body (black oval), both of which are necessarily outside of consciousness. One part of the physical body is the physical brain (circle; shaded and unshaded). It contains the "reality space" of consciousness (unshaded), separated from other, nonconscious (shaded) functional domains by a heavy black line, signifying their exclusion from consciousness. Arrows mark interfaces across which neural information may pass without entering consciousness. The designation ego-centre is a sensorimotor construct unrelated to the concept of self-consciousness. See text for further details.

vance to the tripartite determination of what to do next, as outlined in the foregoing discussion. A striking illustration of this principle is afforded by respiratory control (Merker, 2005). It is automatic and unconscious as long as partial pressures of blood gases stay within normal bounds, yet intrudes most forcefully on consciousness in the form of an acute sense of panic when they go out of bounds. Extreme blood gas values are an indication that urgent action on the environment – such as removing an airway obstruction or getting out of a carbon dioxide filled pit – may be imperative. That is what suddenly makes action selection and target selection relevant to respiratory control, which accordingly "enters consciousness" in the form of a powerful feeling of suffocation.

This example further illustrates the lack of any necessary connection between cognitive sophistication and the reason for something to enter consciousness. Even quite elementary functions may benefit from the efficiency provided by the triangular action-target-motivation interface of consciousness. It serves optimal decision-making in real time, on the broad front of its tripartite information base, concisely packaged in its multivariate simulation space. Such a utility is particularly valuable when a moment's hesitation may make a big difference in outcome, as in the suffocation example (but also in, say, agonistic encounters), quite apart from anything to do with advanced cognition. The evolution of such a utility could accordingly proceed independently of cognitive capacity, to crown the optic brain with its tectal machinery at the very outset of the vertebrate lineage.

In its peculiar nesting of a body inside a world, around an ego-centre in a shared coordinate space subject to motivational bias, this interface possesses the essential attributes of phenomenal consciousness. As implemented in the midbrain and diencephalon, the arrangement is proposed to have served as the innate scaffolding supporting all further elaboration of conscious contents in phylogeny. Centred on the colliculus extending into periaqueductal gray, it will be further defined in Section 4.5. A felicitous term for the functional state supported by the basic (mesodiencephalic) arrangement would accordingly be "primary consciousness" (Hodgson, 1878; Petty, 1998; Trevarthen and Reddy, 2006).

4.4 Coherent, motivated behaviour under sensory guidance in the absence of the cerebral cortex

The superordinate functional position attributed to mesodiencephalic mechanisms in previous sections of this article is supported by a number of empirical findings that receive a unified interpretation in this light. When the behavioural effects of local brain stimulation are systematically surveyed by means of depth electrodes, it is common to find that the most coherent, integrated, and natural-looking (whole, or "molar") behavioural reactions – be they orienting, exploration, or a variety of appetitive, consummatory, and defensive behaviors – are evoked by stimulation of diencephalic and midbrain sites, whereas stimulation at more rostral or caudal levels tends to evoke more fragmentary or incomplete behaviors (Adams, 1979; Bandler and Keay, 1996; Bard, 1928;

Brandao *et al,* 1999; Carrive *et al,* 1989; Fernandez de Molina and Hunsperger, 1962; Hess, 1954; Hess and Brugger, 1943; Holstege and Georgiadis, 2004; Hunsperger, 1956, 1963; Hunsperger and Bucher, 1967; Kaada, 1951; Orlovsky and Shik, 1976; Schaefer and Schneider, 1968; Schuller and Radtke-Schuller, 1990).

All of the behaviors just mentioned have also been exhibited by experimental animals after their cerebral cortex has been removed surgically, either in adulthood or neonatally. Best studied in this regard are rodents (Whishaw, 1990; Woods, 1964). After recovery, decorticate rats show no gross abnormalities in behaviour that would allow a casual observer to identify them as impaired in an ordinary captive housing situation, although an experienced observer would be able to do so on the basis of cues in posture, movement, and appearance (Whishaw, 1990; what follows relies on Whishaw's study, supplemented by additional sources as indicated). They stand, rear, climb, hang from bars, and sleep with normal postures (Vanderwolf *et al,* 1978). They groom, play (Panksepp *et al,* 1994; Pellis *et al,* 1992), swim, eat, and defend themselves (Vanderwolf *et al,* 1978) in ways that differ in some details from those of intact animals, but not in outline. Either sex is capable of mating successfully when paired with normal cage mates (Carter *et al,* 1982; Whishaw and Kolb, 1985), though some behavioural components of normal mating are missing and some are abnormally executed. Neonatally decorticated rats as adults show the essentials of maternal behaviour, which, though deficient in some respects, allows them to raise pups to maturity. Some, but not all, aspects of skilled movements survive decortication (Whishaw and Kolb, 1988), and decorticate rats perform as readily as controls on a number of learning tests (Oakley, 1983). Much of what is observed in rats (including mating and maternal behaviour) is also true of cats with cortical removal in infancy: they move purposefully, orient themselves to their surroundings by vision and touch (as do the rodents), and are capable of solving a visual discrimination task in a T-maze (Bjursten *et al,* 1976; see also Bard and Rioch, 1937).

The fact that coherent and well-organized molar behaviors are elicited by local stimulation in the mesodiencephalic region of intact animals and that coherent motivated behaviour under environmental guidance is displayed spontaneously by animals lacking a cerebral cortex means that the neural mechanisms required to motivate, orchestrate, and provide spatial guidance for these behaviors are present in the parts of the brain that remain after decortication. Some aspects of these behaviors are dependent upon basal ganglia and basal forebrain functions remaining after the loss of their principal (cortical) source of afference (Wishaw, 1990), whereas the basic competences of decorticate animals reflect the capacity of upper brainstem mechanisms to sustain the global patterning, emotional valence, and spatial guidance of the postures and movements of orienting, defence, aggression, play, and other appetitive and consummatory behaviors (Adams, 1979; Holstege and Georgiadis, 2004; Masino, 1992; Maskos *et al,* 2005; Panksepp, 1982; Sakuma and Pfaff, 1979; Swanson, 2000). The particulars of the dependence of these behaviors on key structures located in the mesodiencephalic region has been repeatedly reviewed (Bassett and Taube, 2001; Behbehani, 1995; Groenewegen, 2003; Haber and Fudge, 1997; Horvitz, 2000; Houk, 1991; Jurgens, 1994; Mouton, 1999; Padel, 1993; Panksepp, 1998a; Prescott *et al,* 1999; Swanson, 1987,

2000; ten Donkelaar, 1988; Watt, 2000; Watt and Pincus, 2004; Winn, 1998; Zahm, 2006).

It is into the premotor circuitry of these ancient and highly conserved upper brainstem mechanisms that a wide range of systems place their bids for "where to look" and "what to do," irrespective of the level of sophistication of any one of these "bidding" systems. Each of them has independent access to effectors, and their upper brainstem interactions are not infrequently mediated by collaterals of such projections. The cerebral cortex is one prominent input to this system through the direct and indirect fibre projections emphasized in the foregoing discussion and sketched in Fig. 9.3 (see also Swanson, 2000). This relationship is, however, not a one-way affair. In fact, the manner in which the telencephalon is interfaced and integrated with the mesodiencephalic control system adds further definition to the central role of upper brainstem mechanisms in conscious functions.

4.5 Including the forebrain

Three cortical regions figure repeatedly and prominently in studies of cerebral mechanisms related to attention, neglect, and consciousness: namely, the posterior parietal cortex, the prefrontal cortex, and a medial territory centred on the cingulate gyrus (Baars *et al,* 2003, Fig. 1; Blumenfeld and Taylor, 2003; Clower *et al,* 2001; Corbetta, 1998; Han *et al,* 2003; Lynch *et al,* 1994; Mesulam, 1999; Posner and Petersen, 1990; Raz and Buhle, 2006; Rees and Lavie, 2001). A special connective and functional relationship exists between these three cortical territories and the mesodiencephalic system outlined in the foregoing discussion. It is most easily approached by considering their mutual interface in the nuclei of the dorsal thalamus. The latter can be divided into first-order (largely sensory relay) and higher-order ("association") thalamic nuclei (Sherman and Guillery, 2001), and it is with the latter, *higher-order* nuclei, that the mesodiencephalic system maintains an intimate and complex relationship.

The two major higher-order nuclei of mammals are the mediodorsal nucleus, whose cortical projections define the prefrontal cortex, and the pulvinar complex related to a set of posterior cortical areas, including extrastriate visual areas such as those of the posterior parietal cortex. Though proposed to serve as thalamic relays for cortico-cortical interactions (Sherman and Guillery, 2001), these nuclei are not devoid of extra-telencephalic input, and both receive prominent input from the superior colliculus (Benevento and Fallon, 1975; Harting *et al,* 1980; Lyon *et al,* 2005). Afferents to the pulvinar originate largely from the superficial collicular layers, whereas those destined for the mediodorsal nucleus are predominantly of intermediate layer origin. The latter projection targets a zone at the lateral edge of the mediodorsal nucleus related to the frontal eye fields (see Sommer and Wurtz, 2004), the cortical territory most directly implicated in unilateral neglect of frontal origin (see Mesulam, 1999, and references therein).

The cingulate gyrus, finally, is related to the mesodiencephalic system by its projections to the intermediate and deep layers of the colliculus (Harting *et al,* 1992;

Sherman *et al,* 1979), the periaqueductal gray matter (An *et al,* 1998; Floyd *et al,* 2000), and by a conspicuously heavy projection to the *zona incerta* (Mitrofanis and Mikuletic, 1999). This latter structure is a mammalian derivative of the ventral thalamus of comparative terminology mentioned in Section 4.1, and has emerged from obscurity only recently (see review by Mitrofanis, 2005). It sends a topographically organized inhibitory projection to the superior colliculus, and reaches up into the thalamus above it to selectively innervate its higher-order nuclei bilaterally, likewise with powerful GABAergic inhibition (Barthó *et al,* 2002; Lavallée *et al,* 2005; Power *et al,*, 1999; Trageser and Keller, 2004).

Collicular input to the higher-order nuclei is excitatory, whereas their incertal input is inhibitory. This implies dynamic competition between colliculus and zona incerta for influence over the two principal thalamic dependencies of the prefrontal and the posterior parietal cortex. In this competition the inhibitory incertal element stands under cingulate cortex influence and is also in a position to inhibit the colliculus directly and with topographic specificity (Ficalora and Mize, 1989; Kim *et al,* 1992; Ma, 1996; May *et al,* 1997). These circumstances cannot but profoundly affect the functional dynamics of the three cortical territories with which we are concerned. The principal pathways relating them to the mesodiencephalic control system and the higher-order thalamic nuclei are depicted schematically in Fig. 9.6.

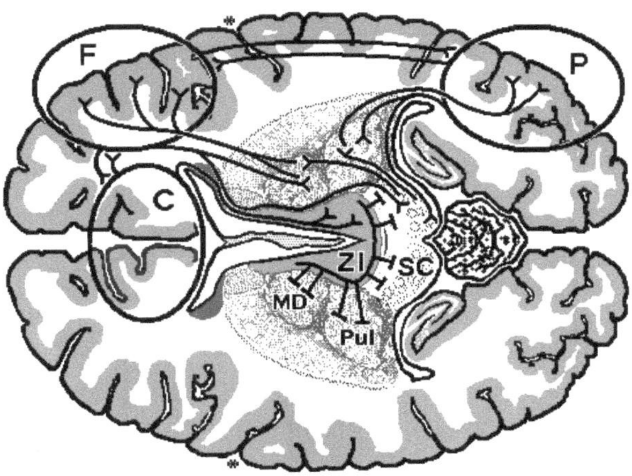

Fig. 9.6 Composite diagram illustrating the interface between the mesodiencephalic system and the thalamocortical complex. Principal pathways by which the superior colliculus and the zona incerta relate to one another, as well as to the dorsal thalamus and the cerebral cortex, are indicated in black heavy lines. Excitatory connections end in a "Y", inhibitory connections in a "T". Abbreviations: **P**: parietal; **F**: frontal; **C**: cingulate cortex; **SC**: superior colliculus; **ZI**: zona incerta; **Pul**: pulvinar complex; **MD**: mediodorsal nucleus of the thalamus. The central sulcus is marked by an asterisk. See text for further detail.

Supplying a key node in the relations depicted in Fig. 9.6, the zona incerta is monosynaptically (and often reciprocally and bilaterally) connected with on the order of 50 separate structures along the entire length of the neuraxis from spinal cord to olfactory bulb (my own conservative inventory of the literature, not counting connections with individual cortical areas separately). Internally, the zona incerta features profuse mutual connectivity in a setting of cytoarchitectonic and cytological heterogeneity in which GABAergic cells are prominent (Benson *et al,* 1991, 1992; Nicolelis *et al,* 1992; see Power and Mitrofanis, 1999, 2001; and Bartho *et al,* 2002, for connective details). A combination of reciprocal external connectivity with internal mutual inhibition is the theoretically optimal solution for implementing global competitive interaction among structures separated by long distances (for background, see McFarland, 1965; Snaith and Holland, 1990; Prescott *et al,* 1999). The zona incerta accordingly may implement such a scheme, and is hereby proposed to do so, as schematically illustrated in Fig. 9.7.

The zona incerta – or the ventral thalamus of non-mammals – thus supplies the integrative machinery of the optic brain with a connective hub that seems designed to conduct mutually inhibitory trials of strength among a truly diverse set of afferents. They include, but are not limited to, visual, auditory, somatosensory, vestibular (Horowitz *et al,* 2005), cerebellar, striatal, collicular, motor, and limbic ones. The outcome of the competition – a neural decision – is conveyed to the intermediate and deep layers of the superior colliculus by a topographically organized inhibitory projection, as already mentioned. The collicular return projection to the zona incerta – like that of many incertal afferents – is non-topographic, implying greater specificity of incertal influence over the colliculus than the reverse. At the same time, incertal inhibitory output ascends into the association nuclei of the dorsal thalamus, establishing the zona incerta as a connective bridge straddling the mesodiencephalic and the thalamocortical systems.

Coupled with the scope of its connectivity along the neuraxis, this nodal position of the zona incerta lends it a potentially strategic role as an arbiter of moment-to-moment decision-making "in the light of all available evidence." As in the case of collicular target selection, the loss of such a high-level function need not generate conspicuous behavioural deficits, and does not appear to do so in rats with incertal lesions (Thompson and Bachman, 1979). Rather, it would be expected to issue in suboptimal levels of resource allocation relative to shifting patterns of multiply interacting opportunities and needs. Preliminary indications regarding the great diversity and complexity of neuronal response properties in the zona incerta are worthy of note in this connection (Crutcher *et al,* 1980; Kendrick *et al,* 1991; Ma, 1996; Mungarndee *et al,* 2002; Nicolelis *et al,* 1992).

Finally, the zona incerta lies in immediate anterior continuity with the prerubral field and rostral interstitial nucleus of the medial longitudinal fasciculus, that is, with the rostral-most pole of the intermediate control system for orienting organized in spherical coordinates, mentioned in Section 4.3. This rostral pole is specialized for vertical movement, whereas the system's horizontal components are found farther caudally, in paramedian reticular structures extending into the pons. Could it be that the zona

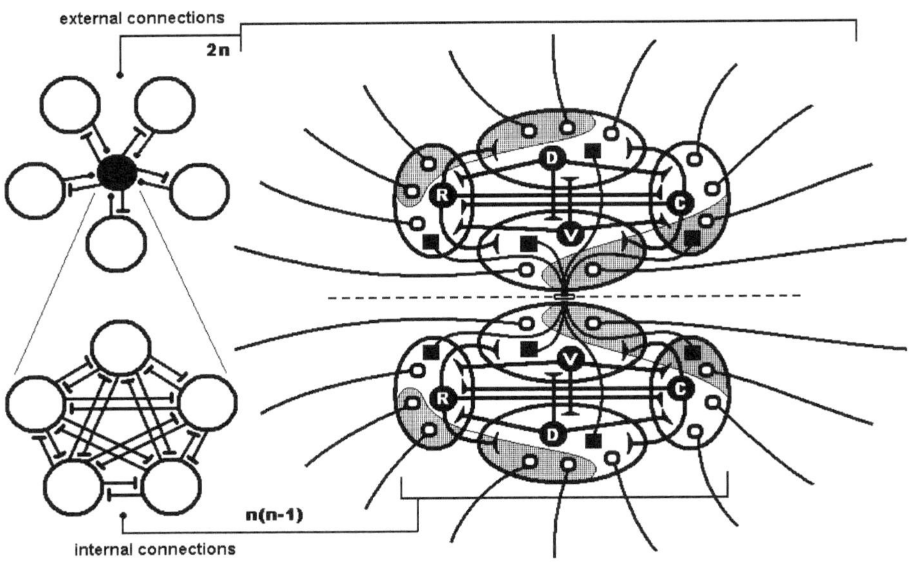

Fig. 9.7 Schematic diagram illustrating zona incerta connectivity to the rest of the brain and of its subdivisions to one another, depicted bilaterally to emphasize the prominent commissural, "midline-straddling," nature of incertal connectivity. Ovals represent the principal incertal subnuclei, shown physically separated for graphical clarity. Labels **R**, **D**, **V**, and **C** mark the rostral, dorsal, ventral, and caudal subnuclei, respectively, and connections among them. Filled squares indicate that each subnucleus projects to all subnuclei on the opposite side. Open circles stand for incertal connections with the rest of the brain (typically reciprocal in nature). Shaded regions stand symbolically for the fact that functionally defined subregions of the zona incerta (sensory modalities, motor, limbic, etc.) often cut across its nuclear subdivisions. Inserts on the left summarize the two connective schemes that appear to be combined in the zona incerta. Long-distance (external) connections, for which wiring-efficiency is at a premium, connect n entities to a central hub (the zona incerta itself) by 2n (i.e., reciprocal) connections (expandable to new targets at the low cost of 2). Internal connectivity within the zona incerta (wiring efficiency *not* at a premium), by contrast, appears to follow the n(n – 1) connective principle (lower diagram, expanding the filled central circle of the upper diagram). The scheme as a whole idealizes evidence supplied by the pioneering studies of Mitrofanis and colleagues (references in the text), and awaits refinement in the light of further detailed studies.

incerta supplies a kind of origin for this coordinate system, a midline-straddling point of unity connected directly and via the colliculus to the rest of the coordinate space (Giolli *et al,* 2001; Kolmac *et al,* 1998; Leichnetz *et al,* 1987)? Incertal omnipause neurons are at least compatible with such an eventuality (Hikosaka and Wurtz, 1983; Ma, 1996). Nothing would be more elegant than to entrust the final arbitration of "what to do next" to a self-inhibitory "winner-take-all" or other decision network (Richards *et al,* 2006) lodged at the origin of the coordinate system that controls the orienting movements which execute that decision once made. As a primary perspectival viewpoint charged with changing motives, it would possess the essential attributes of a self (see sect. 4.3). Prominent incertal afference from cingulate cortex would fit such a role (cf. Northoff *et al,* 2006 for medial cortex and self), but short of further evidence, the suggestion must remain speculative.

4.5.1 Collicular gamma oscillations and cortical "binding."

The superior colliculus is the only place outside of the cerebral cortex in which fast oscillations in the gamma range have been shown to occur and to behave in a manner paralleling in all significant respects that of the cortex (Brecht *et al,* 1998, 1999, 2001). At the cortical level such oscillatory activity has been proposed to serve a "binding" function for consciousness (in the sense of integrating disparate elements of unitary conscious percepts) on circumstantial grounds (Engel *et al,* 1999; Engel and Singer, 2001; Singer, 2001). As we shall see, one need not, however, ascribe a unique role to gamma oscillations in either binding or consciousness to recognize that they may have consequences for cortico-collicular integration nevertheless.

Though sometimes portrayed as "the" problem of consciousness, the acuteness of the cortical binding problem must not be exaggerated. The pyramid architecture of point-to-point interareal connectivity within topographically organized cortical sensory domains ensures that corresponding points on areal topographies featuring different functional content (e.g., contour and colour) are *connectively* and thus coherently related, even though the areas themselves occupy separate locations in the cortical sheet (Felleman and Van Essen, 1991; cf. Merker, 2004a).

The laminar superposition of numerous cortical areas in the colliculus takes this principle further. Here the joining of corresponding points on different cortical maps takes place by direct laminar superposition of topographic projections of different cortical areas within a unified collicular topography. Thus, the output of different cortical areas are brought within the compass of the dendritic trees of single collicular neurons, which often straddle collicular laminar boundaries (Albers and Meek, 1991; Laemle, 1983; Langer and Lund, 1974; Ma *et al,* 1990). Tight temporal synchrony of neuronal firing in separate cortical loci (through coupling to gamma oscillations) increases the probability that their joint activity will fall within the temporal window of integration of any neuron – whether cortical or subcortical – to which they project convergently (Abeles, 1982; König *et al,* 1996). Synchronous activation of corresponding loci on separate cortical maps would accordingly assist such activity in crossing collicular thresholds by summation via the dendritic trees of convergently innervated collicular cells.

In crossing the collicular threshold – whether assisted by gamma synchrony or not – cortical activity would gain access to the mesodiencephalic system in all of its ramifications, projections to the cortex included (see Fig. 9.6). This, according to the present account, would be a principal step by which such activity enters awareness. If so, it follows that *one conscious content will not be replaced by another without involvement of the mesodiencephalic system (centred on the superior colliculus) as outlined here, even when that change is unaccompanied by eye movements.* This prediction is specific to the present perspective, and accordingly renders it testable. The means for doing so are exemplified by a recent functional imaging study of a visual-auditory illusion in humans (Watkins *et al,* 2006). That study revealed collicular activation associated with awareness of the illusion, though stimuli were identical on trials in which the illusion was not perceived, and central fixation was maintained throughout, confirming the prediction just made, in this particular instance.

This, then, would be the identity of the so far unidentified threshold featured in a recent programmatic proposal regarding conscious function (Crick and Koch, 2003). Its identification with the threshold for access to the mesodiencephalic system centred on the colliculus (Figs. 9.4 and 9.6) is reinforced by the fact that layer V pyramidal cells supply the sole cortical projection to the colliculus. These cells exhibit a number of notable specializations: they do *not* give off collaterals to the thalamic reticular nucleus on passing through it (Jones, 2002), their local intra-cortical connectivity appears stereotyped (Kozloski *et al,* 2001), and their apical dendrites branch in cortical layer I and carry specialized conductance mechanisms activated by top-down (feedback) connections in the superficial cortical layers (Larkum *et al,* 2004). This may ensure that activation of both the feedforward and feedback cortical system is typically required for the cortico-mesencephalic threshold to be crossed, such concurrent activation having been proposed as an essential condition for cortical information to reach awareness (Lamme and Spekreijse, 2000; see also Merker, 2004a).

4.5.2 Consciousness and cortical memory.

Penfield and Jasper proposed a role for the centrencephalic system in both consciousness and the laying down of cortical memories across the life span. A rationale for such a memory role is suggested by the present perspective. The perpetual and cumulative nature of cortical memory recording (Merker, 2004a,b; Standing, 1973) puts a premium on economy of storage, that is, on concentrating memory recording to significant information (Haft, 1998). A criterion for doing so is available in the system of integration for action as outlined here: Information that is important enough to capture control of behaviour (i.e., by triggering an orienting movement placing its target in focal awareness) is also important enough to be consigned to permanent cortical storage. The focal presence of the target obviously will be the greater part of ensuring such an outcome, but it is likely to be actively supported as well by the system of dual colliculo-thalamic relays to cortex (cf. Fig. 9.6). From its parietal and frontal target areas, accessed in part via so-called matrix cell projections from the thalamus to the superficial cortical layers (Jones, 1998), the mesodiencephalic influence would then propagate and spread through the cortex via intracortical top-down feedback connectivity.

The evidence for a "general learning system" (which includes the superior colliculus: Thompson, 1993), mentioned in the introduction to Section 3, would seem to bear on this proposal, as well. In fact, the severe capacity limitations of so called working memory (Baddeley, 1992; Cowan, 2001; Mandler, 1975) are likely to derive in large part from the mesodiencephalic bottleneck which all attended (i.e., conscious) information must access according to the present proposal, just at the point where the parallel distributed data format of the forebrain *requires* conversion to a serial, limited capacity format to serve behaviour.

4.5.3 The zona incerta and the seizures of absence epilepsy

It is to be noted, finally, that the Penfield and Jasper postulation of a centrencephalic system *symmetrically related to both cerebral hemispheres* was motivated in part by observations on the generalized seizures of absence epilepsy. The zona incerta sends a rich complement of commissural fibres across the midline not only to itself, but also to the association nuclei of the dorsal thalamus (Power and Mitrofanis, 1999, 2001). It is also a prime locus for the induction of generalized epileptic seizures, being more sensitive than any other brain site to their induction by local infusion of carbachol (Brudzynski *et al,* 1995; see also Gioanni *et al,* 1991; Hamani *et al,* 1994). A number of phenomena that may accompany absence seizures can be readily related to the zona incerta. Thus, a forward bending or dropping of the head (or bending of the whole body to the ground; Penfield and Jasper, 1954) may relate to the already mentioned fact that the transition between the zona incerta and midbrain contains mechanisms for vertical control of eyes and head (Holstege and Cowie, 1989; Waitzman *et al,* 2000; cf. sect. 4.2). The fluttering of the eyelids that often occurs in the same situation is also easily accommodated by the functional anatomy of this region (Morcuende *et al,* 2002; Schmidtke and Buttner-Ennever, 1992).

The Penfield and Jasper definition of their proposed centrencephalic system always included explicit reference to the midbrain reticular formation. The zona incerta resembles a forward extension of the midbrain reticular formation beneath the thalamus (Ramón-Moliner and Nauta, 1966), and much of the functional anatomy of the diencephalon needs to be re-examined in light of its unusual connectivity. As noted by Barthó *et al,* (2002), the identification of a second, incertal, source of GABAergic innervation of the dorsal thalamus, in addition to that of the thalamic reticular nucleus, necessitates a re-evaluation of the entire issue of the nature of thalamic involvement in seizure generation and oscillatory thalamocortical activity (McCormick and Contreras, 2001; Steriade, 2001). This is all the more so since the even more recent discovery of a third source of powerful GABAergic thalamic inhibition, originating in the anterior pretectal nucleus (Bokor *et al,* 2005). One need not, however, await the outcome of such re-examination to identify the zona incerta as the perfect anatomical centre-piece for the Penfield and Jasper centrencephalic hypothesis, though its obscurity at the time kept it from being recognized as such.

5. Consciousness in children born without cortex

Anencephaly is the medical term for a condition in which the cerebral hemispheres either fail to develop for genetic-developmental reasons or are massively compromised by trauma of a physical, vascular, toxic, hypoxic-ischemic, or infectious nature at some stage of their development. Strictly speaking, the term is a misnomer. The brain consists of far more than cerebral hemispheres or prosencephalon, yet various conditions of radical hemispheric damage are historically labelled anencephaly. When the condition is acquired, for example, by an intrauterine vascular accident (stroke) of the fetal brain, the damaged forebrain tissue may undergo wholesale resorption. It is replaced by cerebrospinal fluid filling otherwise empty meninges lining a normally shaped skull, as illustrated in Fig. 9.8. The condition is then called hydranencephaly (Friede, 1989), and is unrelated to the far more benign condition called hydrocephalus, in which cortical tissue is compressed by enlarging ventricles but is present in anatomically distorted form (Sutton *et al,* 1980).

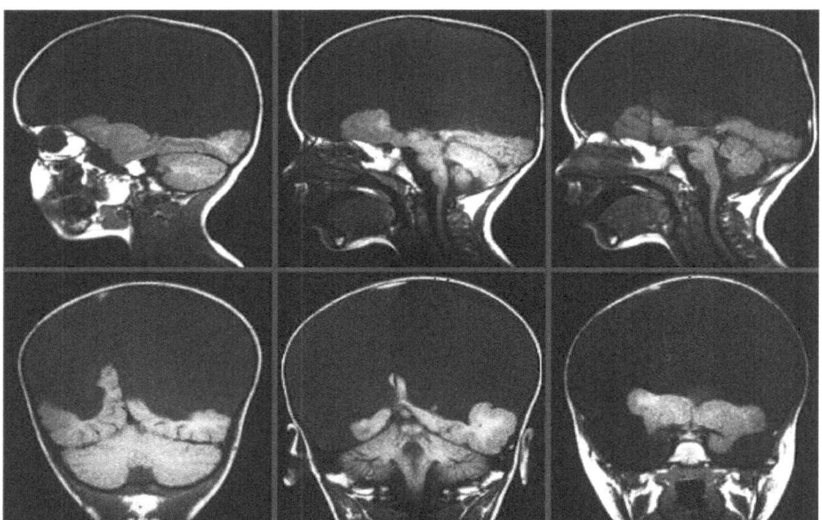

Fig. 9.8 Saggittal and frontal magnetic resonance images of the head of a child with hydranencephaly. Spared ventromedial occipital and some midline cortical matter overlies an intact cerebellum and brainstem, whereas the rest of the cranium is filled with cerebrospinal fluid. (Reprinted with the kind permission of the American College of Radiology; ACR Learning File, Neuroradiol., ed. 2, 2004).

The loss of cortex must be massive to be designated hydranencephaly, but it is seldom complete (see Fig. 9.8). It typically corresponds to the vast but somewhat variable forebrain expanse supplied by the anterior cerebral circulation (Myers, 1989; Wintour *et al,* 1996). Variable remnants of cortex supplied by the posterior circulation, notably inferomedial occipital, but also basal portions of temporal cortex, and midline cortical tissue along the falx extending into medial frontal cortex, may be spared. The physical presence of such cortical tissue, clearly visible in Fig. 9.8, need not mean, however, that it is connected to the thalamus (white matter loss often interrupts the visual radiations, for instance) or that it is even locally functional. On autopsy, such tissue may be found to be gliotic on microscopic examination or to exhibit other structural anomalies indicating loss of function (Marin-Padilla, 1997; Takada *et al,* 1989). As Fig. 9.8 shows, most cortical areas are simply missing in hydranencephaly, and with them the organized system of corticocortical connections that underlie the integrative activity of cortex and its proposed role in functions such as consciousness (Baars *et al,* 2003; Sporns *et al,* 2000).

An infant born with hydranencephaly may initially present no conspicuous symptoms (Andre *et al,* 1975), and occasionally the condition is not diagnosed until several months postnatally, when developmental milestones are missed. In the course of the first year of life, which is often though not invariably difficult, these infants typically develop a variety of complications that always include motoric ones (tonus, spasticity, cerebral palsy), and often include seizures, problems with temperature regulation, reflux/aspiration with pulmonary sequelae, and other health problems occasioning medical emergencies and attended by a high mortality rate. Were one to confine one's assessment of the capacities of children with hydranencephaly to their presentation at this time – which for natural reasons is the period in the lives of these children to which the medical profession has the most exposure – it would be all too easy to paint a dismal picture of incapacity and unresponsiveness as the hydranencephaly norm. When, however, the health problems are brought under control by medication and other suitable interventions such as shunting to relieve intracranial pressure, the child tends to stabilize and with proper care and stimulation can survive for years and even decades (Counter, 2005; Covington *et al,* 2003; Hoffman and Liss, 1969; McAbee *et al,* 2000).

When examined after such stabilization has taken place, and in the setting of the home environment upon which these medically fragile children are crucially dependent, they give proof of being not only awake, but of the kind of responsiveness to their surroundings that qualifies as conscious by the criteria of ordinary neurological examination (Shewmon *et al,* 1999). The report by Shewmon and colleagues is the only published account based upon an assessment of the capacities of children with hydranencephaly under near optimal conditions, and the authors found that each of the four children they assessed was conscious. For detail, the reader is referred to the case reports included in the Shewmon *et al,* (1999) publication. Anecdotal reports by medical professionals to the same effect occasionally see print (Counter, 2005), but compared to its theoretical and medical importance the issue remains woefully underexplored.

To supplement the limited information available in the medical literature on the behaviour of children with hydranencephaly, I joined a worldwide internet self-help group formed by parents and primary caregivers of such children. Since February of 2003 I have read more than 26,000 e-mail messages passing between group members. Of these I have saved some 1,200 messages containing informative observations or revealing incidents involving the children. In October 2004 I joined five of these families for one week as part of a social get-together featuring extended visits to Disney-World with the children, who ranged in age from 10 months to 5 years. I followed and observed their behaviour in the course of the many private and public events of that week, and documented it with four hours of video recordings.

My impression from this first-hand exposure to children with hydranencephaly confirms the account given by Shewmon and colleagues. These children are not only awake and often alert, but show responsiveness to their surroundings in the form of emotional or orienting reactions to environmental events (see Fig. 9.9 for an illustration), most readily to sounds, but also to salient visual stimuli (optic nerve status varies widely in hydranencephaly, discussed further on). They express pleasure by smiling and laughter, and aversion by "fussing," arching of the back and crying (in many gradations), their faces being animated by these emotional states. A familiar adult can employ this responsiveness to build up play sequences predictably progressing from smiling, through giggling, to laughter and great excitement on the part of the child. The children respond differentially to the voice and initiatives of familiars, and show preferences for certain situations and stimuli over others, such as a specific familiar toy, tune, or video program, and apparently can even come to expect their regular presence in the course of recurrent daily routines.

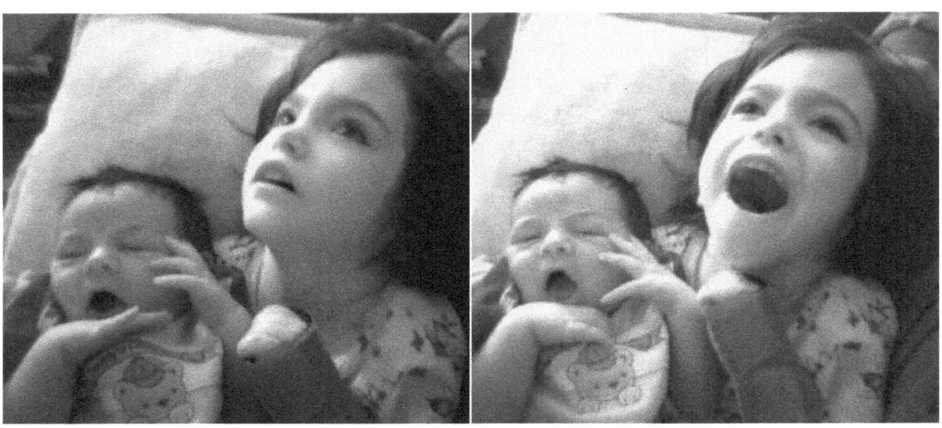

Fig. 9.9 The reaction of a three-year-old girl with hydranencephaly in a social situation in which her baby brother has been placed in her arms by her parents, who face her attentively and help support the baby while photographing.

Though behaviour varies from child to child and over time in all these respects, some of these children may even take behavioural initiatives within the severe limitations of their motor disabilities, in the form of instrumental behaviors such as making noise by kicking trinkets hanging in a special frame constructed for the purpose ("little room"), or activating favourite toys by switches, presumably based upon associative learning of the connection between actions and their effects. Such behaviors are accompanied by situationally appropriate signs of pleasure or excitement on the part of the child, indicating that they involve the kind of coherent interaction between environmental stimuli, motivational-emotional mechanisms, and bodily actions for which the mesodiencephalic system outlined in this article is proposed to have evolved. The children are, moreover, subject to the seizures of absence epilepsy. Parents recognize these lapses of accessibility in their children, commenting on them in terms such as "she is off talking with the angels," and parents have no trouble recognizing when their child "is back." As discussed earlier, episodes of absence in this form of epilepsy represent a basic affliction of consciousness (cf. Blumenfeld and Taylor, 2003). The fact that these children exhibit such episodes would seem to be a weighty piece of evidence regarding their conscious status.

In view of the functional considerations reviewed in the foregoing, none of these behavioural manifestations in children with hydranencephaly ought to occasion any surprise, and no special explanations such as neural reorganization based on plasticity are needed to account for them. Rather, they are what the nodal position of mesodiencephalic mechanisms in convergent neural integration, along with the comparative evidence regarding the behaviour of mammals in the absence of cerebral cortex, would lead us to expect. Nor is there much warrant for attempting to attribute these behaviors to remnant cortical tissue. Besides the questionable functional status of spared cortex already alluded to, a significant functional asymmetry speaks directly against it. As common as it is for some occipital cortex to remain in these individuals, so is it rare for any auditory cortex to be spared. Yet, sensory responsiveness in hydranencephaly shows the opposite asymmetry: hearing is generally preserved, whereas vision tends to be compromised (Hydranencephaly Group Survey, 2003). The pattern is easily accounted for by the intactness of the brainstem auditory system in these children (Lott *et al,* 1986; Yuge and Kaga, 1998), crowned by a projection from inferior to superior colliculus. By contrast, vision in these children is liable to be compromised already at the level of the optic nerve. The latter's blood supply through the anterior cerebral circulation exposes it to damage in hydranencephaly, and its status varies widely in affected children (Jones and France, 1978).

What is surprising, instead, is the routine classification of children with hydranencephaly into the diagnostic category of "vegetative state" (Multi-Society Task Force, 1994), apparently in conformity with a theoretical identification between the cortex as an anatomical entity and consciousness as a function. It is this very identification which has been under critical examination in the present target article. To the extent to which the arguments and the evidence presented here have any merit, such an identification is not tenable, and the routine attribution of a lack of awareness to children lacking cortex from birth would accordingly be inadmissible. The extent of awareness and other capacities in these children must be based on assessment in its own right, by appropriate neurological tests, and not by reference to the status of their cortical tissue (Shewmon, 2004). Moreover, considering the medically fragile status of many of these children, such behavioural assessment must be performed under optimal circumstances.

Properly assessed, the behaviour of children with early loss of their hemispheres opens a unique window on the functional capacities of a human brainstem deprived of its cerebral cortex early in intrauterine development. They tell us, for one thing, that the human brainstem is specifically human: these children smile and laugh in the specifically human manner, which is different from that of our closest relatives among the apes (Provine and Yong, 1991; van Hooff, 1972). This means that the human brainstem incorporates mechanisms implementing specifically human capacities, as shown long ago by the neurologist Gamper on the basis of his detailed cinematographically documented account of a congenitally anencephalic girl entrusted to his care (Gamper, 1926). In her case, there is no possibility that remnant hemispheric tissue might account for her human smile, since detailed postmortem histology disclosed that she had no neural tissue above the level of the thalamus, and even her thalamus was not functional.

The implication of the present account is that unless there are further complications, such a child should be *expected* to be conscious, that is, possessed of the primary consciousness by which environmental sensory information is related to bodily action (such as orienting) and motivation/emotion through the brainstem system outlined in the foregoing. The basic features of that system evolved long before the cerebral hemispheres embarked on their spectacular expansion in mammals to supply it with a new form of information based upon cumulative integration of individual experience across the lifetime (see Merker, 2004a). Now as then, this brainstem system performs for the cortex, as for the rest of the brain, a basic function: that of integrating the varied and widely distributed information needed to make the best choice of the very next act. That function, according to the present account, is the essential reason for our being conscious in the first place. The integrated and coherent relationship it establishes between environmental events, motivation/emotion, and actions around the pivotal node of an egocentre would seem to offer a definition of a "being" in biological terms.

6. Implications for medical ethics

Needless to say, the present account has ramifying implications for issues in medical ethics. One of these concerns pain management in children with hydranencephaly and similar conditions. It is not uncommon for parents to encounter surprise on the part of medical professionals when requesting analgesia or anesthesia for their crying child during invasive procedures, a situation in some ways reminiscent of what was found in the case of neonates only a few decades back (Anand and Hickey, 1987). They also extend to more general issues pertaining to the quality of care appropriate to these children, and ultimately to questions such as the meaning of personhood and even medical definitions of death (see, e.g., Shewmon *et al,* 1989, and references therein). Such questions are decidedly beyond the scope of the present article, which is meant only to raise those issues of a theoretical and empirical nature which are prior to and essential for finding reasoned and responsible answers to the ethical ones. Suffice it to say that the evidence surveyed here gives no support for basing a search for such answers on the assumption that "awareness," in the primary sense of coherent relatedness of a motivated being to his or her surroundings, is an exclusively cortical function and cannot exist without it.

7. Conclusion

The evidence and functional arguments reviewed in this article are not easily reconciled with an exclusive identification of the cerebral cortex as the medium of conscious function. They even suggest that the primary function of consciousness – that of matching opportunities with needs in a central motion-stabilized body–world interface organized around an ego-centre – vastly antedates the invention of neocortex by mammals, and may in fact have an implementation in the upper brainstem without it. The tacit consensus concerning the cerebral cortex as the "organ of consciousness" would thus have been reached prematurely, and may in fact be seriously in error. This has not always been so, as indicated by the review of the Penfield and Jasper (1954) "centrencephalic" theory of consciousness and volitional behaviour with which we began. As we have seen, their proposal has not only been strengthened by certain findings accumulating since it was first formulated more than half a century ago, but, suitably updated, it still appears capable of providing a general framework for the integration of a vast array of diverse facts spanning from the basics of the vertebrate brain plan to evidence for awareness in children born without a cortex. Whether such a framework can be developed into a comprehensive account of the neural organization of consciousness will depend upon resolving a number of the empirical and theoretical questions left unanswered in the foregoing discussion. Preliminary though it may be, that discussion suggests that part of the endeavour to resolve these questions will require close scrutiny of conserved and convergently innervated upper brainstem mechanisms as potential key components of a neural mechanism of consciousness.

Acknowledgements

My gratitude goes first to all the members of the hydranencephaly internet group founded by Barb Aleman. To them and their children I owe not only this article, but a precious enrichment of my life. To Karen Krueger my thanks for permission to use photos of her daughter Heather Joy. I am grateful as well to Bernard Baars for his personal encouragement, and to him, Douglas Watt, Alan Shewmon, Jaak Panksepp, and Sunny Anand for most helpful comments on an earlier version of this manuscript. The comments and suggestions provided by five referees helped me to further improve this article. Finally, my intellectual debt to the late Eugene Sachs is gratefully acknowledged.

Notes

1. In what follows, the term "cortex" will always be taken to mean all or part of the cerebral cortex along with its associated dorsal thalamic and claustral nuclear apparatus. The thalamic reticular nucleus, being functionally intrinsic to this thalamocortical complex is regarded as being part of it despite its embryological and phylogenetic origin in the ventral thalamus (it is directly continuous with the lateral margin of the zona incerta). Unless otherwise indicated, "subcortical" will refer to all central nervous system tissue that is not thalamocortical complex in this sense, and "brainstem" will refer to diencephalon and the rest of the entire neuraxis caudal to it.

2. To avoid possible misunderstanding of this key point, note that the analogue "reality simulation" proposed here has nothing to do with a facility for simulating things such as alternate courses of action by, say, letting them unfold "in imagination," or any other version of an "inner world," "subjective thought," "fantasy," or the like. Such capacities are derivative ones, dependent upon additional neural structures whose operations presuppose those described here. The purpose of the "analogue simulation" defined here is first and foremost to veridically reflect states of the world, the body, and needs at whatever level of sophistication a given species implements those realities. It is thus most directly related to the model of Philipona and colleagues (2003, 2004), as well as to the "situation room analogy" developed by Lehar (2002).

3. Note that in some of the animal and human studies cited in this passage the term "Cartesian" occurs as a misnomer for "spherical." They all refer to a system organized in terms of "azimuth" and "elevation," that is, a system of spherical coordinates.

References

A complete reference list (with 380 entries) is published in the original article Merker B (2007), Consciousness Without a Cerebral Cortex: A Challenge for Neuroscience and Medicine. *Behavioral and Brain Sciences* **30**:63-81.

Chapter 10

Three Types of State Transition Underlying Perception

Walter J. Freeman

1. Introduction

The biology of brains should be approached first and foremost in terms of how they construct intentional behaviour, by which humans and other animals know what they are doing, and why. The term 'intentional' is used in preference to 'voluntary', because it broadly includes behaviour that is classed as instinctive, innate, emotional, and unconscious, as well as conscious and ethical, only requiring that it originate within the subject, not as stereotypic reactions to particular stimuli, but as self-directed actions based in the context of all past experience. The behaviourist stimulus-response and cognitivist rule-driven logical paradigms fail to address the most basic properties of biological intelligence and control, which are its autonomy and its creative powers. The theory of nonlinear dynamics has greatly expanded our understanding of neural mechanisms by which large-scale patterns of brain activity are self-organized. The new concepts give us fresh insight into the neurodynamics of intentional behaviour, how consciousness emerges within the brain, and how brains regulate their own sensory inflows into the several sensory cortices (Freeman, 1995, 2000/2006).

Evidence has been summarized (Freeman, 2000; Tsuda, 1992, 1996, 2001) that brains are chaotic dynamic systems that not only can destroy information (in the Shannon-Weaver sense); they also can create information. Brain chaos is not deterministic chaos, which is low dimensional, stationary, autonomous, and noise-free. To the contrary it is stochastic chaos; it is *constrained noise* (Freeman, 2006), which emerges from mutual excitation among neurons, and which is modulated during engagement of brains with the environment through the body. Hence the stochastic chaos is noisy, infinite-dimensional, non-stationary and non-autonomous, yet indistinguishable from deterministic chaos by conventional statistical tests (Rapp, 1993; Freeman, 2001). The source of the noise being mutual excitation among pyramidal cells (Freeman, 2004a,b, 2006),

Consciousness Transitions: Phylogenetic, Ontogenetic, and Physiological Aspects
Edited by Hans Liljenström and Peter Århem

the chief constraints are provided by the neuronal thresholds and refractory periods and by the net feedback gain among the neurons that evaluates the strength of mutual interdependence in the regulation of mean firing rates. The reduction in disorder by interaction is expressed in Haken's (1983; 1999) 'slaving principle' by which a collective generates its activity serving as an 'order parameter'. Oscillations in the gamma range result from negative feedback through inhibitory interneurons; aperiodic oscillations stem from interactions among three major components of the olfactory system, the bulb, anterior olfactory nucleus, and prepyriform cortex, each with its characteristic limit cycle frequency that is incommensurate with the other frequencies (Freeman, 2000/2006).

These aperiodic, unpredictable, chaotic oscillations are revealed by recordings of local field potentials (LFP) from depth electrodes, electrocorticogram (ECoG) from surface electrodes and electroencephalogram (EEG) from scalp electrodes. All three derivations arise from summed dendritic synaptic currents that accompany axonal action potentials (nerve impulses) as both cause and effect. Spatial patterns of amplitude modulation (AM) of the aperiodic oscillations that are correlated with conditioned stimuli show that brains, as chaotic systems, do not merely filter and process sensory input. By self-organization in generating order parameters, they actively seek sensory stimuli as raw material from which to create perceptual patterns, with or without awareness, that replace stimulus-induced activity (Freeman 1991).

This review sketches (2) the neurodynamics of the prototypical olfactory sensory lobe of the brain, (3) some ways in which its state changes are controlled by the forebrain in the mechanisms of attention and motivation, (4) the development and evolution of its state space through bifurcations during learning, (5) the formation and maintenance of an attractor landscape that mediate the discrimination of classes of stimuli by generalization and abstraction, (6) the role of chaos in creating new attractors in cerebral cortex, and (7) the clarification of the meanings of intentionality in the context of embodied cognition.

2. Neural activity in the olfactory system

The creative property of chaos is not readily apparent in the nonlinear homeostatic feedback mechanisms of brain reflexes, which insure the stability of brain function by keeping the internal environment of the brain at an optimal level. Brain creativity becomes clear in modelling perception, which requires interaction with the external environment. An example comes from study of the neural activity in the olfactory system. This is a semi-autonomous lobe that interacts with other parts of the forebrain, both receiving sensory input from the nose and regulatory centrifugal controls from other parts of the forebrain, and transmitting its self-organized patterns widely through the forebrain, especially from the prepyriform cortex to the entorhinal cortex. Its endogenous activity persists after it has been surgically isolated from the rest of the brain, which shows that its basic functions are self-organizing. However, its chaotic activity disappears when its parts have been surgically disconnected (Freeman, 1975/2004), showing that its chaos is a global property that is not due to the entrainment of the whole system by single neurons acting as chaotic generators (Aihara, Takabe and Toyota, 1990). This distinction is important, because the stimuli that are recognized in perception are spatiotemporal patterns. They are characterized by spatial relationships between each part and every other part of the patterns, so that the determinants of sensory cortical function must also be global, not local as in 'feature detector' neurons (Singer and Gray, 1995).

The clearest demonstrations of the self-organization of complex spatiotemporal patterns of neural activity are based in simultaneous recordings of EEG activity from arrays of 64 electrodes placed on the olfactory bulb and cortex of rabbits. In each case the recordings reveal a common oscillation in field potential over the entire array (Freeman 1991) constituting repeated bursts of oscillation in a cinematographic process. Each burst manifests a 'wave packet' having a measurable duration, diameter, and location in time and space (Freeman, 2004a,b). The fluctuation serves as a carrier wave for perceptual information in its spatial AM pattern (Freeman, 2005). The waveform of the carrier is aperiodic and unpredictable, reflecting the chaotic dynamics of sensory cortices. The spatial AM patterns of wave packets emerge following stimulation of the olfactory receptors by the inhalation of air carrying odorant substances, but only when the substances are accompanied by reinforcement (Freeman, 2001). During an act of perception the spatial pattern of excitation in the nasal receptors is replaced by a spatial pattern of action potentials that represents the stimulus to the olfactory bulb when the pulses are transmitted on the primary olfactory nerve. The topographic mapping of the axons in the nerve onto the bulbar surface preserve the representation as a set of localized discharges of bulbar mitral cells that are driven by the afferent volley.

The action potentials from the receptors for odorants are not merely those representing a stimulus designated by the experimenter. The volley with inhalation also includes the action potentials from receptors responding to whatever is in the background or control air, generally unknown (Freeman, 2001). The bulbar response to massive background input is tuned by habituation to enhance bulbar sensitivity not in the gamma range, as for the foreground input, but in the theta range. Inhalation causes a strong wave of depolarization phase-locked to the respiratory frequency (Freeman, 1975/2004). Thereby the bulbar population is destabilized, leading to the emergence by a state transition of a burst of gamma oscillation constituting a wave packet. Unlike the spatial pattern of the action potentials of single cells, the AM pattern of the gamma wave packet does not literally represent the stimulus. It cannot do so, because the AM pattern lacks invariance with respect to the stimulus. When the context, significance, value or meaning of the stimulus is changed, the AM pattern changes (Freeman, 1991). Moreover, the AM pattern is distributed; no channel is any more or less important or effective than any other channel in classification of AM patterns with respect to conditioned stimuli. By inference the microscopic pattern of action potentials that represents the stimulus in sensation has high information density on few neurons, whereas the AM pattern of the wave packet that embodies perception has low information density on very many neurons.

The AM patterns are transmitted to the prepyriform cortex over the divergent-convergent lateral olfactory tract, in which the transmitting neurons have widely branching axons, and from which each receiving neuron has synaptic input from widely spaced transmitting neurons. This output tract does not map like the topographic mapping of the input pathway. Instead the tract performs a spatiotemporal integral transformation that transmits only the activity having the shared carrier waveform. The sensory-evoked microscopic activity lacks spatial coherence in frequency, so it is attenuated by frequency and phase dispersion on summation over large populations. The result is that only the endogenous, self-organized AM pattern is effectively transmitted to the targets of bulbar transmission. Thereby the bulbar outputs transmit some relevant portion of the knowledge from experience that has been accumulated in synaptic modifications with learning from stimuli during periods of training to respond to them, in brief, the meanings of the stimuli that are unique for each of the subjects. Having done the work of selecting an attractor and updating its defining synaptic connections, the sensory evoked activity is expunged. The sensory information about which receptors a stimulus activated among equivalent receptors has no value. The deletion enacts abstraction; the selection enacts generalization.

3. The role of preafference in attention

The dynamics of the three-layered cortices of the olfactory sensory system is a prototype for the dynamics of six-layered neocortex. In both types of sensory cortex the response to conditioned stimuli is the formation of intermittent cinematographic frames of activity, in which a mesoscopic aperiodic shared wave carries AM patterns that are classified with respect to the inducing discriminated stimuli (Viana Di Prisco and Freeman, 1985; Barrie *et al*, 1996). Behavioural studies show unequivocally that humans and other animals alike successfully predict and therefore selectively attend to expected stimuli, indeed actively search for them, knowing what they are searching for. The question considered here is: What are the principal neural mechanisms of selective attention? A key concept in models of perception and motor control is that of preafference (Kay *et al*, 1995; Kay and Freeman, 1998). When a macroscopic brain activity pattern that is generated by chaotic dynamics initiates action toward a goal, that pattern has two components. One is a motor command that activates the descending motor systems. The other is a set of messages to all of the central sensory systems, which prepares them for the impending changes in sensory input that will be caused by the motor action. This corollary discharge (Sperry, 1950) in preafference primes and updates the attractor landscapes (Section 4) in the several primary sensory cortices, constituting the ongoing construction of the brain's knowledge base.

Recordings of EEG in rabbits trained in a classical aversive paradigm and cats trained in an operant appetitive paradigm have shown that in the first 300 ms after stimulus onset, a brief succession of AM patterns emerges in the sensory cortex to which the stimulus is directed, having a carrier frequency in the gamma range (Fig. 10.1 (A) (Freeman, 2005). In the next 300-500 ms before the onset of the conditioned response there is a brief succession of AM patterns in the beta range (B) (Gaál *et al*, 2003; Freeman and Burke, 2003; Freeman and Rogers, 2003). These beta AM patterns have contributions from all senses (Fig. 10.2), so they are multimodal (Freeman and Burke, 2003), and they display phase locking of shared oscillations over the visual, auditory, somatic and olfactory sensory areas and as well the entorhinal cortex (Fig. 10.3).

Multimodality is shown by measuring the goodness of classification after the EEG data from each of the five cortices contributing to the classification are removed. Deletion of any of them reduces the classification level (Fig. 10.2), which indicates that these are macroscopic patterns covering much or possibly the whole hemisphere of recording. The inference drawn from these data is that the destabilization of the several sensory areas by input results in the transmission to the superficial layers of the entorhinal cortex of a set of 4 gamma bursts, followed by convergence and formation of a response that is sent to the hippocampus via the perforant path. Transmission through the hippocampus establishes a multimodal percept that is assigned a time and place of occurrence by the hippocampal cognitive map and short term memory. The hippocampus sends its output through the subiculum mostly to the deep layers of the entorhinal cortex, which diverges its output to all of the primary sensory cortices. This divergent transmission is the proposed vehicle for preafference, appearing in the olfactory bulb as bursts in the beta range between the gamma bursts induced by inhalation (Bressler, 1988; Kay *et al*, 1995).

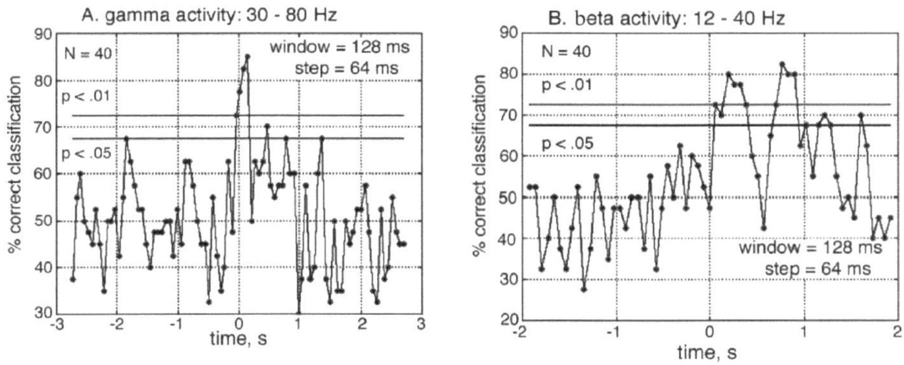

Fig. 10.1 Probability of % correct classification occurring by chance. **A.** Gamma band. **B.** Beta band. AM patterns were detected by application to sets of 40 trials of a moving window 64-128 ms in duration, which was stepped along the sets of 6 s trials at 16 ms intervals. The 20 frames from CS+ vs. CS- trials were divided into 10 training and 10 test trials. At each step the % correct classification was calculated and plotted with trial time on the abscissa. The probabilities of those % values occurring by chance were calculated with the binomial distribution (Barrie *et al*, 1996). As shown by the example here, peaks of high classification occurred just after the onsets of CSs and thereafter at varying intervals. The level of % classification declined through the onsets of the CRs, due to variation in the latencies of later classifiable wave packets called "jitter" (Tallon-Baudry *et al*, 1998). (From Freeman and Holmes, 2005).

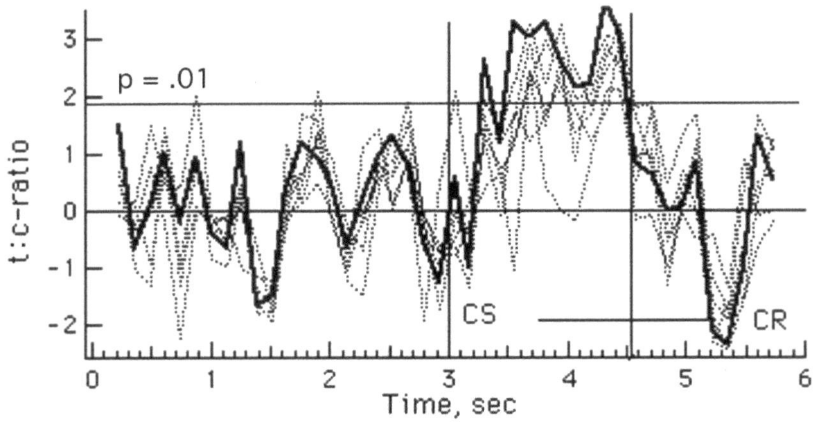

Fig. 10.2 Beta ECoG in a moving window gave 64x1 multimodal feature vectors from 64 electrodes that served to classify AM patterns from cats trained to discriminate conditioned stimuli. Removal of the channels from either visual, auditory, somatic, olfactory or entorhinal cortex reduced the goodness of classification. (From Freeman and Burke, 2003).

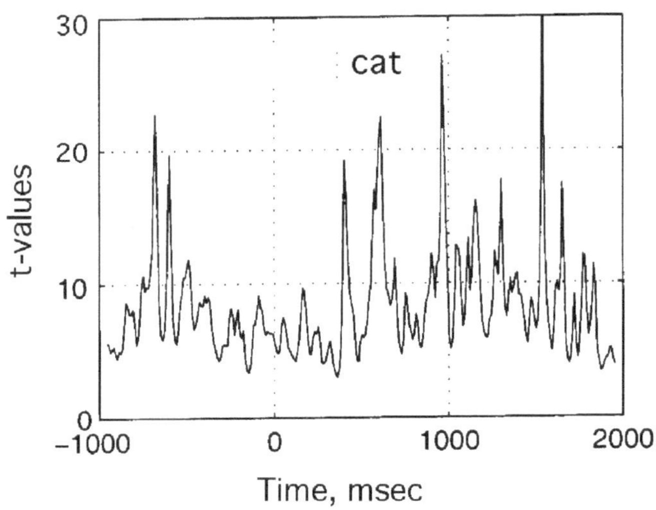

Fig. 10.3 An index of phase locking across the ECoG of 64 electrodes from visual, auditory, somatic, olfactory, and entorhinal areas showed intermittent phase locking of beta activity at times corresponding approximately to classifiable AM patterns. (From Freeman and Rogers, 2003).

By this mechanism within half a second of sample onset every cortex is updated with the contents of the mesoscopic wave packets that have been constructed in every other sensory cortex. They all contribute to a macroscopic AM pattern, for which the carrier wave is synchronized beta oscillation in all sensory areas and the entorhinal cortex. That unity in diversity makes this macroscopic state a candidate for the neural correlate of consciousness of a multisensory percept, the Gestalt (Freeman, 1986). This inference is supported by the apparent loss of consciousness in *absence*. When the lateral olfactory tract is tetanized by a train of electrical pulses, the normal chaotic dynamics is suppressed by a state transition to the abnormal epileptic state of complex partial seizure, which is characterized by the replacement of beta and gamma activity by a stereotypic 3/s spike and wave, loss of autonomic control and the abilities to track, attend, and learn or remember new stimuli presented during the seizure. It is remarkable how abruptly the state of the brain changes, first in the catastrophic onset, and then in the equally sudden return to normal function. In humans there is often no awareness of time lapse or a gap in the flow of experience.

4. Associative (Hebbian) learning requires structural bifurcation

A major question in the study of neural networks is: Where do the synaptic changes take place when a network learns a new input? The sites of olfactory synaptic modification in animals were found by giving an electrical impulse to axons in the lateral olfactory tract, which runs from the bulb to the olfactory cortex, and training the animal to recognize the impulse as a conditioned stimulus. The impulse responses of the prepyriform cortex and olfactory bulb were measured by fitting damped cosines to them, before and after the animal had been trained to recognize and respond to the stimulus as a meaningful input. The form of the impulse responses had the shape of a damped cosine wave (Basar, 1980), and it changed in a characteristic way as the animal learned to identify and respond to the electrical stimulus (Freeman 1975/2004). The initial amplitude did not change or decreased slightly, which showed that the site of modification was not at the synapses between the input axons and the excitatory neurons, as it is in back-propagation networks. Instead, the duration of the initial half cycle increased (Fig. 10.4, B), which showed that the modified synapses were situated between the excitatory neurons. Compared with the naïve state (A) the frequency decreased slightly, but the duration of the oscillation increased substantially. This decrease in the decay rate of the impulse response showed that the modified population of excitatory neurons was sensitized and made more likely to become unstable in the oscillatory mode. The prolongation was not apparent in the mean (dotted curve) but in the standard deviation (SD, solid curve), in which the peaks and cancellations reflected the frequency modulation (FM) of the oscillation that degraded the mean. This synaptic change with associative learning was irreversible, though its effects could be over-ridden by habituation or newer learning (C).

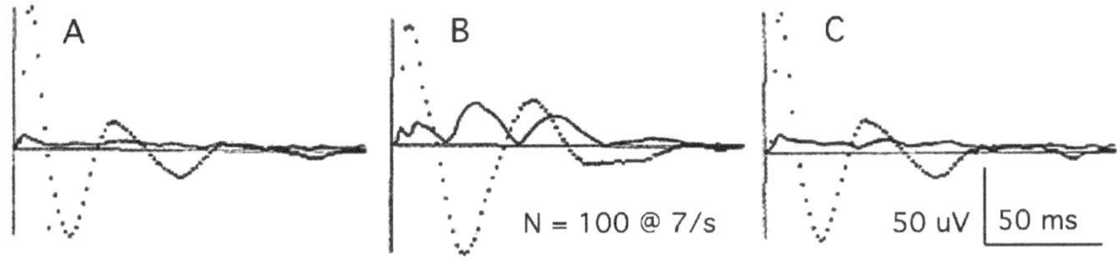

Fig. 10.4 Associative learning to recognize as a conditioned stimulus the electrical pulse driving the evoked potential increased the duration and not the amplitude of the first peak of the impulse response, showing that the site of synaptic change was not monosynaptic at the input to the cortex but disynaptic between excitatory neurons. (From Freeman, 1975/2004).

Learning in animals occurs when a reward or punishment accompanies a novel stimulus. The reinforcement is mediated by the release of norepinephrine into the olfactory system from the locus coeruleus triggered by an unconditioned stimulus (Gray *et al*, 1986). If there is no reinforcement, then habituation takes place, by which the neural system decreases its sensitivity to the stimulus on repeated presentations. The decay rate of the impulse response increases with foreshortening of the oscillation, showing an increase in stabilization. This process is easily reversible with dishabituation. Learning is always a combination of processes for association and habituation. Input that is wanted is reinforced, and input that is unwanted is habituated. The stimulus to be learned is by definition an unknown figure in a complex background. Habituation attenuates cortical neural responses to the background by shifting the energy of cortical activity away from the gamma range (30-80 Hz), where cognitive content is located, and into the theta range (3-7 Hz), which is the common range of repetition rates for wave packets. As noted earlier the surge of background input is necessary to induce state transitions in sensory cortices. The neural input from specific stimuli that selects attractors is too weak to actualize the latent attractor landscape by inducing a state transition. Hence there can be no perception without background input, and every attractor and its basin are drenched in contextual cues that contribute strongly to associations on recall.

In modelling these data from studies of the changes in evoked potentials with learning, the simulated synaptic strengths are increased or decreased in a model of olfactory dynamics (Freeman, 1987) at the sites corresponding to modifiable synapses, until the simulated impulse responses change their wave forms in conformance with the experimental observations. Pairs of neurons that are co-active have their interconnecting synapses strengthened in Hebbian learning. A 2-D layer of 64 coupled oscillators is constructed in software to simulate the olfactory bulbar dynamics, with simulated regulatory input from other parts of the brain, so as to show that, with training by examples under reinforcement, the spatial AM pattern of the common carrier changes from a pretraining pattern to a new pattern, whenever an example is given as an input of the newly learned class. With each new category of stimulus to which the model is trained to respond, there is a distinctive new spatial pattern of output. Each spatial pattern consists of 64 scalar values of the amplitude of the carrier wave, so that each output is a vector that defines a point in 64-space. The degree of similarity between patterns is given by a Euclidean distance in 64-space (Kozma and Freeman, 2002; Li *et al*, 2006). Classification is based on clustering of points derived from measurement of individual responses. The centre of gravity of a cluster gives the average AM pattern of the category. These properties replicate the performance of the olfactory system during learning (Freeman, 2005).

5. Modifiable synapses provide for generalization over categories of stimuli

The identification of the modifiable synapse as the connection between excitatory neurons in the bulb shows that in this respect the olfactory system closely resembles a class of associative memory systems exemplified by the self-organizing feature map of Kohonen (2001), the W weight matrix of Amari (1977), the brain-state-in-a-box of Anderson *et al* (1977), the spin glass model of Hopfield (1982) and the recurrent network of Aihara *et al* (1990) in the establishment and maintenance of stored information in a symmetric matrix of connection strengths. The simplest description of each main part of the olfactory system is a double layer formed by interconnecting a Hopfield excitatory net from spin glass with a mutually inhibitory net from the Limulus eye. Connection strengths change only in the excitatory net in accordance with a modified version of the Hebb rule. The system does not learn single events but assigns inputs into categories (Yao and Freeman 1990; Ohl *et al*, 2001), each class having an attractor and its own basin of attraction corresponding to its generalization gradients (Kozma and Freeman, 2002). The basin of attraction solves the problem known as "Lashley's dilemma", from the description by Karl Lashley (1942):

> "Generalization is one of the primitive basic functions of organized nervous tissue (p 302). ... Here is the dilemma. Nerve impulses are transmitted ... from cell to cell through definite intercellular connections. Yet all behaviour seems to be determined by masses of excitation. ... What sort of nervous organization might be capable of responding to a pattern of excitation without limited specialized paths of conduction? The problem is almost universal in the activities of the nervous system (p 306)."

The convergence property implied by the basin reflects the nature of the operation that the olfactory system performs, which is to generalize over its specific inputs. The very large number of receptor neurons in the nose, on the order of 10^8 with 10^3 types and 10^5 responsive to any one odorant, enables detection at extremely low concentration. On any one inhalation not all of the available receptors actually receive the odorant, but because of turbulence in the nasal passage, the selection of 10^2 - 10^3 from the 10^5 receptors that are capable of responding is different on each trial. A nerve cell assembly forms during training by strengthening synapses among the bulbar excitatory cells that are co-excited by the specific odorant input, which thereafter responds in a stereotypic way when any of its members are excited. The Hebbian nerve cell assembly provides for generalization from examples to significant categories of input. This operation constitutes the first step in the processes of perception (Skarda and Freeman, 1987) prior to consciousness of the percept.

The necessity for this process is infinite complexity of the olfactory environment of odorant chemicals, while the olfactory mechanism for classification of perceptual odours is finite. A stimulus is not admitted, filtered, and compared. It triggers the construction of a store of knowledge about its context and significance. The particular receptors that receive a stimulus are members of an equivalence class, are not reproducible, and are not knowable. This categorization takes place in the olfactory system at the first synapse of the pathway, where the enormous load of information from the receptors is washed away and replaced by a categorical signal. The same modes of category formation and selection have been found in the visual, auditory, and somatosensory cortices (Barrie *et al*, 1996; Ohl *et al*, 2001; Freeman *et al*, 2003). The conclusion is that the store of knowledge in brains emerges from pair-wise changes in large numbers of synapses cumulatively and continually in evolving attractor landscapes. The landscape in each sensory area is latent in the form of selective sensitivity to classes of input. Each basin of attraction embodies a possible expected outcome of an act of observation. At any moment in a given context an animal expects a limited number of possibilities, only one of which can be realized with each act of sampling the environment: odorants A, B, C, ..., none of the above, or some unexpected combination or unknown odorant. Each realization brings an opportunity for an additional increment of learning, if the stimulus activates receptors of the equivalence class not previously selected by an event of that class. Notably this incremental change at the microscopic level cannot be described as formation of a representation of a stimulus or as an extraction of one of its features.

6. The role of stochastic chaos in Hebbian learning

If the inhaled air does not contain a known odorant but a novel chemical, then a gamma burst fails to form, and the spatiotemporal response is unpatterned, disordered. The neural activity does not conform to any pre-existing spatial AM pattern. This suggests an important role for chaos in Hebbian learning. Strengthening of a Hebbian synapse requires concomitant activity in both the presynaptic and postsynaptic neurons. When this activity is organized in an already existing basin of an attractor, then the synaptic change merely reinforces an existing spatial AM pattern of output. If a new pattern is to be created, then the activity that is required to drive the Hebbian synapses to change must have no previous spatial pattern. A chaotic generator appears to be an optimal way that the bulb can provide novel spatial patterns of broad-spectrum neural activity, which can then be consolidated by the learning process.

The origin of the *constrained noise* lies in the mutual excitation of pyramidal cells, which is stabilized at a mean steady state by the refractory periods, which limit the feedback gain of the positive feedback loop formed by the interacting neurons (Ch. 8 in

Freeman, 2000/2006). The role of inhibitory neurons is not to stabilize the noise level but to introduce gamma oscillations by negative feedback (Ch. 6 in Freeman, 1975/2004). As the amplitude of oscillations induced by sensory input increases, the interaction strength and the negative feedback gain increase, which can destabilize the cortical dynamics, leading to narrow band bursts of gamma in relation to selected attractors, which are seen as phase-locked oscillations in wave packets correlated with behaviour. The broad-band oscillations that are seen with novel stimuli for which no attractors exist do not show phase locking, but they do reveal the existence in cortex of a chaotic attractor that has been called an "I don't know" basin of attraction (Skarda and Freeman, 1987). Once an attractor has emerged under reinforcement, the mechanism for emergence of a wave packet on selection of the attractor by a stimulus is predominantly the reduction in disorder with phase-locking and emergence of an AM pattern, owing to the enhanced constraints imposed by the synaptic interactions and expressed in the ECoG as an order parameter (Haken, 1983, 1999).

These properties support the inference that, in the operations of associative memory systems, a response to a stimulus is freshly created with each cinematographic engagement with the environment. The construction of each pattern is guided by a basin of chaotic attraction, which was formed during learning. Perceptions are triggered by stimuli, but they are shaped by connectivity patterns that were laid down during the past learning, and by neural messages from the limbic system that modulate the attractor landscapes of the sensory cortices that embody the knowledge base of the subject. The images of memory are not stored and retrieved as in computer systems. The AM pattern constructed by a stimulus-induced state transition in a sensory cortex is not a representation of the stimulus but of the meaning of the stimulus. It is a partial actualization of the history of past experience, the present context, and the future significance of the stimulus for the subject receiving it.

7. Interpretations of intentionality in neurodynamics

Thomas Aquinas (1272/1952) described perception during intentional action as having two stages: first the replacement of the stimulus by the *phantasm,* a product of the imagination that was as unique, ephemeral and unknowable as was the stimulus, and then the replacement of the phantasm by the *intelligible species*. In terms of neurodynamics the phantasm in stage one corresponds to the volley of action potentials generated by the sensory receptor cells that replaces the energy of the stimulus, and to the activity of cortical neurons driven by sensory input that centrally represent the stimulus, the raw sense data of psychologists that is inaccessible to consciousness. The intelligible species in stage two corresponds to the frame containing the wave packet that is created by state transition under the guidance of the attractor selected by the input. The important distinction here is not between object and subject, or neuron and quale, or brain and mind, but between the specific and the generic, the concrete and the abstract, the event and the class, which captures the difference between sensation and perception. In Aquinian terms the material events in the world are unique and unknowable, and they can only be approximated by assimilation, which is by bringing the lived body into conformance with the situation. All knowledge is first imagined by the intelligible species detected in the wave packet.

Intentionality, phantasms and imagination were dismissed by Descartes in his program for mechanization of the body and mathematization of the mind. When in the 19th century Franz Brentano reintroduced the concept, he did so in order to distinguish between the human performing a task while knowing of doing so versus the machine performing the same task without understanding. The "Scholastic" meaning of intentionality was dismissed in the widely accepted interpretation of the concept of intentionality derived by analytic philosophers from Brentano and Husserl (1913/1960), which has been entitled "aboutness" (Searle, 1983): a thought or belief is about something in the world, and its form in the brain is a pattern of activity serving as a representation of that something. This interpretation was contested by Heidegger (1975/1988), who reintroduced what he called "the enigmatic phenomenon of intentionality" in forms close to those of Aquinas, addressing what he called "the central problem of philosophy": in his terms, "... the 'transposition' (transcendence) of the Dasein over to things" (p. 162). In the present context 'transposition' conforms to the transmission of the wave packet through the spatial integral transform done by the lateral olfactory tract simultaneously with the deletion of the raw sense data driven by the stimulus. Merleau-Ponty (1942/1963), likewise rejected the Cartesian subject-object dichotomy and conceived the Aquinian intentional arc as "to render oneself present to something through the body," and assimilation as the search for and achievement of "maximum grip". He continued:

> "Once the prejudice of sensation has been banished, a face, a signature, a form of behaviour cease to be mere 'visual data' whose psychological meaning is to be sought in our inner experience, and the mental life of others becomes an immediate object, a whole charged with immanent meaning. More generally it is the very notion of the immediate which is transformed: henceforth the immediate is no longer the impression, the object which is one with the subject, but the meaning, the structure, the spontaneous arrangement of parts." pp 58-59 (Merleau-Ponty, 1945/1962).

Other phenomenologists, existentialists and pragmatists in the past century have included the body as essential for perception, most clearly among the American pragmatists by John Dewey (1914) in his critique of the conditioned reflex: "Actions are not reactions to stimuli; they are actions into the stimuli." Embodied cognition was further developed by Gestalt psychologists, who studied the impact on perception of objects embedded in environmental contexts, leading to the conception of a field of force:

> "Our present knowledge of human perception leaves no doubt as to the general form of any theory which is to do justice to such knowledge: a theory of perception must be a field theory. By this we mean that the

> neural functions and processes with which the perceptual facts are associated in each case are located in a continuous medium." (Köhler, 1940, p 55)

Koffka expanded this to include interaction between fields of force in the environment and in the brain:

> " ... let us think of the physiological processes not as molecular, but as molar phenomena. ... Their molar properties will be the same as those of the conscious processes which they are supposed to underlie." (Koffka 1955, p 57) "The task of psychology...is the study of behavior in its causal connection to the psychophysical field." (p 67) "...things in our environment tell us what to do with them. ... Their doing so indicates a field of force between these objects and our Egos... which...leads to action." ... "A handle wants to be turned,... chocolate wants to be eaten, ..." (p 353)

Gestaltism is inconsistent with Aquinian self-organization, because the source of the ordering is still conceived as the structuring done by the importing of forms from exogenous fields, not as the internal creation of structure that is then tested against raw sense data in trial-and-error learning by hypothesis testing. J. J. Gibson (1979) further conceived the role of behaviour that is generated within brains in the definition of objects by means of *affordances*:

> " ... the affordance, being invariant, is always there to be perceived. An affordance is not bestowed upon an object by a need of an observer and his act of perceiving it. The object offers what it does because it is what it is. ... But this does not in the least imply separate realms of consciousness and matter, a psychophysical *dualism*. It says only that the information to specify the utilities of the environment is accompanied by information to specify the observer himself. ... exteroception is accompanied by proprioception ... to perceive is to coperceive oneself." (op. cit., p 139)

In Gibson's conception, "information" derives from objects in the world and is incorporated by resonance of Gestaltist fields that "in-form" the brain. His conception at first glance appears to be related to the Aquinian intentional act of "stretching forth" and conforming to the environment by learning, and to the "intentional arc" of Merleau-Ponty, but the difference is that an affordance constitutes information in an object, and the resonance transfers that information into the meaning. Gibson's conception has the difficulty that the affordance is different for every user, so that the information content of objects must be infinite, and the organization of the resonant filter is unique for each subject unspecifiable. Contrastingly in the Aquinian view, the intentional brain state defines the object in term of an emergent goal with its unity of inner context, and there is no transfer of information, only a modification of the self as the result of action. The Gestaltist and Gibsonian approaches essentially retain the Cartesian subject-object distinction, by virtue of the source of organization being attributed to objects and events in the world, whereas the phenomenological approach attributes the structure of behaviour to the active self-organizing dynamics within embodied brains engaged with the environment.

Situated cognitivists (Clancey, 1993) and enactionists (Tani, 1996; Clark, 1997) have also emphasized the basis of cognition in actions of the body into the world, but they continue to postulate the existence of symbols (numbers, images, firing patterns, etc.) in support of a distinction between "internal" representations derived from “external” representations by the transfer of information in perception, contrasting with assimilation toward maximal grip by assimilation without representation (Merleau-Ponty, 1942/1963; Freeman, 1995).

8. Conclusions

The critical link between past and future in sensation transposing to perception occurs at the moment of selection of gamma attractors followed by up-dating of the limbic pattern of beta activity, which incorporates the immediate result as a fresh small step along a trajectory extending into the future. The formulation of the next succeeding step is shaped by the entire body of past experience at play within each step, including the projection from the past by the presently existing threads extending into the future. With all its limitations of perspective and short-field emphasis, this field of activity is the basis for decisions and choices, and for the awareness likely mediated by that field, which is a moment of consciousness. An important limitation imposed by the requirement for passage through the limbic system is the time required for neurodynamics to construct the gamma patterns and reorganize the beta patterns following each definable search frame. Libet (1994) has shown that there is a delay approximating 0.5 sec between the arrival of a stimulus and the onset of awareness of the stimulus, though that onset is subjectively backdated close to the actual time of arrival. Popper and Eccles (1977) described this as having no physiological explanation. However, the process is analogous to the two-threshold technique in common use by physiological psychologists for identifying *that* a response has occurred with a high threshold, and detecting *when* it occurred with a preceding low threshold. Libet (1994) has extended his studies to the decision process, with indication of similar delays in awareness of actions.

In these terms, the intentional, dynamic, public, limbic self continually constructs the neural activity patterns that seek sensory input. The global updating that sets the matrix for each next step lags by half a second, both the initiating actions and the arrival of sense data de-stabilizing cortices and initiating fresh constructions. In this view the private experiential awareness of self, the ego, is invariably half a second behind, always justifying, explaining, rationalizing, and claiming credit by virtue of the ability to back-date, which was designed by evolution of the mammalian neocortical sensory system to provide a mechanism for keeping the foundational self in near synchrony with the unfolding real world. This is the reverse of the Cartesian view; the phenomenologist says, "I act, therefore I think." Mahayana Buddhists and Lacanian psychoanalysts have written alike about the "illusion of the self." In the intentional view the illusion is not that the conscious ego exists, but that ego is in control of the self. Fortunately only a small fraction of intentional behaviour is conscious, because that fraction is much too small a sample to be trusted with the management of intent and assimilation

References

Aihara K, Takabe T, Toyoda M (1990) Chaotic neural network. *Phys Lett A* **144**:333-340.

Amari S (1977) Neural theory of association and concept formation. *Biol Cybern* **26**:175-185.

Anderson JA, Silverstein JW, Ritz SR, Jones RS (1977) Distinctive features, categorical perception, and probability learning: Some applications of a neural model. *Psychol Rev* **84**:413-451.

Aquinas, St Thomas (1272/1952) *The Summa Theologica of Saint Thomas Aquinas.* Volume 19 in the Great Books Series. Chicago: Encyclopedia Britannica.

Barrie JM, Freeman WJ, Lenhart M (1996) Modulation by discriminative training of spatial patterns of gamma EEG amplitude and phase in neocortex of rabbits. *J Neurophysiol* **76**:520-539.

Basar E (1980) *EEG - Brain Dynamics*. Amsterdam: Elsevier.

Bressler SL (1988) Changes in electrical activity of rabbit olfactory bulb and cortex to conditioned odor stimulation. *J Neurophysiol* **62**:740-747.

Clancey WJ (1993) Situated action: A neuropsychological interpretation response to Vera and Simon. *Cogn Sci* **17**:87-116.

Clark A (1997) *Being There – Putting Brain, Body, and World Together Again.* Cambridge MA: MIT Press.

Dewey J (1914) Psychological doctrine in philosophical teaching. *J Philos* **11**:505-512.

Freeman WJ (1975/2004) *Mass Action in the Nervous System.* New York: Academic Press.

Freeman WJ (1986) Petit mal seizure spikes in olfactory bulb and cortex caused by runaway inhibition after exhaustion of excitation. *Brain Res Rev* **11**:259-284.

Freeman WJ (1987) Simulation of chaotic EEG patterns with a dynamic model of the olfactory system. *Biol Cybern* **56**:139-150.

Freeman WJ (1991) The physiology of perception. *Sci Am* **264**:78-85.

Freeman WJ (1995) *Societies of Brains. A Study in the Neuroscience of Love and Hate.* Mahwah NJ: Erlbaum.

Freeman WJ (2000) A proposed name for aperiodic brain activity: Stochastic chaos. *Neural Networks* **13**:11-13.

Freeman WJ (2000/2006) *Neurodynamics. An Exploration of Mesoscopic Brain Dynamics*. London UK: Springer. http://sulcus.berkeley.edu/eBook:Neurodynamics:Mesoscopic Brain Dynamics.

Freeman WJ (2001) The olfactory system: odor detection and classification. In: *Frontiers in Biology*, Volume 3 Intelligent Systems, Part II Brain Components as Elements of Intelligent Function. New York: Academic Press, pp 509-526.

Freeman WJ (2004a) Origin, structure, and role of background EEG activity. Part 1. Analytic amplitude. *Clin Neurophysiol* **115**:2077-2088.

Freeman WJ (2004b) Origin, structure, and role of background EEG activity. Part 2. Analytic phase. *Clin Neurophysiol* **115**:2089-2107.

Freeman WJ (2005) Origin, structure, and role of background EEG activity. Part 3. Neural frame classification. *Clin Neurophysiol* **116**:1118-1129.

Freeman WJ (2006) Origin, structure, and role of background EEG activity. Part 4. Neural frame simulation. *Clin Neurophysiol* **117:**572-589.

Freeman WJ, Burke BC (2003) A neurobiological theory of meaning in perception. Part 4. Multicortical patterns of amplitude modulation in gamma EEG. *Int J Bifurc Chaos* **13**:2857-2866.

Freeman WJ, Gaál G, Jornten R (2003) A neurobiological theory of meaning in perception. Part 3. Multiple cortical areas synchronize without loss of local autonomy. *Int J Bifurc Chaos* **13**:2845-2856.

Freeman WJ, Rogers LJ (2003) A neurobiological theory of meaning in perception. Part 5. Multicortical patterns of phase modulation in gamma EEG. *Int J Bifurc Chaos* **13**:2867-2887.

Freeman WJ, Holmes MD (2005) Metastability, instability, and state transition in neocortex. *Neural Networks* **18**:497-504.

Gibson JJ (1979) *The Ecological Approach to Visual Perception.* Boston: Houghton Mifflin.

Gray CM, Freeman WJ, Skinner JE (1986) Chemical dependencies of learning in the rabbit olfactory bulb: acquisition of the transient spatial-pattern change depends on norepinephrine. *Behav Neurosci* **100**:585-596.

Haken H (1983) *Synergetics: An Introduction*. Berlin: Springer.

Haken H (1999) What can synergetics contribute to the understanding of brain functioning? In: C Uhl (Ed) *Analysis of Neurophysiological Brain Functioning*. Berlin: Springer-Verlag, pp 7-40.

Heidegger M (1975/1988) *The Basic Problems of Phenomenology*. Bloomington IN: Indiana University Press.

Hopfield JJ (1982) Neuronal networks and physical systems with emergent collective computational abilities. *Proc Natl Acad Sci USA* **81**:3058-3092.

Husserl E (1913/1960) *Ideas: General Introduction to Pure Phenomenology*. New York: MacMillan.

Kay LM, Freeman WJ (1998) Bidirectional processing in the olfactory-limbic axis during olfactory behavior. *Behav Neurosci* **112**:541-553.

Kay LM, Freeman WJ, Lancaster L (1995) Limbic markers during olfactory perception. In: I Gath and G Inbar G (Eds) *Advances in Processing and Pattern Analysis of Biological Signals*. New York: Plenum.

Koffka K (1935) *Principles of Gestalt Psychology*. New York: Harcourt Brace.

Köhler W (1940) *Dynamics in Psychology*. New York: Grove Press.

Kohonen T (2001) *Self-Organizing Maps*. Berlin: Springer.

Kozma R, Freeman WJ (2002) Classification of EEG patterns using nonlinear dynamics and identifying chaotic phase transitions. *Neurocomputing* 44:1107-1112.

Lashley K (1942) The Problem of Cerebral Organization in Vision. *Biol Symp* **7**:301-322.

Li X, Li G, Wang L, Freeman WJ (2006) A study on a bionic pattern classifier based on olfactory neural system. *Intern J Bifurc Chaos* **16**:2425-2434.

Libet B (1994) *Neurophysiology of Consciousness: Selected Papers and New Essays*. Boston MA: Birkhauser.

Merleau-Ponty M (1942/1963) *The Structure of Behavior*. Boston: Beacon Press.

Merleau-Ponty M (1945/1962) *Phenomenology of Perception*. New York: Humanities Press.

Ohl FW, Scheich H, Freeman WJ. (2001): Change in pattern of ongoing cortical activity with auditory category learning. *Nature* **412**:733-736.

Rapp P (1993) Chaos in the neurosciences: Cautionary tales from the frontier. *Biologist* **40**:89-94.

Searle JR (1983) *Intentionality*. Cambridge UK: Cambridge University Press.

Popper KR, Eccles JC (1977) *The Self and Its Brain*. Berlin: Springer.

Singer W, Gray CM (1995) Visual feature integration and the temporal correlation hypothesis. *Ann Rev Neurosci* **18**:555-586.

Skarda CA, Freeman WJ (1987) How brains make chaos in order to make sense of the world. *Behav Brain Sci* **10**:161-195.

Sperry RW (1950) Neural basis of the spontaneous optokinetic response. *J Comp Physiol* **43**:482-489.

Tallon-Baudry C, Bertrand O, Peronnet F, Pernier J (1998) Induced gamma-band activity during the delay of a visual short-term memory task in humans. *J Neurosci* **18**: 4244-4254.

Tani J (1996) Model-based learning for mobile robot navigation from the dynamical systems perspective. *IEEE Trans Systems, Man, and Cybern* **26B**:421-436.

Tsuda I (1992) Dynamic link of memory - Chaotic memory map in nonequilibrium neural networks. *Neural Networks* **5**:313-326.

Tsuda I (1996) A new type of self-organization associated with chaotic dynamics in neural networks. *Int J Neural Syst* **7**:451-459.

Tsuda I (2001) Towards an interpretation of dynamic neural activity in terms of chaotic dynamical systems. *Behav Brain Sci* **24**:793-810.

Viana Di Prisco G, Freeman WJ (1985) Odor-related bulbar EEG spatial pattern analysis during appetitive conditioning in rabbits. *Behav Neurosci* **99**: 962-978.

Yao Y, Freeman WJ (1990) Model of biological pattern recognition with spatially chaotic dynamics. *Neural Networks* **3**:153-170.

Chapter 11

Conscious Contents Provide Coherent, Global Information†

Bernard J. Baars

Consciousness is what we might expect to find in an organ, added for the sake of steering a nervous system too complex to handle itself.
William James (1890)

1. Introduction

In recent years, psychologists have begun to approach the issue of consciousness quite pragmatically, largely free from the theoretical obstacles that restricted the scope of earlier attempts. Recent authors agree reasonably well on the phenomena that a theory of consciousness must explain, and it appears that some theoretical consensus is beginning to emerge as well. For example, authors like Posner and Warren (1972), Mandler (1975a,b), and Straight (1979) agree that consciousness is closely associated with short-term memory and the limited-capacity components of the nervous system. Others are impressed by the very large range of the phenomena that are consciously accessible (e.g., Smith, 1969; Hilgard, 1976, 1977; Tart, 1975), and still others relate consciousness to the control of action, executive functions, and automaticity (e.g., Shallice, 1972, 1978; Norman, 1976; Shiffrin and Schneider, 1977; Norman and Shallice, 1980). This paper describes a unified theoretical approach to a large set of phenomena associated with consciousness, and it suggests that a rather simple set of principles can account for many of these phenomena. This approach is further extended in Baars and Mattson (1981), and Baars and Kramer (1982).

Whenever the words *consciousness* or *awareness* are used in this essay, they refer to "consciousness of some content," as opposed to other meanings such as "waking

---

† "Conscious Contents Provide the Nervous System with Coherent, Global Information" by Bernard J Baars, from *Consciousness and Self-Regulation – Advances in Research and Theory*, Vol. 3, edited by RJ Davidson, GE Schwartz and D Shapiro, Plenum Press, New York, 1983. Reprinted with kind permission of Springer Science and Business Media.

Consciousness Transitions: Phylogenetic, Ontogenetic, and Physiological Aspects
Edited by Hans Liljenström and Peter Århem

consciousness" (e.g., Natsoulas, 1979), To focus on this sense of consciousness, the empirical constraints shown in the tables are stated in terms of *contrasting pairs of statements* about some conscious content compared with a similar unconscious one (Tables 11.1 and 11.2),

For example, one would like to know why we tend to lose consciousness of predictable stimuli, such as the pressure of our clothing, the ambient light, or background noise. Or we would like to know why we are not usually conscious of certain higher-level constancies, such as discourse presuppositions. It is important to note that at times, we can become quite conscious of discourse presuppositions, and we can often make habituated patterns conscious as well,

There are many more examples of our being conscious of something at one point but not another. In the cocktail-party effect, we are largely unconscious of nonattended information, but we can make any stream of information conscious "at will". Further, we are unaware of subthreshold energies, although under the right conditions, these same energy patterns are consciously available. Long-term memory is typically unconscious, but an extraordinary number of things in memory can be made conscious, There are many other such empirical conscious-unconscious contrasts, which generally evoke very little controversy, The job of theory is to find an explanation that will fit all the empirical constraints simultaneously. This approach can lead to an explanation of conscious and unconscious functioning in perception, voluntary action, problem solving, memory retrieval, and attention.

The tables show the empirical constraints on consciousness that this paper attempts to explain. These tables contrast each conscious event or property with a corresponding unconscious one, as described above. Table 11.1 shows a number of empirical *capability constraints*, which contrast the capabilities of conscious and unconscious processes.

Table 11.2 shows some *boundary constraints,* specifying the synchronic and diachronic boundaries of conscious contents. Together, these sets of constraints serve to exclude a very large number of possible theories of awareness. Note again that the constraints on conscious phenomena described in the tables are entirely empirical: they derive from consistent evidence regarding the experiences of very many people. The claims made in the tables regarding unconscious processes are naturally more inferential, but they seem to generate little disagreement. Thus, the empirical constraints seem to be quite firm.

These empirical constraints fit a small number of rather simple ideas that form the conceptual core of this paper. The following sections explore these ideas in detail and give specific arguments in favour of applying them to a theory of consciousness.

1.1 The need for a central information exchange in a distributed processing system: some theoretical definitions

Following are some theoretical definitions that will prove useful in exploring this perspective. These definitions are largely conceptual, and the remainder of this chapter shows how they can be used to illuminate the issue of consciousness.

The nervous system resembles in many respects a "distributed" information-processing system, in which highly complex and detailed processing is performed by specialized subsystems in a relatively independent way. There is extensive psychological and neurophysiological evidence for such specialized and relatively autonomous processors (e.g., Geschwind, 1979; Hilgard, 1976, 1977; La Berge, 1974; Shiffrin and Schneider, 1977). Of course, there is also sound evidence for molar, nonspecialized activity (e.g., John, 1976; Mandler, 1975b). In the present point of view, consciousness is that which unites specialized and nonspecialized processes.

Perhaps the key concept is the idea of distributed information processing, a kind of systems organization that is currently being studied intensively in computer science. In such a system, a large number of specialized processors may be "data-driven"; that is, they may themselves retain the processing initiative. The specialists themselves decide by their own criteria what is and what is not of interest, so that a central mechanism does not necessarily have executive power over the special-purpose systems. However, these specialists do require some means of interacting with each other.

It is somewhat difficult conceptually to think of our nervous system as a distributed system because we tend to attach great importance to executive processes. Indeed, we tend to identify with some core executive component of our system, and we tend to believe that this component is in charge of all the others. No doubt, there is some truth to this commonsense view, but the idea of a distributed system can account for a great deal of the evidence. Actually, these ideas are not contradictory. One can speak of executive systems that operate in a domain defined by distributed operations, much as a government may be viewed as an executive operating in a domain defined by multitudinous interactions between individual human beings. But we are in the habit of thinking hierarchically about the nervous system, and it takes some reorganization to think distributively.

If we accept the analogy of the nervous system as a distributed society of specialized processors, some of which try to act as a governmental executive toward the others, then consciousness is much like a publicity organ in this society (see Fig. 11.1). Consciousness seems to be closely associated with a mechanism that permits interaction between specialized, dedicated processors. This mechanism behaves remarkably like a global data base – a "central information exchange" used by artificial intelligence workers to permit any set of specialized processors to cooperate or compete in order to solve some central problem (Kaplan, 1973; Reddy and Newell, 1974; Erman and Lesser, 1975). A global data base is essentially a memory to which all processors in the system have potential access, and from which all can potentially receive input (see Section 1.2). Any representation in the global data base is distributed to the specialized processors much as a television broadcast is distributed to a viewing audience. But unlike an ordinary television audience, some subset of the specialized processors can act on the global data base in return, to propose hypotheses that can then be broadcast to any of the others.

Each of the specialists in the audience can decide on the relevance of the global representation for its own domain. Specialists are assumed to be triggered by the mismatches between the global representation and their own internal representation of their domain. Thus, syntactic specialists are sensitive to linguistic input, and spatial specialists to visual input. However, if the input is already known, if there is no mismatch, the

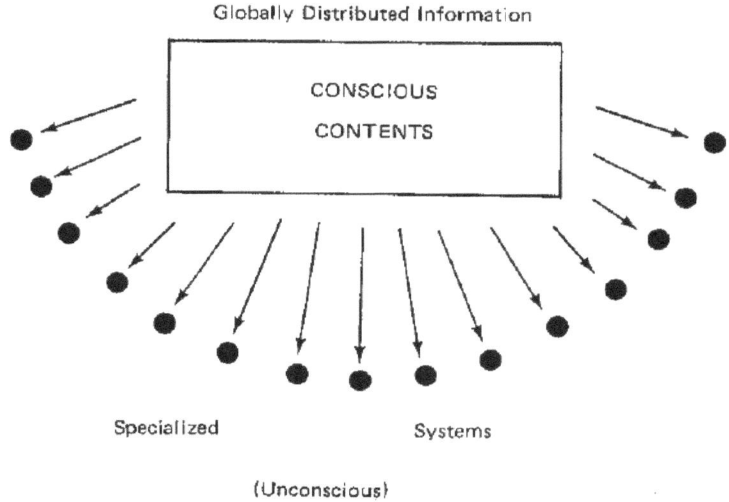

Fig 11.1 This diagram provides a first approximation to a theory of consciousness. The rectangle symbolizes a global data base, a memory whose contents are available to any specialized processor in the system, and that is in turn controlled by some subset of specialists. Only the global information is conscious – the operation of the specialists is not normally conscious, unless it is displayed by the global data base.

global representation is simply redundant. Further, if the global information is in some other domain that is irrelevant to a specialist, it simply fails to respond. In this sense, any specialist decides by its own criteria whether to process the global representation.

Thus, all specialists are potentially responsive to global input, but the word *global* does not mean that all processors must always accept all global information. It would be useless for a syntactic processor to accept visual input, and vice versa. Rather, *global* implies that a representation is available to *any* processor that has a relevant domain of specialization, providing that it finds the global input to be informative. The word *global* applies whenever we can make this "any" argument, that is, whenever we can plausibly say that some representation is available to any processor, or that any processor can act on it. Generally speaking, the ability to distribute information globally is especially useful if one cannot decide ahead of time which one of the specialized processors needs some information.

What constitutes a specialized processor? This point is rather tricky, because the extent to which a processor behaves autonomously may depend on the task (e.g., Hyde and Jenkins, 1973; Jenkins, 1974). Thus, in riding a bicycle with perfect automatic skill, one may wish to speak of a "bicycle-riding system" with some justification. Certainly, in the beginning of learning to ride a bicycle, this would not be justified. And certainly, when the skill runs into problems – if the right pedal falls off – it must be decomposed, so that control systems for the right foot can learn to behave differently. Yet, we would be in trouble if we had to recruit all the components of bicycle riding each time we leaped on a bicycle; normally, it is better to access the skill as a whole, and for this reason, automatic bicycle riding may be treated as a unified, relatively autonomous specialized processor. This kind of flexibility may discomfit a builder of neat minitheories, who would like to have building bricks that remain stable in all circumstances. But the nervous system probably profits by this potential for flexibility.

As we noted above, global representations are distributed to specialized processors much as a television program is distributed to a large number of viewers. Each viewer has the option of processing or not processing the television program. If the viewer already knows the information, or if it is irrelevant, it may be ignored. If a global representation is neither redundant nor irrelevant to some specialist, it will attempt to adapt to the global information. Adaptation is defined as an attempt by the specialized processor to match the global information in its domain, to reduce the mismatch that triggered it in the first place. At a physiological level, there is extensive evidence for processes like this: both neurons and systems of neurons habituate selectively to input, ceasing to fire when the input is absorbed. But if a change occurs in the habituated pattern (that is, in conditions of mismatch with the previous adaptation), these systems activate again until the new input has become redundant, equilibrium is restored, and

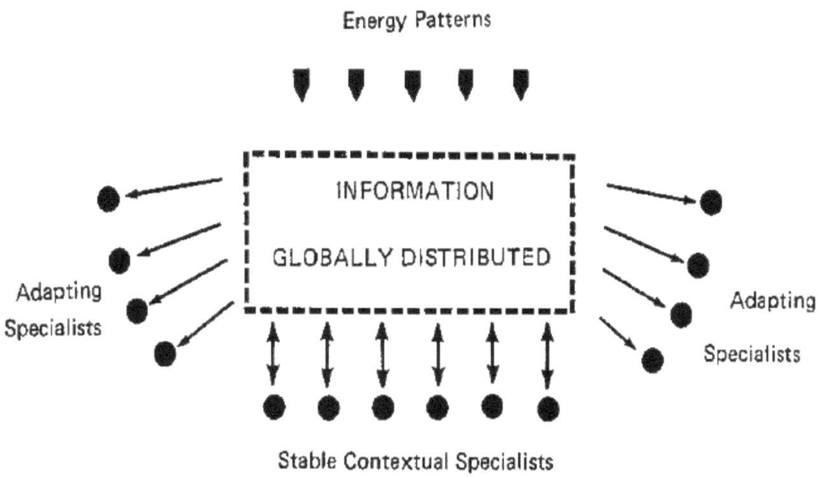

Fig 11.2 A somewhat more complete diagram, showing that the global information is, in fact, the result of an interaction between incoming energy patterns and a stable set of specialized systems, which provide a context. The resulting global representation, when it has become stable, triggers adaptation in the remaining specialists in the system. Each of the specialists attempts to reduce mismatch between global information in its specialized domain and its own model of the input. Since the global data base exists in a "distributed" system, the processing initiative is left to the specialists themselves.

they cease firing (Sokolov, 1963; Asratyan, 1965). Note that selective habituation to current input is, in fact, a way in which neurons can store information about any current state of the input. (See Fig. 11.2).

Any global representation that triggers widespread adaptation can be said to provide global information (MacKay, 1969). Neurophysiological evidence on this point is quite good: new events in the environment, those that are psychologically most likely to be conscious, cause extensive neural activity, far beyond the anatomical pathways of the sensory modality involved (John, 1976). But when this same input is presented repeatedly, the pattern of activity becomes much more localized and limited to special pathways. Redundant input ceases to be global.

Thus, global information is defined as being information to the system as a whole – and here again, one may use the television analogy. If everyone in a country tuned in to a television news program at the same time, one might similarly speak of the content of the news program as providing global information to the country as a whole. Global information is to be distinguished from local information, which is information that is processed w\within a single specialist without the benefit of the global data base.

1.2. The system in operation

A number of different processors may cooperate or compete in sending hypotheses to the global data base by acting to confirm or disconfirm global hypotheses until ail competition is resolved. If some global hypothesis proposed by one specialized processor is immediately contradicted by another, the hypothesis w\ill have only a transient existence. In order to establish a stable global representation, a number of processors must cooperate; that is, they must create a context[1]. A context is defined as being a set of stable constraints on a global representation, provided by a set of cooperating processors. That definition implies, of course, that these constraints are consistent with each other; if this were not so, the processors providing the constraints would begin to compete, and the global representation would lose stability. One can view a context as consisting of the set of constraints that is relevant to some particular process to which the system as a whole has *already* adapted.

A set of stable global constraints (a context) is not necessarily complete; there may be many degrees of freedom left, so further constraints can be added to the context. Indeed, some of the constraints that define a context may be changed by incoming information. Finally, incoming information may be able to fill in certain parameter values in dimensions that are specified by the context. Thus, the existence of a stable, coherent set of contextual constraints is not incompatible with the acquisition of new information.

So far, all these definitions have been purely conceptual. Our psychologica1 argument wm be that in the nervous system, a stable, global representation becomes conscious *if it provides global information* (Section 3). Conscious representations provide information to the system as a whole, or alternatively, one may say that they trigger adaptation in the system as a whole. Thus, any specialized processor can respond to conscious information relevant to its domain.

It is important to be aware that a global data base is not an executive, though it may be *used by* systems acting in an executive capacity. Indeed, the power and usefulness of distributed information-processing derives from its decentralized organization (Greene, 1972; Turvey, 1977). Again, it is more accurate to compare the global data base with a broadcasting station, which can send information to a vast number of processors, and which can, in turn, be controlled by some processors. In much the same way, a government can control a broadcasting station – but it is the government that acts as an executive, while the broadcasting facility is merely a medium. Consciousness is viewed in this paper as a certain operating mode of this medium, and consciousness can likewise be used by processors acting as executives, without itself being an executive.

In many ways, this approach is not new. The global data base has been used by a number of researchers in artificial intelligence (Kaplan, 1973; Erman and Lesser, 1975), and it has a clear similarity to the idea of working memory. Mandler (1975b) has pointed out the dose relation of consciousness to short-term memory. Furthermore, Shallice (1972, 1978) has proposed that action systems may dominate the limited-capacity system in much the way that this chapter suggests that consistent sets of processors may dominate the global data base. Others have discussed this kind of a system in a number of different contexts (Lindsay and Norman, 1976; Arbib, 1982; Tart, 1975; Hilgard, 1977; Baars and Kramer, 1982). Nonetheless, the present approach does suggest new ways of viewing the psychological implications of such a system.

Given this perspective, a large amount of evidence falls into line. In Section 2, detailed evidence is discussed for associating consciousness with the psychological equivalent of a global data base, and Section 3 presents evidence concerning the boundaries of conscious contents.

1.3 Advantages of the global data base

The special advantages and disadvantages of global data bases can be enumerated. First, the advantages:

(1) Global information is distributed to all relevant processors, so that if there exists some specialist able to handle it in a fast, efficient way, it can be found immediately.

(2) In a problem space that is uncertain or badly understood, a global data base can unite information from many incomplete sources to produce greater certainty than any individual specialist could produce by itself (Erman and Lesser, 1975).

(3) A distributed processing system with a global data base would seem to be an ideal learning device. In our version, it is inherently an adapting system since global information is displayed to many different specialists, which are assumed to adapt to those new aspects of the global display that are within their purview. Indeed, we argue in Section 3 that we have an experience of some event whenever the system as a whole is adapting to a stable, global representation of the event.

(4) A global data base can optimize the fundamental trade-off between structure and flexibility. This is a general problem for large systems: on the one hand, it is vital to have specialized, structured solutions available for standard problems, and on the other hand, such structured solutions can be a drawback when the system is faced with really

new conditions. In new situations, *flexibility* is at a premium. A global data base permits one to change from a highly structured approach to a highly flexible one. One can have the advantage of structure if the problem is in the province of specialization of a particular processor, along with the advantage of flexibility in choosing between alternative processors, and in the possibility of having a number of specialized processors cooperate in solving some problem.

(5) Though a global data base is not an executive, it can be used by executive systems. Executive systems can use the global data base to distribute information to control other systems, and to receive feedback from subordinate systems.

(6) New processors can be added in a modular fashion, The system can grow without serious disruption, since specialized processors can be added without having to change the previous set of processors, Indeed, the entire configuration may be used to *develop* new specialized processors: as joint information from partial knowledge sources becomes more and more determinate, a new rule-set is defined. This new ruleset may become autonomous and may begin to behave as a specialized processor in its own right. Karmiloff-Smith (1979) has observed a process very much like this in the acquisition of language and other representational systems by children,

(7) The same processor may be used in different tasks, For instance, speech perception and speech production have many components in common; thus, it may be that speaking and listening, in fact, involve many of the same processors, which are merely organized differently for speech input and output. Along these lines, Geschwind (1979) claimed that "the primary motor and sensory areas are specialized in the sense that each one is dedicated to a specific function, but the functions themselves are of general utility, and the areas are called on in a great variety of activities."

(8) In terms of *content addressability*, one does not need to call any particular processor from a global data base: it is necessary only to present the conditions that the processor finds unambiguously informative. The "name" of any processor is, in a real sense, the information to which it is responsive. This property has considerable advantages and corresponds well to what is known about human cognition (e.g., Norman, 1976).

(9) A global data base can solve the need for a mental *lingua franca* (Dennett, 1978). In Western thought, the idea that there must be a common mental code so that one sense modality can communicate with the others goes back to Aristotle's notion of a "common sense". The problem is, of course, that visual information is to some extent unique and different from auditory information, which differs, in turn, from the motor code that controls speech articulation (e.g., Norman, 1976). The unique properties associated with the efficient control of speech are simply not directly translatable into a visual code. Thus, on the one hand, a *lingua franca* is desirable; on the other, it would vitiate the advantages of specialized languages.

A global data base operating in a distributed system obviates this need for a *lingua franca*. Only those systems will respond to a global representation that already "speak the language" of the global representation. Other systems simply don't look at this information, because they do not speak the language. This is perhaps a poor metaphor because the system has not an arbitrary code like natural language, but a content-addressable code more like a semantic network. If the content of the global data base changes, or if in the process of adaptation the specialists change so that they become sensitive to new dimensions of the global information, one might say that they are "learning to speak" the language of the global representation.

There may be one common code able to access all processors, and that is temporal simultaneity. Work on evoked potentials (John, 1976) shows that for a brief time, a prominent new event in any sensory modality reverberates widely throughout the nervous system, far beyond the special anatomical areas and pathways associated with the sensory input. Further, we know from biofeedback research (Schwartz, 1975) that an enormous variety of specialized processors can respond intelligently to those widespread events that affect the whole nervous system simultaneously for a short period of time. It is well established that conscious feedback can be associated with remarkably specific events in the nervous system, events that are presumably controlled by specialized processors. Note that we are not claiming that these events – EEG rhythms, autonomic functions, or single motor neurons – are controlled consciously. In fact, we do not control anything consciously in the literal sense of knowing precisely how we do things (Baars and Mattson, 1981). Rather, the idea is that in biofeedback, specialized systems that control EEG rhythms, or single motor units can independently decide to respond and adapt to widely broadcast information if there is effective temporal simultaneity.

(10) A final advantage of this kind of system has been pointed out by Hayes-Roth and Lesser (1976): it consists of flexibility of access to the global data base. Given the same set of specialists, it is possible to experiment with various strategies to control access to the global data base. Certain specialized processors may be given a higher priority than others. This "focus-of-attention" problem has intriguing parallels to the psychological issue of attention. Furthermore, it provides a theoretical mechanism whereby certain potential conscious contents can be avoided, so that in principle, one could incorporate ideas of repression and the dynamic unconscious.

1.4 Disadvantages of the global data base

No system design is without drawbacks, and the global data base has some obvious ones. For one, it uses a large number of processing resources because all specialists must continually monitor the central information relevant to their domain. Further, global problem-solving is relatively slow, certainly when compared with the fast and efficient action of a specialist that knows how to solve a specific problem. Many different processors must learn to cooperate in order to produce a solution to the global problem. Whenever possible, the global data base should relegate some problem to a specialist (i.e., as soon as a determinate solution is found).

As we point out below (Section 2.1), all these disadvantages have parallels in the "computational inefficiencies" of consciousness: consciousness, too, seems to demand a very great number of resources; it, too, is slow compared with unconscious information-processing. When a conscious solution to a problem is discovered, it is also quickly relegated to unconscious processors: it becomes habituated or automatic.

This paper does not claim that consciousness is identical to the operation of a global data base in the nervous system. Rather, it seems that we are conscious of some content when there exists an internal representation that meets three criteria: it must be global, stable, and informative (Section 3).

We turn now to a set of arguments to show that conscious processes are closely associated with a system that acts very much like a distributed system with a global data base.

2. Capability constraints: arguments for associating consciousness with a global data base

How do we know when someone is conscious of something? Most obviously we are willing to infer that someone is conscious of an object if the observer can describe it. But consciousness cannot be limited to verbal description – first, because that would confound the construct of consciousness with the evidence that is used to infer it, and second, because that would include talking parrots and computers while excluding babies, aphasics, and ourselves when we are not talking! Adequate measures for any construct *result* as much *from* good theory as they *lead* to it. Thus, any initial definition may need to be changed as the theory is developed. But as a first approximation, we may say that we are willing to infer consciousness when someone can potentially *act discriminatively* toward some internal representation, especially when the internal representation is nonroutine. This inference would include babies, aphasics, at least some animals, and ourselves in our more silent moments; it would exclude even ourselves if we were engaged in extremely routine tasks or were processing very routine representations of things. It would include the objects of perception, which can reasonably be thought of as represented in the nervous system because one can do sensitive memory tests afterwards, for example, and find quite remarkable recognition memory for these perceptions. Also, it would include images among conscious experiences, since all cognitive measures of imagery ask the subject to behave discriminatively toward the image (e.g., Paivio, 1975). Note that a subject does not have to act discriminatively at all times toward some internal representation to be considered conscious of the object. As long as the subject can potentially do this, we may consider him or her to be conscious because of this potential.

In practice, we capitalize on the fact that people can consistently answer questions like, "Are you conscious of the words in front of you?", "Before you read this question, were you conscious of the feel of your chair, of the presuppositions of this question, of the breakfast you ate yesterday?" Such questions are answered so consistently by so many people that when it comes to collecting empirical constraints on a theory of consciousness (Tables 11.1 and 11.2), no practical obstacles arise.

An initial definition like this may not help us to decide on difficult cases like trance states, automatic writing, cases of multiple personality (e.g. Hilgard, 1976, 1977), or reports of "consciousness without content" (e.g. Naranjo and Ornstein, 1971; Globus and Franklin, 1980). Further, it is possible that people can be conscious of some things so fleetingly as to be unable to answer these questions accurately, as suggested, for example, by the well-known Sperling (1960) phenomenon. But in theory building, as in law, hard cases make bad laws. The great bulk of phenomena we wish to capture can be incorporated in the definition. Perhaps further insight can be gained by considering the

Table 11.1 Capability constraints on a theory of consciousness[a]

	Conscious processes	Conscious processes
1.	Computationally inefficient.	Highly efficient in special tasks.
2.	Great range and relational capacity.	Limited domains and relative autonomy.
3.	Apparent unity, seriality, and limited capacity.	Highly diverse, can operate in parallel, and together have great capacity.

[a]The capability constraints provide one set of conscious-unconscious contrasts that is quite well established and uncontroversial. Yet, these constraints place considerable limits on possible theories.

more typical phenomena first so that we can then approach these other fascinating questions more intelligently.

Table 11.1 compares the capabilities and limits of conscious and unconscious processes. These comparisons are purely relative. For example, when we claim that entirely conscious processes are computationally inefficient, this is only in comparison to the evident efficiency of unconscious processors. Note, by the way, that the first column refers to unitary conscious processes, while the second column refers to multiple unconscious processors. This is itself, of course, a theoretical claim that is supported by some of the arguments made below.

In this section, we discuss each of these points in detail and show how, together, the capability constraints lead to the notion that consciousness reflects the functioning of a global data base in a distributed processing system.

2.1 Computational inefficiency

If by *computational efficiency* we mean the ability to compute some standard algorithm quickly and without error, then it is dear that conscious processes lack computational efficiency, while unconscious processors are often remarkably efficient.

2.1.1 Some limitations of conscious phenomena

Virtually all authors have noted that the vast preponderance of truly effective human information processing is not normally open to awareness (e.g., Miller, 1962; Mandler, 1975a,b; Shallice, 1972, 1978; Straight, 1979). In recent years, we have grown increasingly aware of the awesome complexity of processes needed in the comprehension of even a single sentence, the analysis of a visual scene, or the control of a single motor gesture. People can point to correct and incorrect instances of these events, but they cannot specify the rules involved or explain how they work. This lack of awareness of processing details is universal: it applies to perception and memory retrieval, to most of

Table 11.2 Boundary constraints on the contents of consciousness[a]

Conscious events	Events that do not become conscious
Synchronic phenomena:	
1. Percepts.	Context required to organise percepts.
2. Input consistent with context.	Input inconsistent with context.
Diachronic phenomena:	
3. Percepts.	Preperceptual processes.
4. Any change in a habituated stimulus	Habituated percepts.

[a]The boundary constraints suggest that conscious contents are coherent (and hence stable) and also informative.

problem solving, to the control of speech and action, and so on. Where we are able to carry out some mental algorithm in a largely conscious way (as in mental arithmetic), the process is often slow and prone to error. Indeed, as conscious processes become more and more proficient, they also become less and less consciously available.

Some people may be tempted to conclude from these observations that consciousness is unimportant or even "epiphenomenal", that is, that it has no functional role to play in the nervous system. That is not the solution advanced here. However, it is clear that a functional role for consciousness cannot be found in its computational efficiency.

2.1.2 Comparison with unconscious processors

The claim that unconscious processors are *highly efficient and specialized for routine tasks* is clearly inferential, but not unreasonable. We may be aware of the sound of a sentence, and of the words, but we are certainly not aware of the fast, complex, and generally error-free processes that mediate between the awareness of sound and the awareness of words. We get some inkling of this complexity when we first acquire some knowledge or skill, but once the new ability is learned to the point of proficiency, it drops out of consciousness. It is then no less complex than before, and it is processed a good deal more efficiently than when it was first acquired – yet, it *seems* easier, presumably because the processing is handled by specialized systems that make little demand on our conscious processing capacity.

There is well-known physiological evidence for independent special-purpose processors in speech, spatial analysis, emotion, metabolic control, and even music (e.g., Geschwind, 1979). On the other hand, there is also a large amount of well-established evidence to show that neural activity in response to new or significant information is extremely widespread, going far beyond the classical neuroanatomical pathways of each sensory modality (John, 1976). (Notice that consciousness of some content is also associated with new or significant events; see Table 11.2). But there is no contradiction between localized, dedicated processors and global activities: the tasks performed by the brain require both specialization and global coordination .

It makes sense to suppose that all the truly efficient processors are specialized. The rule systems for spatial analysis are different from those involved in moral judgments, which, in turn, differ from the rules of syntax. Different, yes, but not absolutely autonomous. If we want to understand property law, we must understand how spatial relationships such as boundaries, thoroughfares, and surface features interact with considerations of morality; and, of course, the law has its own syntactic devices to make these considerations explicit. Thus, these three highly cohesive rule systems – morality, spatial relations, and syntax – must interact in some cases. More generally, one can claim that for any two apparently separate and internally cohesive rule-systems one can discover a context in which they must interact. Consciousness is presumably involved in facilitating the interaction between such rule systems, until their interaction becomes routine. Once it has become routine, one could speak of a single moral–spatial–syntactic processor with a considerable degree of internal coherence.

2.1.3 Relevance of these points

All of these points are consistent with the view that consciousness is associated with a global data base in a distributed processing system. In such a system, purely global processes would be slow and inefficient, because they require the cooperation of a number of otherwise separate processors. By comparison, once some specialist is able to compute a standard algorithm for some known problem, it can do so with very great speed and efficiency.

2.2. Some advantages of conscious processes

While completely conscious processes are computationally inefficient, the contents of consciousness have extraordinary *range, relational capacity,* and *context sensitivity.* By contrast, unconscious processors by themselves have *limited domains* and are *relatively autonomous* (Table 11.1).

2.2.1 Conscious phenomena

Range

Consciousness seems to participate in all known mental processes at some time. This extraordinary range of conscious contents is one of those "obvious" facts that become puzzling only on further examination. If, for instance, perceptual experience were merely the result of energy transduction, we might explain the vast variety of conscious percepts quite simply: we need only suppose that many different kinds of energy are transduced into some common electrochemical form. But sensation and perception seem to require highly intelligent algorithms, which are probably so complex that each different form of perception demands a specialized set of processors. And *if* perceiving one kind of object or event demands a specialized processor, how many processors do we need to account for the vast range of percepts, thoughts, feelings, and intentions to which we have conscious access?

Indeed, we seem to have access to an astonishing variety of events in the nervous system. Under optimal conditions, sensory sensitivity approaches the theoretical minimum in which a single retinal receptor may be stimulated by a single photon, or (in the case of hearing) a hair cell may be stimulated by the Brownian motion of molecules in the ear canal. Electrical stimulation of cortical neurons can sometimes be experienced, and by means of conscious feedback, the firing patterns of a single spinal motor unit can come under voluntary control (Schwartz, 1975). This does not mean that the biofeedback subject is conscious of what she or he is doing; rather, it means that the system controlling the motor unit behaves as a specialized distributed processor, able to look at global information. The conscious feedback signal presumably represents such global information. Further, stimulation to which we have become habituated can become conscious (as when we are reminded of background noise, of the effects of gravity, of the feel of a chair, and so on). The unconscious presuppositions of perception, comprehension, and action can become conscious when they are strongly violated (e.g., Offir, 1973; Hornby, 1974). And memories we had long thought lost can reappear in a variety of circumstances (Williams and Hollan, 1981; Hilgard, 1977).

Contrary to widespread opinion, the contents of consciousness are not limited to the so-called higher mental functions. Indeed, there is a striking ability to range far and wide between the most abstract conceptual representations and supposedly more "concrete" sensory-perceptual elements.

What kind of a system could model this extraordinary range of contents? Certainly, a single task-specific processor could not do it. Insofar as a processor is good at some particular task, it is likely to be limited in doing other tasks. No such limitation appears for consciousness.

Relational Capacity

We can relate the contents of consciousness to each other almost without limit. Many decades of research on conditioning indicate that people and even animals have a remarkable capacity for learning arbitrary relationships between different stimuli and between stimuli and responses, though it is easier, of course, to learn nonarbitrary relations.

Although there seems to be evidence that humans can sometimes absorb input without awareness (Dixon, 1971), there is little or no evidence that we can acquire new relationships between inputs without awareness (Brewer, 1974). Indeed, several decades of experimental attempts have shown how difficult it is to demonstrate this. Clearly, the acquisition of new relationships generally involves awareness. Indeed, Smith (1969) has proposed that the capacity to arbitrarily relate any stimulus to any other, or to any response, is the criterion-like property of consciousness.

In humans, there is one relational capacity that is of special importance, and that is, of course, our ability to operate on conscious contents with highly specialized linguistic processors. It is important to note again that the existence of some conscious content does not depend on our capacity to express it in words. Verbal report is usually good evidence of some conscious content, but it is not the same as the conscious content.

Context sensitivity

This is a special kind of relational capacity, and it is of fundamental importance to the present discussion. It differs from the kind of relational capacity that is shown in cases like conditioning. During classical conditioning, for example, we learn that a tone signals the coming of a shock; both of these events are experienced as events. They each "stand out as figure from ground". But when we speak of context sensitivity in this chapter, we mean that an experience is affected by factors that do not stand out in this manner. It is a truism that experimental tasks are always affected by a host of variables of which the subjects are not aware. Many of these variables may have been conscious at one time, but they often have their effect long after they have become unconscious. This kind of context sensitivity is basic to the arguments made in this chapter (see Section 3).

While context sensitivity implies that conscious contents are affected by a variety of unconscious factors, one can equally well make the complementary point that conscious contents have widespread effects that are themselves not conscious. Both of these observations make sense if we think of consciousness as an operating mode of a global data base in a distributed processing system. Specialized processors sensitive to contextual factors can affect conscious representations, and, in turn, the conscious representations can have a widespread effect throughout the system.

2.2.2 Corresponding unconscious phenomena

Compare the wide range, the relational capacity, and the context sensitivity of conscious events to what we claim to be the relatively *limited domain of unconscious processors and their relative autonomy*. What is the evidence for these claims?

In this view, unconscious processors *by themselves* (i.e., without the intervention of the global data base) are relatively limited and autonomous. The only problem with this claim is that the people we study are generally conscious, so that the limitations of unconscious processors are seldom exhibited overtly. But in the case of involuntary slips of speech or action, we are privileged to see some unconscious processors in a relatively uncontrolled way. Involuntary slips can apparently violate any level of linguistic control (Fromkin, 1973, 1981), and the same observation applies to slips of complex action (Norman, 1981). We can define involuntary slips as those actions that are surprising to the actor: they are not consistent with his or her own previous plans. In extreme cases, slips like this can violate rules that the actor is highly motivated to follow. But in all cases, one can argue that *some specialized rule system, which should have anticipated and prevented the slip, was momentarily decoupled.* Since slips can violate any kind of rule system, it follows that any particular system can be momentarily decoupled from the others.

Presumably, if the actor had only a little more time to think about the action, or if his or her attention had been drawn to the relevant dimension, he or she would have anticipated the problem and prevented it (e.g., Baars et al., 1975). People are invariably surprised at having made a slip *when they become aware of it*; but this surprise implies that after the slip has been committed, the straying processor must have become coupled again (otherwise the slip would not be recognized as a slip). Indeed, it seems that the process of becoming conscious of an action has the effect of relating the action to its proper context. This again makes sense from the viewpoint of a global data base: if some representation is globally distributed, all the relevant factors can operate on it and respond to it. Conversely, as long as a processor is isolated from the global data base, it can violate rules imposed by other specialized processors.

That unconscious rule systems are relatively autonomous also follows from our frequent inability to exert lasting voluntary control over undesired habits. Most people seem to have automatisms that they would like to eliminate but that seem to be quite autonomous and resistant to external considerations. The more overlearned they are, the less they are conscious, and the harder it may be to exert voluntary control over them. They seem to appear especially when we are consciously distracted or overloaded.

For another example, consider the apparent autonomy of inputs to which we have become habituated. Suppose we have an air conditioner that emits a constant hum of which we rapidly lose awareness. If we need to leave the house and want to shut off the air conditioner (i.e., if the context changes so that we need to operate on the source of the habituated stimulus), we need to become aware of the fact that the air conditioner is on. If we fail to bring this fact to awareness, we are likely to leave the air conditioner on, because the habituated representation of this information is not sensitive to changes in context (i.e., to the fact that we are leaving the house).

Like consciousness, a global data base in a distributed processing system has enormous range, relational capacity, and context sensitivity. By comparison, each specialized processor has a relatively limited domain and is relatively autonomous.

2.3 Apparent unity, seriality, and limited capacity

2.3.1 Conscious phenomena

The impressive relational capacity and context sensitivity of conscious contents should not suggest that there are no limits on our conscious relational ability. However, these limits are of a very interesting kind and seem to depend exclusively on the mutual informational compatibility of the mental contents. There are many well-known demonstrations suggesting that we cannot simultaneously experience two mutually exclusive organizations of input (Gregory, 1966; Bransford and Johnson, 1973; Bransford and McCarrell, 1974).

Along very similar lines, there is an extensive lore in the history of science regarding the inability of scientists working within one paradigm to understand a competing paradigm (Kuhn, 1970). Comparable demonstrations of "fixedness" in problem solving go back to Luchins (1942) and Duncker (1945), illustrating the very general fact that a problem cannot be solved if approached within a framework that resists the correct solution. Similarly, Levine (1971) has demonstrated that an extremely simple discrimination task cannot be solved, even under "ideal S-R reinforcement contingencies", if subjects approach it with the wrong set. In the area of conditioning, Dawson and Furedy (1976) have shown that human galvanic skin response (GSR) conditioning will not take place if people are given an explanation of the conditioned-stimulus – unconditioned-stimulus relation that "masks" the contingency between these events. Similar conclusions emerge from work on ambiguous stimuli (e.g., MacKay, 1970) and on brain damage (e.g., Gazzaniga and LeDoux, 1978). If one can safely generalize over a large literature involving such disparate experimental techniques, one might say that any two pieces of information can be consciously related to each other, provided that they can coexist within a single, coherent framework. Facts like the above suggest that conscious organization demands unity, even if the unity is spurious.

The "unity of consciousness" fits quite well with the global-database notion. Any global representation that is not consistent with some processor will quickly encounter competition, so that it will be very unstable. Stable global representations must be coherent at anyone time, though they may be contradicted at some later time by another stable global representation.

If the contents of consciousness must be coherent, this requirement also implies that incompatible contents must become conscious serially, and that there is a limited capacity for competing contents. Thus, the apparent unity, seriality, and limited capacity of conscious contents seem to belong together as a set of related phenomena,

Shaffer (1976) has pointed out that people can do a number of tasks (such as conversing and playing the piano) simultaneously, though we would ordinarily consider them contradictory. It is to be noted that Shaffer's subject is extremely well practiced at these tasks, so that her conscious capacity is not likely to be taxed. In general, it appears that tasks that compete when they are new stop competing after enough practice has been gained (i.e., after the tasks are taken over by efficient and relatively autonomous specialized processors). Thus, it may well be that two otherwise "incompatible" tasks can go on in the nervous system, provided that they are not conscious at the same time, Hilgard (1977) provided some spectacular examples of such apparently incompatible processes.

2.3.2 Comparison with unconscious processes

Compared with the unity, seriality, and apparent limited capacity of conscious processes, it appears that unconscious processors are highly diverse, that they can operate in parallel (unless they need to interact in some way), and that together, the set of specialized unconscious processors has a very great processing capacity. What is the evidence for these claims?

The diversity of processors follows immediately from the idea that they are highly specialized, combined with the incontestable observation that the nervous systems does an enormous number of different things, Parallelism and the idea of a very large unconscious processing-capacity can be viewed in the following way.

Capacity

Consider the physiological facts. There are some 10^{10} neurons in the cerebral cortex alone, firing at an average rate of perhaps 40 Hz (Eccles, 1973). Thus we have about 40 x 10^{10} events taking place each second, or roughly one-half trillion. This certainly seems like a system with very great capacity – yet we know that conscious capacity seems extraordinarily limited. We can store only 7 plus or minus 2 isolated items even with conscious rehearsal, we can process only one stream of speech at a time, and it takes us at least 100 msec to react to a conscious stimulus. Unless we assume that most neurons are firing away "epiphenomenally", so that their activity has little effect on psychologically interesting variables, we must somehow reconcile this picture of frenetic activity and relatively ponderous conscious processing. The idea of a global data base appears to reconcile this apparent conflict, As a distributed system, it is no surprise that the great amount of processing activity is not global but is relegated to dedicated processors. Because a global representation requires the cooperation of a number of specialists, it must change much more slowly than any single specialist. This view helps to resolve what seems to be a contradiction between the limited capacity of consciousness and the enormous amount of processing activity that we observe at the same time.

Parallel processing

The neurophysiology also suggests that "the organization of the brain implies parallel processing" (Thompson, 1976; Anderson et al., 1974). However, we well know that conscious processes seem to be quite serial (e.g., Newell and Simon, 1972). How can these different impressions be reconciled?

Consider some of the general properties of parallel information processing, Superficially, it would seem that one could get much more accomplished by parallel than by serial processing, However, parallel processors are restricted if there is some contingency between one process and another-and intelligent processes often involve a series of contingent decisions, In particular, if processors operating in parallel need to interact, there exists a contingency between otherwise independent systems, The result is a bottleneck, which behaves in an apparently serial fashion, These considerations are especially relevant to this discussion, because *a global data base can be vie-wed precisely as a device that facilitates interaction between otherwise independent, parallel systems.* Marslen-Wilson and Welsh (1978) have provided evidence that some components of speech perception are, in fact, mediated by a parallel interactive system of this kind.

2.3.3 Summary: How the global data base fits the capability constraints of Table 11.1

The relevance of this theoretical metaphor for consciousness is now beginning to emerge. A global data base is not itself an efficient computational device; rather, it permits a multitude of efficient processors to communicate in some commensurable way. Hence, it must be able to display an enormous range of representations, and to relate any two arbitrary representations to each other, so that distributed processors can help to specify the relationship between any two global representations. At anyone time, the global data base can display only a single *coherent* content; incoherent representations will swiftly decay because of internal competition. Indeed, a *context* has previously been defined as a stable, coherent set of constraints on a global representation. This need for unity at anyone time will make it appear that the global data base has a very limited capacity, and for this reason, competing contents must appear serially. In short, all of the capability constraints of Table 11.1 very naturally "fall out of" the concept of a global data base.

By contrast, the specialized processors in a distributed system are highly efficient in their special tasks; necessarily, their domains of specialization are also limited and relatively autonomous. They are highly diverse, they can operate in parallel (provided they do not need to interact), and together, they possess great processing capacity.

Thus, there is a close association between consciousness and the kind of systems configuration discussed here, but not an identity. In the following section, arguments are given that conscious representations must be global, and also stable and informative.

3. The boundaries of conscious contents

What does it mean to experience something? What are the boundary conditions of conscious *contents*? More is known about perception than about any other kind of conscious content, and most detailed examples will be perceptual. However, the conclusions to be drawn from these examples should suggest an approach to such conscious contents as images, which are not dependent on external input. The arguments given here depend on a detailed consideration of the boundary constraints (Table 11.2) which show under what conditions conscious events become unconscious, and vice versa. Following is a short summary of the arguments, followed by more detailed considerations.

Two kinds of boundary conditions may be called synchronic, since they exist at the same time as any conscious experience, though they ale not themselves conscious. First, we know that there must be internal representations of the context within which a percept is defined, but that this contextual representation is not conscious. In Section 1, a context was defined as a stable set of constraints on any global representation. Thus, the claim is made that those properties of a global representation that are entirely constrained are not conscious. Until we encounter an Ames trapezoidal room, we are not conscious of the fact that we interpret trapezoidal shapes as rectangles in our carpentered world, and until we encounter someone of a different ancestry and culture, we are not conscious of the fact that we have certain assumptions about people's appearance, dress, and mannerisms. These contextual assumptions are clearly used constantly to make sense of the world, but they are not conscious. Second, it is also the case that sensory input that is *not interpretable within the current context* is not conscious. When we hear a word in a meaningful sentence, we are typically conscious of only one meaning, even though a glance at a dictionary should convince anyone that all words have more than one meaning. We are not conscious of any other meaning until we are in the right context. When we are confronted with a foreign language, especially one with a very different phonology, the sounds of the language do not become conscious until the phonology is spelled out, or until we practice it, or until we hear minimally contrastive pairs of words. We can listen to rock music many times without understanding the words, until someone tells them to us; after that, the words seem limpidly clear. The examples can be multiplied indefinitely. All the phenomena discussed under the heading of "Apparent unity, seriality, and limited capacity" (Section 2.3), represent cases where some input is perfectly interpretable within one context but incoherent in another one.

These two points can be summed up as follows: Context, taken by itself, is unconscious, and input, taken by itself and in the absence of the appropriate context, is also unconscious. Only when *both* of these conditions exist – that is, when there is input that can be organized within a current context – are we conscious of some percept.

Next, there are two kinds of unconscious representation that exist diachronically, before and after a conscious representation. The first of these diachronic representations involves *preperceptual* processes, which are clearly representational in nature but not conscious. Second, and of very great importance to the present argument, conscious percepts habituate rather quickly, if the input remains predictable. Habituation in all its forms is something of a stepchild in the psychological literature on learning and memory. It is often treated as a rudimentary kind of learning, but one that is not very interesting because it is not associative. Perhaps it is a result of "neural fatigue", etc. In any case, habituation is viewed as something of a by-product, of limited interest.

But in the present approach, habituation is thought to be an intelligent matching of input by any neural system – perhaps the fundamental form of learning, therefore. Neural systems stop working when they complete an internal match of the input, as suggested by Sokolov (1963), and this decrement of responding may resemble a kind of fatigue from the outside. But the decrement simply reflects the fact that after the system actively matches the input, the input becomes redundant with respect to the system. Habituation of consciousness of some object is treated in this chapter as global redundancy, indicating that the system as a whole has adapted to some stable global representation (see also Nadel, 1981). In the same sense, actions that have become automatic can be considered globally redundant, though they do require local information processing (see Section 1). The notion of global redundancy emerges most dearly from a consideration of habituation of awareness, but once established, it can be seen in a number of other cognitive phenomena.

What follows now is a detailed presentation of these arguments.

3.1 Contexts

It is a very general fact that the perception of some object or event requires a stable context or framework (e.g., Minsky, 1975; Asratyan, 1965) and, indeed, that without this stable context a percept cannot be established. A change in the relevant context will produce a different experience or render the input pattern incoherent. This is true whether we speak of linguistic presuppositions, of assumptions made about space in order to process the visual world, or of the set of assumptions that produce fixedness in problem solving.

Contextual constraints of any experience must be represented with great precision, *yet they are invariably unconscious*. That does not mean that we cannot become aware of some contextual assumption, but that as we become aware of it, it ceases to be context and requires some other unconscious assumptions to be comprehended. The context as context always escapes our awareness.

One way to bring a contextual constraint to awareness is by violating it strongly (e.g., Offir, 1973; Hornby, 1974). In the Ames trapezoidal room, such a violation occurs when the observer tosses a ball against the wall, and the ball bounces back in an unexpected way. As the observer becomes aware of the trapezoidal shape of the room, his or her contextual assumptions – the stable constraints on his or her conscious representation of the room – go through a transformation. Indeed, one may argue that contextual factors can become conscious *only* when they are challenged.

Cognitive psychologists can often avoid dealing with stable contextual factors by operating *within* a given experimental situation. In other disciplines, such as anthropology, developmental psychology, or the history of ideas, this is not possible. As a result, scholars in these fields are often acutely aware of the effects of changes of context. But it seems that in any situation, the invisible context contains the most powerful factors in the situation, so that this contextual frame is well worth the attention of cognitive psychologists.

Are we defining *context* circularly? Ultimately, we need to work with well-defined theories that make explicit exactly the content of contextual frameworks. Fortunately, over the past 20 years, workers in artificial intelligence have made great strides in the direction of specifying the knowledge needed to understand everyday situations, It seems likely that over the next few decades, this work will yield theories rich enough, and explicit enough, so that we can follow someone around a supermarket, perhaps, talking about being conscious of this and that while taking for granted active but unconscious contextual knowledge.

3.2 Undefined inputs

There is a second synchronic boundary to a conscious experience, which becomes especially clear in the case of perceptual input. Consider a perceptual demonstration of a hidden figure, such as the well-known "Dalmatian in the park", which shows a spotted Dalmatian in a shadow-flecked park. Since the entire picture is in black and white, it is initially very difficult to spot the hidden figure. Until the right organization is discovered, it does not become conscious. (It is noteworthy that contextual factors, in the sense defined before, can suggest the correct organization). However, once proficiency is gained in spotting the hidden figure, the process of spotting the figure becomes fast and efficient to the point where it becomes very difficult to *avoid* seeing the dog. Now the situation is reversed; it becomes difficult to see the spots *as* spots, unless the image of the dog is further obscured (for instance, by turning the picture upside down). Thus the hidden figure is bistable, much like an ambiguous stimulus, but it is nonreversible; once the hidden dog is discovered, the newer, more coherent representation tends to prevail.

This example is really a paradigm case for much of this discussion. The idea of a "hidden pattern" may be generalized to cover a multitude of psychological tasks. For example, in word perception, there is, in a reasonable sense, a "hidden figure" in the pattern of sound. Before one makes a scientific discovery, there is, so to speak, a hidden pattern in the evidence. The reader can no doubt supply further examples of the search for hidden patterns. The analysis of this example may be applied to many similar cases.

The input pattern specifying the Dalmation is unconscious *until* we find the right context – the right set of stable constraints – within which to interpret the input. Thus, we can now state a second very general boundary condition on consciousness of input: available energy patterns are unconscious if they are not defined coherently within the current context. The problem of organizing a pattern in the right way is the problem of finding the right set of stable contextual constraints for that pattern.

The first and second boundary conditions are really different sides of the same coin. The first claims that, to be conscious, any input demands some contextual constraints that are themselves unconscious. The second one states that any pattern is unconscious unless the right framework is available to organize it in a coherent fashion. In the absence of the right context, any input pattern is unconscious. *Thus, context alone is unconscious and input alone is unconscious.* Only when these two unconscious components interact in the right way are we conscious of some event. This is a fundamental claim.

The third and fourth boundary conditions on conscious contents add another requirement. Not only must a conscious representation be coherent and stable, but it must also provide *global information*; that is, it must make a difference to the system as a whole. Arguments for this requirement are discussed next.

3.3 Global information and global redundancy

Consider the very general phenomenon of habituation of awareness, the third boundary condition. When some stimulus is repeated or continued past a certain point, it is no longer experienced. This effect is as general as perception itself: it occurs in all sensory modalities, with any kind of stimulation.[2] Indeed, at any time, there are a large number of predictable energy patterns impinging on us, from gravity to the ambient light, sound, and temperature, to the pressure of our clothing. All these energy patterns are typically unconscious. It may be that the visual system is especially protected against premature habituation to constant inputs by means of physiological nystagmus (the constant high-frequency tremor of the eyes), which causes light edges in the visual field to fall always on a slightly different part of the retina. Without this feature, awareness of the visual input is lost within seconds. Clearly in all other sensory modalities, awareness of some redundant input tends to habituate quite rapidly.

Habituation of neural structures occurs at all levels, from single cells to complex structures. As we noted above, habituation has not been thought to be a full-fledged kind of learning in the literature on learning and memory, even though, as Sokolov (1963) pointed out, habituation cannot be a form of fatigue because it is stimulus-specific, and because habituated animals will dishabituate to the *absence* of the repeated stimulus. That is, dishabituation is a response not to energy input but to new information. And if dishabituation (orienting, etc.) is a response to information, it is not too great a leap to suppose that habituation is a response to redundancy (Asratyan, 1965; Nadel, 1981). In the present perspective, it is assumed that all specialized processors attempt to model global input that is relevant to them, and that having done so, they cease responding to this input. Thus, habituation for these processors is a sign that learning has occurred; that is, the input matches the local representation sufficiently so that no further adaptation is required. In this view, *habituation of awareness* means that the system as a whole has adapted to the input, so that the input no longer provides global information. (Recall that *global information* has been defined as a global representation that triggers adaptation in the system as a whole, so that any relevant processor can adapt to it). Thus, habituation of awareness to any input may be considered a sign of global redundancy, which is simply the absence of global information.[3]

But now we can go one important step further. By definition (see Section 1), a context is a set of stable, global constraints – constraints to which the system as a whole has adapted. But this is, of course, equivalent to saying that these constraints are globally redundant. Anything that is globally redundant can therefore become part of a new context, able to affect the way relevant new stimuli will be experienced. We may say, then, that conscious representations that are lost from awareness because of habituation do not disappear: they continue to provide a context within which future related representations are defined.

This is not surprising if we consider well-established facts on gross perceptual readaptation (e.g., Köhler, 1962). Consider an everyday example. When we first step onto a small sailboat, we are very much aware of the movements of the boat, but most of us adapt fairly rapidly, so that the movements of the boat become highly predictable and are lost from awareness. They become globally redundant, in the sense suggested above. What previously constituted information has habituated and is now a part of the context. It is easy to show that this habituated information is not completely lost; it merely becomes the framework within which new information is defined. Thus, on returning to dry land, what is usually in the background now becomes information. That is to say, it now seems as if the *world* is reeling drunkenly: we make false predictions regarding our orientation to gravity and attribute the source of this information not to ourselves but to the world. What was globally redundant (and unconscious) at sea now becomes globally informative and hence conscious, until we once again adapt to land, and the spatial context once again fades into the background.

Thus, habituation of conscious contents can create new contextual constraints that can affect the way conscious information is structured. This is true not only of relatively gross properties, such as our orientation to gravity. Rather subtle properties of our perceptual experience can also be affected by a habituation phenomenon like selective adaptation (e.g., Eimas and Corbit, 1973). In the remainder of this chapter, the idea that habituated conscious contents can create the context for future conscious contents will be considered a general and very important property of the system.

Note that even though a certain process may be globally redundant, it can still require some local information-processing. Consider automaticity, which is the counterpart of habituation in the case of a proficient skill (LaBerge, 1974). As we walk around the world, we are largely unaware of the fast, complex, and subtle details of balancing and moving. That is, the action of walking is largely globally redundant, in the sense defined above. Yet we cannot claim that these fast-moving details of walking are nowhere computed; rather, we may say that they do not require global information-processing, because they are essentially routine and predictable at a global level. This is presumably not true for babies just beginning to walk, nor is it true for bed-ridden individuals who are readapting to walking, nor for people just stepping off a small sailboat. For all these individuals, walking requires global information-processing.

Again, the question may be raised whether we are defining *information* and *redundancy* in a circular way, by reference to the phenomena they are supposed to explain. We know that when we can control stimulation, we can repeat a stimulus over and over again, and it will disappear from consciousness. Further, we know that if we change only one dimension of the stimulus – its amplitude, energy envelope, quality, or temporal parameters – the stimulus will become conscious again. These are not circular claims if we have experimental control. But this explanation is still far from satisfactory. We cannot predict the informativeness of a new stimulus in many situations, and until we have adequate theoretical representations of these situations, we will not be able to specify what is informative and what is not. Work in artificial intelligence does seem to be moving rapidly in the direction of such explicit representations for everyday situations.

So far, the argument applies to entirely predictable inputs, such as the ambient light or temperature. However, the notion of global redundancy can also apply to highly predictable *components* of input. For example, when we hear a series of paraphrases of a single sentence, there is little doubt that we will begin to ignore the meaning rather quickly, even though the physical input is continuously changing. Indeed, it will be difficult to attend to the meaning at all: it has become globally redundant. More generally, it seems likely that as soon as we fully understand the meaning of any sentence, any repetition of the meaning will be globally redundant. One may interpret the "click" of comprehension as that moment in which the meaning of a sentence becomes globally redundant, so that now it can be used to interpret new, incoming information. Thus, the notion of global redundancy can be extended beyond the pure case of completely predictable input.

The idea that globally redundant constraints are not conscious adds a very important qualification to the discussion so far. Such things as predictable stimuli and automatic skills must be represented in the nervous system in a coherent fashion, but of course they are not conscious. But this means that the coherence of a representation is a necessary but not a sufficient condition for the representation to be conscious. A conscious representation must be new or *globally informative* as well as coherent. We can now add this to the set of boundary conditions specified so far and claim that *conscious experience of some content involves an interaction between an energy pattern and contextual constraints, resulting in a coherent representation that provides global information to the system.* This is one of the major conclusions of this chapter.

We now have three sets of constraints along with a reasonable theoretical interpretation. Conscious contents always seem to involve an interaction between some energy pattern and a set of contextual assumptions, so that the interaction results in a coherent representation. But if this coherent representation provides no news, if it is not informative in some sense, it is not conscious. This observation suggests that after habituation of awareness, contextual processors have adapted to the representation so that the news has been absorbed and has indeed become a part of the system itself.

3.4 Preperceptual processing

Consider now the fourth boundary condition of perception. The idea is widely accepted that input information is preprocessed for a few tenths of a second before it becomes conscious (e.g., Neisser, 1967). Preperceptual processing is usually viewed as a kind of hypothesis testing, in which many different hypotheses are brought to bear on the problem of representing the input. Hypotheses are representations, of course, and we must explain why this kind of input representation is not conscious. This is the fourth boundary condition for conscious experience, and it is "diachronic" because it refers to a stage of stimulus representation that comes before the conscious experience.

Indeed, in practice, the global-data-base configuration has been used primarily for the kind of hypothesis testing that presumably takes place before the establishment of a percept (Erman & Lesser, 1975). The global data base is useful when the processors needed to analyze the input are unknown, so that *any* relevant processor can be brought to bear on any global hypothesis. Because the global data base is very useful in broadcasting information to relevant but unpredictable processors, it seems plausible to assume that the nervous system makes use of something like a global data base during preperceptual processing. But this means that there must be global hypotheses that are *not conscious*. Now explain the difference between conscious hypothesis-testing and the unconscious hypothesis-testing that presumably takes place before perception?

That explanation is really already available. It was previously shown that input patterns not defined within the current context are not conscious. But preperceptual processes involve precisely a set of hypotheses that are undefined within the current context, because they are unstable and mutually competitive. By the time they cooperate sufficiently to establish a coherent context, they become conscious. It is therefore not surprising that the preprocessing of input is not conscious.

If we look at the hidden-figure example again, we can observe this process in slow motion. In the beginning, it is difficult to find the Dalmatian at all, but given the right context (which can be induced by external hints, etc.), the input can be interpreted in the right way. On repeated exposure, it is indeed difficult to avoid seeing the dog, to access the less-coherent interpretation of the input. But even when the pattern is analyzed quite efficiently, we can presume that preprocessing still takes place. It is then no longer conscious, presumably because the processing has speeded up and the lower-level hypotheses are fleeting and unstable.

This point is further supported by a set of examples that show the opposite effect. That is to say, what would happen if we were to *slow down* preperceptual processing? Would the previously unconscious process of hypothesis testing become conscious? This has in fact been tried (Norman, 1976; Kolers, 1978; Bruner and Potter, 1964). For instance, when we read a sentence such as this one, fast and efficient unconscious processes take care of letter and word recognition. But try reading a sentence upside down: suddenly we begin to test *conscious* hypotheses about letters and words. Similarly, when we see an unfocused slide that is gradually brought into focus, we begin "spontaneously" to generate conscious hypotheses. These conscious hypotheses are probably very similar to the unconscious ones that occur preperceptually, although they may be more coherent than the unconscious hypotheses.

There is thus a rather thin dividing line between unconscious hypothesis-testing and conscious hypothesis-testing. It is not clear whether the global representation becomes conscious because it is coherent or because it is stable, because stability and coherence tend to covary perfectly. Certainly, in the model developed so far, specialists that compete with each other can display a global hypothesis for only a very short time before it is contradicted. Therefore, any hypothesis that is coherent will also be stable, and vice versa. Thus, it is safer to restate the conclusion reached above as follows: "Conscious experience of some event always involves an interaction between some energy pattern and a contextual framework that results in a coherent *and stable* representation that provides global information to the system".

This is the fundamental perspective on conscious contents that is advanced here. It seems to account for the empirical constraints in an economical fashion. The model is still a thinking tool – not a permanent position to cling to. But it appears to be both adequate and economical. For instance, it is theoretically pleasing to see that the four general boundary conditions actually reduce to only two properties of conscious contents. First, preprocessing and undefined input turn out to be unconscious for the same reason. Preperceptual processes are unconscious because they lack a stable and coherent context, and similarly, undefined inputs lack such a context. Further, context and highly predictable input patterns are both unconscious because they both involve stable, global representations to which the system as a whole has *already adapted*. Thus, four classes of phenomena reduce to two theoretical terms: events are unconscious if they are undefined in the current context, or if they are so stable as to be part of the context. This theoretical economy is encouraging and suggests that the analysis is on the right track.

4. Summary and conclusions

The metaphor of a global "information exchange" in a distributed processing system helps to explain a number of psychological phenomena. It appears that we are conscious of something when there is an interaction between input and context, resulting in a stable and coherent global representation that provides information to the nervous system as a whole. This description fits the empirical constraints of Tables 11.1 and 11.2 and makes a good deal of functional sense as well. When we are conscious of something, we are adapting to it in a global way.

4.1 Extensions of the theory

Because of limitations of space, we can only suggest several extensions of this theoretical framework to incorporate further empirical constraints (Baars, 1980; Baars and Mattson, 1981). For example, the theory can incorporate the classic "insight" phenomena in problem solving, and it suggests a role for the conscious components of intentional action. These points are briefly summarized.

In both spontaneous problem-solving and intentional action, a global context serves to guide specialized processors able to complete the context. Although we are not directly conscious of this global context, any of its components can become conscious when contradicted in some way, by either internal or external influences. In problem solving, one accumulates a set of constraints that are at first fully conscious, and as the system adapts to these constraints, they become components of the problem context. As part of a context, these constraints are, of course, not conscious unless they are violated in some way.

An *intention* may be considered a special kind of problem context, one that serves as a global goal to mobilize and organize a large, diverse set of action specialists to prepare and execute an action. Further, the intention has a timing component that permits the action to run off when it is ready.

When there is a conflict between different intentions, or when some action violates a component of its intention, the appropriate component of the intention seems to become fully conscious. All these cases involve changes in the intention context and therefore fit our previous characterization of the conditions under which components of a context become conscious. The "internal monologue" that we often use to command ourselves can be viewed as a way in which a processor able to broadcast a conscious command can trigger the creation of an intention by other processors. The intention then can serve to prepare and execute an action. It is to be noted that such conscious commands are never as complete as the intention: they seem to involve only what is new and different about an intention. Most of the content of the command is tacitly understood, just as the bulk of our communications with each other are tacitly understood (viz., Bam and Mattson, 1981).

Finally, we can deal with the closely related issue of attention. Attention involves a set of systems able to *select and maintain* some particular conscious content, either voluntarily or involuntarily. In voluntary control of attention, we may give a conscious command that triggers an intention that can control future conscious contents. In the involuntary case, specialized processors act to control the contents of consciousness. The experience of mental effort is thought to result from conflict between voluntary and involuntary means of controlling consciousness. Finally, the ideas of the dynamic or "affective" unconscious fit in naturally with the notion of attention, with the difference that some specialized processors may exercise control of access to the blackboard in order to avoid certain conscious contents, while others may seek to display certain other contents.

4.2 Conclusion

A great deal of work needs to be done to expand and clarify this approach, and to test it for theoretical adequacy and consistency. There are very many empirical implications that have not been discussed in this paper and that must be considered in detail elsewhere. Nevertheless, the theory sketched here fits a large number of facts about consciousness.

No theory at this stage can be more than a thinking tool, to be falsified and changed as our understanding grows. If the present paper serves to define some of the issues with more precision, and if it helps to develop a vigorous and pointed debate about them, a large part of its purpose will have been achieved.

Acknowledgements

The author gratefully acknowledges Michael A. Wapner (CSULA) for a number of the ideas presented in this paper, and Donald A. Norman (UCSD) for his welcome insistence on brevity and clarity, which helped to improve the paper immensely. Others whose comments were most helpful include Robert Buhr (UCSD, Cognitive Science Program), Lynn Nadel (UC Irvine), Michael Posner (University of Oregon), David Cross (SUNY Stony Brook), and Michael T. Motley (UC Davis). The comments of Robert Monk (Chicago), Richard J. Davidson (SUNY Purchase), Daniel Dennett (Tufts University), and Walter Ritter (Einstein College of Medicine) were most helpful as well.

Notes

1. The word context is often rightly criticized for being theoretically and empirically empty. In this paper, however, it has a number of very specific implications. Theoretically, it is defined as a set of stable, global constraints, which serve to guide and define inputs to the global data base. Empirically, contextual factors are defined as being those factors that can change conscious contents without themselves being conscious. Thus, in any experimental situation, it is dear what factors are and are not part of context, though, of course, this definition does not tell us ahead of time which factors will be contextual in any particular case. Nevertheless, it is quite possible to make some rather strong empirical claims, which include the following: Any conscious content can become a contextual constraint when the system is globally habituated to it. As such, the formerly conscious content will constrain the interpretation of future conscious contents. Components of a context must always be mutually consistent, otherwise competition would occur between them. Incomplete contexts serve to mobilize processors able to complete them, and whenever any component of a context is strongly violated, it will tend to become conscious. All these predictions can be tested in appropriate experimental situations. Thus, the word context, as used here, is nonempty both theoretically and empirically.

2. It may be objected that one can have the "same" experience many times without losing awareness of it. For example, one may travel the same road to work each day without a complete loss of awareness of the road. There are a number of answers to this objection. First, we lose complete awareness only of stimuli that are entirely predictable and, in particular, of stimuli with entirely predictable temporal properties. Very often, we do not adapt completely to some conscious content; we simply shift to a different content. Nevertheless, it is easily conceivable that we shift away if some particular dimension of the content has been absorbed, especially if that dimension is relevant to a current task. Second, we are not merely exposed to the same road each day; we interact with it, so that often we can voluntarily override the tendency to lose awareness of the information. Third, whenever some change takes place in a well-known environment to which we have become habituated, the change must be integrated into a larger set of events. If someone is exposed to a regularly repeated burst of noise, for instance, and we change only one aspect of the noise burst, all the other properties of the stimulus must be reevaluated. Thus, if the onset ramp of the habituated noise is changed, the subject will not become aware of the onset ramp *in abstracto* but will become aware of the whole noise burst. Similarly, if one aspect of the road to work changes, other aspects most become conscious as well. In sum, our continued awareness of routine events does not constitute a counterexample to the claim made here.

3. What about the impact of significant stimuli on habituation? It seems obvious that awareness of significant events is lost more slowly than awareness of insignificant ones. In some cases, like chronic pain, one could indeed maintain that awareness is never permanently lost. It seems useful to treat the effect of significance in informational terms. It is highly plausible to think that significant events require more adaptation throughout the system; thus, in a very strict sense, significant events are more "informative" than other events. Significant changes must propagate more widely throughout the system before adaptation can take place. Presumably significant events also demand more problem solving (Baars and Kramer, 1982) before the system achieves complete adaptation.

References

Anderson JA, Silverstein JW, Ritz SA, Jones RS (1973) Distinctive features, categorical perception, and probability learning: Some applications of a neural model. *Psychol Rev* **84**:413-451.

Arbib MA (1982) Perceptual structures and distributed motor control. In: VB Brooks (Ed), *Handbook of Physiology* (Vol. 3). American Physiological Society.

Asratyan EA (1965) *Compensatory Adaptations, Reflex Activity, and the Brain.* Oxford: Pergamon Press.

Baars BJ (1980) What is the role of consciousness? (Unpublished manuscript).

Baars BJ, Mattson ME (1981) Consciousness and intention: A framework and some evidence. *Cognition and Brain Theory* **4**:247-263.

Baars BJ, Kramer DN (1982) Conscious and unconscious components of intentional control. *Proceedings of the Fourth Annual Cognitive Science Conference.* Ann Arbor University of Michigan Program in Cognitive Science.

Baars BJ, Motley MT, MacKay DG (1975) Output editing for lexical status in artificially elicited slips of the tongue. *Journal of Verbal Learning and Verbal Behavior*, **14**:382-391.

Bransford JD, McCarrell NS (1974) A sketch of a cognitive approach to comprehension: Some thoughts about what it means to comprehend. In: WB Weimer and DS Palermo (Eds), *Cognition and the Symbolic Processes.* Hillsdale, NJ: Erlbaum.

Bransford JD, Johnson MK (1973) Consideration of some problems of comprehension. In: WG Chase (Ed), *Visual Information Processing.* New York: Academic Press.

Brewer WF (1974) There is no convincing evidence for operant or classical conditioning in adult humans. In: WB Weimer and DS Palermo (Eds), *Cognition and the Symbolic Processes*. Hillsdale, NJ: Erlbaum.

Bruner JS, Potter MC (1964) Interference in visual recognition. *Science* **144**:424-425.

Dawson ME, Furedy JJ (1976) The role of awareness in human differential autonomic classical conditioning: The necessary-gate hypothesis. *Psychophysiology* **13**:50-53.

Dennett DC (1978) Toward a cognitive theory of consciousness. In: DC Dennett, *Brainstorms.* San Francisco, CA: Bradford Press.

Dixon NF (1971) *Subliminal Perception: The Nature of a Controversy.* London: McGraw-Hill.

Duncker K (1945) On problem solving. *Psychological Monographs* **58**.

Eccles JC (1973) Brain, speech, and consciousness. *Naturwissenschaften* **60**:167-176.

Eimas PD, Corbit J (1973) Selective adaptation of linguistic feature detectors. *Cogn Psychol* **4**:99-1109.

Erman LD, Lesser VR (1975) A multi-level organization for problem solving using many, diverse, cooperating sources of knowledge. *Proc 4th Int Joint Comp Conf*, Georgia, USSR, pp 483-490.

Fromkin VA (Ed) (1973) *Speech Errors as Linguistic Evidence.* The Hague: Mouton,

Fromkin VA (Ed) (1981) *Errors of Speech and Hearing.* New York: Academic Press.

Gazzaniga M, LeDoux JP (1978) *The Integrated Mind.* New York: Plenum Press.

Geschwind N (1979) Specializations of the human brain. *Sci Am* **241**:100-201

Ghiselin B (1952) *The Creative Process*. New York: Mentor.

Globus G, Franklin S (1980) Prospects for a scientific observer of perceptual consciousness. In: JM Davidson and RJ Davidson (Eds), *The Psychobiology of Conciousness.* New York: Plenum Press.

Greene PH (1972) Problems of organization of motor systems. In: R Rosen and FM Snell (Eds), *Progress in Theoretical Biology* (Vol. 2) New York: Academic Press.

Gregory RL (1966) *Eye and Brain: The Psychology of Seeing.* New York: McGraw-Hill.

Hayes-Roth F, Lesser VR (1976) Focus of attention in the Hearsay II speech understanding system. *Proc 5th Int Joint Comp Conf*, Cambridge, MA, pp 27-35.

Hilgard ER (1976) Neodissociation theory of multiple cognitive control systems. In GE Schwartz and D Shapiro (Eds), *Consciousness and Self-Regulation: Advances in Research* (Vol. 1). New York: Plenum Press.

Hilgard ER (1977) *Divided Consciousness: Multiple Controls in Human Thought and Action*. New York: Wiley.

Hornby PA (1974) Surface structure and presupposition. *Journal of Verbal Learning Verbal Behaviour* **13**:530-538.

James W (1890) *The Principles of Psychology.* New York: Holt (Reprinted by Dover, New York, 1950).

Jenkins JJ (1974) Remember that old theory of memory? Well, forget it! *American Psychologist* **29**:785-95.

John ER (1976) A model of consciousness. In: GE Schwartz and D Shapiro (Eds), *Consciousness and Self-Regulation: Advances in Research* (Vol. 1). New York: Plenum Press.

Kaplan RM (1973) A general syntactic processor. In: R Rustin (Ed), *Natural Language Processing.* New York: Algorithmics Press.

Köhler I (1962) Experiments with goggles. *Sci Am* **206**:62-75.

Kuhn TS (1970) *The Structure of Scientific Revolutions* (2nd ed.). Chicago: University of Chicago Press.

LaBerge D (1974) Acquisition of automatic processing in perceptual and associative learning. In: PMA Rabbitt and S Dornic (Eds), *Attention and Performance* (Vol. 4). London: Academic Press.

Levine M (1971) Hypothesis theory and nonlearning despite ideal S-R reinforcement contingencies. *Psychol Rev* **78**:130-1411.

Lewis FL (1970) Semantic processing of unattended messages using dichotic listening. *J Exp Psychol* **85**:225-228.
Lindsay PH, Norman DA (1976) *Human Information Processing* (2nd Ed.). New York: Academic Press.
Luchins AS (1942) Mechanization in problem solving. *Psychological Monographs* **54** (6, Whole No. 248).
MacKay DG (1970) Mental diplopia: Towards a model of speech perception at the semantic level. In: GB Fores d'Arcais and WJM Levelt (Eds), *Advances in Psycholinguistics.* Amsterdam: North Holland.
MacKay DM (1969) *Information, Mechanism and Meaning.* Cambridge, MA: MIT Press.
Mandler G (1975a) Consciousness: Respectable, useful, and probably necessary. In: R Solso (Ed), *Information Processing and Cognition: The Loyola Symposium.* Hillsdale, IL: Erlbaum.
Mandler G (1975b) *Mind and Emotion.* New York: Wiley.
Miller GA (1962) *Psychology: The Science of Mental Life.* New York: Harper & Row.
Minsky M (1975) A framework for representing knowledge. In: PH Winston (Ed), *The Psychology of Computer Vision.* New York: McGraw-Hill.
Nadel L (1981) Cognitive and neural maps. In: P Juscyzk and R Klein (Eds), *Essays on the Nature of Thought: In Honor of D. O. Hebb.* Hillsdale, NJ: Erlbaum.
Naranjo C, Ornstein RE (1971) *On the Psychology of Meditation.* New York: Viking.
Neisser U (1967) *Cognitive Psychology.* New York: Appleton-Century-Crofts.
Newell A, Simon HA (1972) *Human Problem Solving.* Englewood Cliffs, N.J.: Prentice-Hall.
Norman DA (1976) *Memory and Attention: An Introduction to Human Information Processing* (2nd ed.). New York: Wiley.
Norman DA (1981) Categorization of action slips. *Psychol Rev* **88**:1-15.
Norman DA, Shallice T (1980) Attention and action: Willed and automatic control of behavior. (Unpublished paper), UCSD, La Jolla, CA.
Offir CE (1973) Memory for presuppositions of relative clause sentences. *Journal of Verbal Learning and Verbal Behavior* **12**:636-643.
Paivio A (1975) Perceptual comparisons through the mind's eye. *Memory and Cognition* **3**:635-647.
Posner M, Warren RE (1972) Traces, concepts and conscious constructions. In: AW Melton and E Martin (Eds), *Coding Processes in Human Memory.* Washington, O.C: Winston.
Reddy R, Newell A (1974) Knowledge and its representation in a speech understanding system. In: LW Gregg (Ed), *Knowledge and Cognition.* Potomac, MD: Erlbaum.
Schwartz GE (1975) Biofeedback and physical patterning in human emotion and consciousness. *Am Scientist* **63**:314-324.
Shaffer LH (1976) Intention and performance. *Psychol Rev* **83**:375-393.
Shallice T (1972) Dual functions of consciousness. *Psychol Rev* **79**:383-393.
Shallice T (1978) The Dominant action system: An information-processing approach to consciousness. In: KS Pope and JL Singer (Eds), *The Stream of Consciousness: Scientific Investigations into the Flow of Experience.* New York: Plenum Press.

Shiffrin RM, & Schneider W (1977) Controlled and automatic human information processing: II. Perceptual learning, automatic attending, and a general theory. *Psychol Rev* **84**:127-190.
Smith K (1969) *Behavior and Conscious Experience: A Conceptual Analysis.* Athens, OH: Ohio University Press.
Sokolov EN (1963) *Perception and the Orienting Reflex.* New York: MacMillan.
Sperling G (1960) The information available in brief conscious presentations. *Psychological Monographs* **74**.
Straight SH (1979) Consciousness as a workspace. *SISTM Quarterly* **1**:11-14.
Tart C (1975) *States of Consciousness.* New York: Dutton.
Thompson RF (1976) The search for the engram. *American Psychologist* **31**:209-227.
Turvey MT (1977) Prelimenaries to the theory of action with reference to vision. In: R Shaw and JD Bransford (Eds), *Perceving, Acting, and Knowing: Toward an Ecological Psychology.* Hillsdale, MD: Erlbaum.
Williams MD, Hollan JD (1981) The process of retrieval from very long-term memory. *Cogn Sci* **5**:87-119.

Chapter 12

Evolutionary and Developmental Aspects of Intersubjectivity

Peter Gärdenfors

1. Intersubjectivity

Humans have a unique capacity to share and understand others' mental states. The mental states involve not only beliefs, but all forms of consciousness such as emotions, desires, attentional foci and intentions. Representing the mental states of others will here be called *intersubjectivity* (Stern 1985). In the philosophical and psychological debate, the capacity is commonly known as having a "theory of mind" (Premack and Woodruff, 1978). I want to avoid that term since it often presumes that one can understand the beliefs of others, something which, on my account, is but one aspect of intersubjectivity. In this chapter, I shall compare phylogenetic and ontogenetic aspects of intersubjectivity. For this purpose, intersubjectivity will be decomposed into five capacities. In Gärdenfors (2001, 2003), I discussed four: representing the emotions of others (empathy), representing the attention of others, representing the intentions of others and representing the beliefs and knowledge of others (theory of mind). To these I now want to add one more: representing the desires of others. From the analysis of these five factors it will be clear that humans exhibit more intersubjectivity than other animals. In particular, we have a well-developed competence for representing the beliefs of others, but we also excel at forming joint intentions (Tomasello *et al*, 2005).

By studying the development of human children, we have learnt a great deal about when and how the different forms of intersubjectivity emerge. Similarly, from studies of other animal species, in particular primates, a rough picture of the phylogeny of intersubjectivity can be drawn. In this chapter, I shall, in addition to the ontogenetic and phylogenetic themes, speculate about the evolution of the uniquely human forms of intersubjectivity, more precisely about what have been the evolutionary forces that have generated the advanced levels in humans.

Consciousness Transitions: Phylogenetic, Ontogenetic, and Physiological Aspects
Edited by Hans Liljenström and Peter Århem

2. Sensations, perceptions and imaginations

Intersubjectivity is a form of consciousness – the awareness of the consciousness of others. Before discussing the different forms of intersubjectivity, I shall introduce some conceptual tools for studying subjectivity, the most immediate form of consciousness, in different ways. I shall divide the different ways that animals (and humans) represent the world around them and its possibilities into:

(1) *Sensations* that are the immediate sensory impressions.

(2) *Perceptions* that are *interpreted* sensory impressions. Perceptions form one category of representations.

(3) *Imaginations* (or images) that are not directly governed by sensory impressions. I will also call them *detached representations*.

On the first level, consciousness contains sensations. Our subjective world of experiences is full of such experiences: tastes, smells, colours, itches, pains, sensations of cold, sounds, etc (what philosophers of mind call *qualia*). They provide an *awareness of the world*. Humphrey (1993) argues that the biological role of sensations is to tell about what is *happening right now* to the organism.

The sensations often involve an (unconscious) evaluation of whether what is happening is good or bad. This is sometimes referred to as "raw feels." They facilitate a decision about what should be done and thus prepare the organism to act. These decision processes do not by any means have to be conscious. Sensations are often closely aligned with basic choice of action – eat, fight, flee or mate.

But an organism that can not only find out what is happening to its own body but also receives signals about *what is going on out there in the world* has better opportunities to foresee the future and thus survive in an inhospitable world. This is the role of the perceptions. First and foremost, perceptions provide information about the spatial structure of the world and what objects are in it.

The brain is full of mechanisms that contribute new information. In particular, there are many well-studied examples of the visual process. When we see an object, we sense that it has contours, for example. But if we examine the influx of light that hits the retina, we find nothing that corresponds to such contours – they are part of the information that the visual process *constructs*.

Phenomena of this kind are very common and well known from psychophysics. They show that we have a plethora of mechanisms that complement the signals provided by our senses. These mechanisms create the *representations* with which thinking works, since what we experience is not only that which is presented by our sensory receptors but also that which is recreated, that is, represented. The filled-in sensations are what I call perceptions.

It is clear that our senses don't give us information about everything that exists in the world. The evolutionary point is that the richer representations help us *predict* what the world will look like. They create the future in advance. The weaker the sensations, the more important it is for the brain to create its own perceptions. In fact, we are so dependent on perceptions that if someone prevents us from having sensations, we soon begin to hallucinate. Sjölander (1984), pp. 43-57, writes that perceptions can be seen as controlled hallucinations — the content comes from within, but is guided by sensations. Dreams are in a like category, where sensations are similarly turned off. Dreams do not have to be connected to sensations, and hence they become less controlled.

It does not matter much if the information that is added in perceptions has no direct counterpart in the surrounding world as long as the perceptions produce the right results, that is, lead to appropriate actions. In particular, the perceptions become more efficient by exploiting "hidden variables." For example, when an organism observes a moving object, its sensations consist only of the positions of the object, while the *forces* that influence the movement of the object are not sensed. However, if the organism has been able to extract "force" as a hidden variable and relates this to the sensations via something like Newton's second law, then the system would be able to make more efficient and general, if not more accurate, predictions.

It can be argued that much of our reasoning about *causes* depends on this kind of additions (Gärdenfors, 2003). Interestingly enough, there is a marked difference between humans and other animals in their capacities for causal reasoning. It seems that humans are considerably more skilled at perceiving physical causality. In particular, recent experiments indicate that monkeys and apes have great difficulties understanding the causes of physical events and of mental events (see e.g. Povinelli 2000). Tomasello (1999, p. 19) provides the following explanation for why primates cannot understand causal mechanisms and intentionality in others: "It is just that they do not view the world in terms of intermediate and often hidden "forces," the underlying causes and intentional/mental states, that are so important in human thinking."

In contrast, even very small human children show strong signs of interpreting the world with the aid of hidden forces and other causal variables. Gopnik (1998, p. 104) claims that "other animals primarily understand causality in terms of the effects of their own actions on the world. In contrast, human beings combine that understanding with a view that equates the causal power of their own actions and those of objects independent of them."

My interpretation of this crucial difference between humans and other primates is that the perceptions we use to understand the physical world are much better at using hidden variables to make predictions about the future. The causes are not part of our sensations, but our perceptions fill them in. In this way we *perceive* the causes. These causal fillings produce in humans' inner worlds a much more fully formed toolbox for reasoning about the future than exists in other apes.

Apart from physical forces, the perceptions sometimes also involve *mental "forces"* – the emotions, beliefs, desires and intentions – that govern the behaviour of others. The perception of these forces forms the basis for intersubjectivity and they will be the focus of this chapter.

There is nothing that requires that there is any feedback from the senses to make the filling-in mechanisms start working. Actually, it is only an extra step for the evolutionary process to let the mechanisms work without any signals being sent to the body at all. This is what happens when the organism creates for itself an image of what will happen if a certain action is undertaken. Animals' brains, especially humans', utilize *imaginations* of this kind – representations of things that we can move around in our heads before trying to move the things in reality.

The founding father of psychology William James writes: "The commonly received idea [about imagination] is that it is only a milder degree of the same process which took place when the thing now imagined was sensibly perceived" (1890, p. 68). Although James did not have much evidence for this position, there is now quite strong support from brain research that the same mechanisms are involved in controlling the body as in imagining controlling it. The so-called motor imagining ability has been investigated thoroughly by Jeannerod (1994, 2006) and others. It turns out that the same parts of the motor cortex are involved when an action is carried out as when one only imagines the action.

Even simple organisms have sensations – sensory impressions that report what is happening with the body. More advanced animals have perceptions that consist of interpretations of sensations. In particular, perceptions can provide knowledge about what is going on in the animal's immediate surroundings. It is probably only mammals and birds that have imaginations – in their inner world they can direct their thinking towards something that is not immediately present in the surrounding environment. I submit that this is the evolutionary order in which the different functions appear.

3. The inner world

Perceptions and imagination are two types of representations that the brain avails itself of. I call perceptions *cued* representation that stands for something that is present in the surrounding situation (Gärdenfors, 1996, 2003). Imaginations, on the other hand, are here called *detached* representation: something the individual can utilize regardless of whether what it represents is present or not. A detached representation can even stand for something that does not exist at all. For example, our imaginative worlds are full of centaurs, unicorns, elves and trolls – about which we easily communicate although they do not truly correspond to any sensory impressions we have received.

Being able to use an imagination requires that one can *suppress* the sensations one has for the moment; otherwise they will come into conflict with the representation. Glenberg (1997) says that imaginations put reality in quarantine. That places new demands on mental capacities. The suppression of information coming in from reality is probably managed by the frontal lobe of the brain, which is the part that is in charge of planning and fantasizing and the so-called "executive functions" of self-control (Hughes, Russell and Robbins, 1994).

Glenberg (1997) presents a theory of how memory works in which he distinguishes between "automatic" and "effortful" memory. The automatic memory is the one that is used to fill in sensations so that they become perceptions. When I recognise a person in a crowd it is because I blend what I see of the person with my memories. Sometimes there is a mistake – I overlay my memory of one person on top of the sensory impression I receive of another person.

The effortful memory is the one we use when we create images. Images do not appear from nowhere – they build on our previous experiences. What we usually call remembering is only a special kind of image that we think corresponds to something that has actually happened. But memory is also used in fantasies: you cannot imagine a centaur (if you have not seen a picture of one) without memories of horses and people. But images must not be confused with perceptions (except when we pretend (Leslie 1987)). An effort is required to ignore sensations. That is why we often close our eyes or look up at the ceiling when we want to remember something or when we fantasize. This effort can be investigated since, for example, it is more difficult to carry out some action at the same time as we are remembering or imagining.

The collection of perceptions and imaginations I call the *inner world* of an organism (Gärdenfors, 1996). The existence of an inner world has clear evolutionary advantages as was pointed out already by Craik (1943, p. 61):

> "If the organism carries a "small-scale model" of external reality and of its own possible actions within its head, it is able to try out various alternatives, conclude which are the best of them, react to future situations before they arise, utilize the knowledge of past events in dealing with the present and future, and in every way to react on a much fuller, safer and more competent manner to the emergencies which face it."

Under the heading of the "inner environment" this kind of "small-scale model" has been discussed by Dennett (1978). Such an environment is necessary for representing objects (like food and predators), places (where food or shelter can be found), actions (and their consequences), etc, even when these things are not perceptually present. The evolution of such a representational power will clearly increase the fitness of the animal.

It seems that many animal species have inner worlds. For example, the searching behaviour of rats is best explained if it is assumed that they have some form of "spatial maps" in their heads. Evidence for this, based on their abilities to find optimal paths in mazes, was collected by Tolman already in the 1930's (Tolman, 1948). However, his results were swept under the carpet for many years since they were clear anomalies for the behaviourist paradigm. It should be noted I am not assuming that the animal is *aware of* its inner world, nor of the processes utilizing this construct.

One of the main evolutionary advantages of an inner world is that it frees an animal that is seeking a solution to a problem from dangerous trial-and-error behaviour. Jeannerod (1994) says that "actions are driven by an internally represented goal rather than directly by the external world." By exploiting its inner world, the animal can *simulate* a number of different actions in order to "see" their consequences and evaluate them (also compare Grush (1998, 2004), Barsalou (1999), Hesslow (2002), Gärdenfors (2003)). After these simulations, it can choose the most appropriate action to perform in the outer environment.

4. Components of intersubjectivity

A necessary prerequisite for an animal (or a human) to entertain any form of intersubjectivity is that it has an inner world. The crucial issue is whether an individual has any representation of other individuals' inner worlds.

The question of whether an animal or a child exhibits intersubjectivity does not have a simple yes or no answer. I will distinguish between five components of intersubjectivity in order to understand the cognitive capacities of animals and children at different ages.

(1) Representing the *emotions* of others. At this level one can, for example, understand that someone else is in pain. This is what is usually meant by *empathy*. Even though one can understand others' emotions, it does not imply that one understands what they believe or want.

(2) Representing the *desires* of others. This capacity involves understanding, for example, that others don't like the same things as you do.

(3) Representing the *attention* of others. This means that one can understand, for example, what someone else is looking at. However, this ability does not presuppose any conception of other parts of their inner world.

(4) Representing the *intentions* of others. This capacity involves, above all, being able to understand the objective that may lie behind another individual's behaviour.

(5) Representing the *beliefs* and *knowledge* of others. This means, among other things, that one can understand that others don't know the same things as you do.

I have not found any other researcher who grades in this way what is involved in intersubjectivity. Tomasello (1999, p. 179) distinguishes between three levels of a child's development: seeing others as agents; seeing others as intentional agents; and seeing others as mental agents. His second and third levels correspond roughly to my capacities 4 and 5. Proust (1998) presents a longer list of capacities that are part of intersubjectivity, but she does not sort them into stages or discuss how they are related.

Baron-Cohen (1994) proposes that "mindreading" has four components: an intentionality detector (ID); an eye direction detector (EDD); a shared attention mechanism (SAM); and a theory of mind mechanism (ToMM). His ID does not primarily involve representing the intention of others, but it is used to detect whether something moves intentionally (cf. Premack, 1996) which is what Tomasello *et al.* (2005) call acting animately (see below). This component is thus much weaker than my (4). The EDD is a mechanism for following the gaze of others, but does not directly involve representing what others attend to, albeit an EDD is necessary for my (3). The SAM corresponds to what I will call joint attention below, which is a development of my (3). However, the final component ToMM seems to map quite well onto my (5). In sum, there are partial overlaps between Baron-Cohen's account and mine, but he does not include representing the emotions of others. Taken together with the other differences, my account of the different forms of intersubjectivity is arguably more complete than his.[1]

Wellman and Liu (2004) write about a "scaling of theory-of-mind tasks" where they address the "sequence of understandings" in children's developing intersubjectivity. They make a meta-analysis of studies of children's understandings of different forms of intentions, emotions, desires and knowledge. There is thus a considerable overlap with the capacities presented here. They also identify a partial developmental sequence of these understandings. For example, children's understanding of desires seem to precede their understanding of beliefs (e.g. Flavell *et al*, 1990).

Another account that comes close to mine is Stern's (1985) theory of infant development where he discusses intersubjective behaviour as interaffective, interattentional, or interintentional. What is missing from his list is, most importantly, that sharing of beliefs is also a form of intersubjectivity. He defines the three categories as follows: (a) Interaffectivity consists in the infant's matching its own "feeling state as experienced within" with the feeling state "seen 'on' or 'in' another" (1985, p. 132), and he says that still by twelve months, affective exchange is "the predominant mode and substance of communications" (p. 133). It supports the sharing and exchange of affects between sender and observer, and underlies social referencing. (b) Interattentionality means that the infant has "some sense that" persons, including the infant itself, can have individual, different, attentional foci, which can be "brought into alignment and shared" (p. 130). It concerns the exchange of attentions and it can result in mutual interaction (c) Interintentionality "impl[ies] that the infant attributes an internal, mental state" to the adult, namely, "comprehension of the infant's intention and the capacity to intend to satisfy that intention" (p. 131). It prepares for the sharing of intentions, which depends on having states of the mind of different orders that are nested. Brinck (2006) notices that Stern uses mentalistic terms in his definitions and she presents a redefinition of the capacities that does not use such terms.

In order to determine whether an animal (or a child) exhibits a certain form of intersubjectivity, the crucial issue is whether an individual has any representation of other individuals' inner worlds. In the discussions of the intersubjectivity of various apes (for it is mostly about them), ethologists and animal psychologists apply a form of Occam's razor. This means that, for a given cognitive capacity, an animal is not assumed to have it, unless the general behaviour of the animal *cannot* be explained adequately without presuming the capacity. As regards the existence of intersubjectivity, Occam's razor becomes the question of whether their behaviour can be explained solely from the assumption that the animals have an inner world, without assuming that they have some conception of others' inner worlds. Unfortunately, the experiments in this research field have led to a tangle of interpretations from which it is difficult to draw any unambiguous conclusions.

5. Representing emotions

Since we have so much biological history in common with the monkeys and apes, their emotional expressions are obvious to us and it is easy for us to feel sympathy for them. This makes it very difficult to understand how Descartes could claim that animals are just automata; sophisticated automata, admittedly, but with no emotions or moral status. Darwin's theory of evolution, however, changed the status of animals: according to him, animals are basically of the same essence as humans and there is no longer any question that many species, including all mammals and birds, have rich forms of consciousness.

Before Darwin, philosophers regarded it as a category mistake to talk of animals' emotions – it was like saying that a clock is happy or in pain. After Darwin, however, it became a possible field of scientific inquiry, Darwin himself leading the way with his studies of the emotional expressions of apes. In the apes, emotions are often signalled by highly distinct bodily expressions, above all in the face, but also by emitting sounds. Many of these involuntary reactions are still found in humans – it is hard to suppress a cry of surprise or a laugh.

The ability to represent the emotions of others is often called *empathy*. The evolutionary explanation seems to be that the empathy leads to greater solidarity within the group. This reduces the risk of violent conflicts in which individuals thoughtlessly injure one another. The resulting solidarity increases the chances of survival for the individuals in the group.

Bodily expressions of emotions have a communicative purpose. The expressions are most obvious among social animals. Preston and de Waal (2002) have argued that as a mechanism involving the linkage of perception and action, a basic form of empathy is available to most, if not all mammal species. Defining empathy as "any process where the attended perception of the object's state generates a state in the subject that is more applicable to the object's state or situation than to the subject's own prior state or situation" (*ibid.*, p. 4), they see a clear evolutionary motivation for its emergence in the ability to recognize and understand the behaviour of con-specifics. It is characteristic that facial signals, such as the "play-face" expressions of chimpanzees and gorillas (a likely evolutionary precursor to the human smile) carry emotional rather than referential meaning.

As a more advanced understanding of the other's thoughts evolves, the bodily expressions of emotions become less important as signals. Deacon (1997) goes so far as to claim that the evolution of language may have blocked the communicative function of the emotions, since the linguistic message competes with that expressed by the body.

There are several experimental paradigms for studying the empathy of animals. An experiment with rhesus monkeys from Masserman *et al* (1964) is typical. In the experiment, a monkey was trained to choose between two chains: one that delivered large amounts of food and one that delivered half as much. The conditions were then altered so that pulling the chain the gives the larger reward caused another monkey in sight of the first to receive and electric shock. After the subjects saw the shock of the conspecific, a large majority preferred the chain that did not give the other ape a shock, even though it resulted in considerably less reward. Of the remaining monkeys, one stopped pulling the chains altogether for 5 days and another for 12 days after witnessing the shock of the other monkey. These monkeys were starving themselves to prevent the shock to the conspecific, which is a clear indication that they understand the emotions of others.

In many species one also finds examples of *shared emotions* that involve combined experiences of the form "I feel that you feel" and "you feel that I feel". The clearest example is perhaps collective panic, where I am afraid because you are afraid and you are afraid because I am afraid. There are also cases of contagious emotions, like when everybody starts laughing just because everybody else is laughing. Shared emotions are obviously useful a useful basis for collaborative actions, but they are not sufficient for intentional cooperation (see Section 11). An even more advanced form would be *joint emotions* based on even higher order emotions such as "I feel that you feel that I feel", etc. This form presumably occurs among humans, but I know of no evidence from other species.

There is a more advanced form of representing the emotions of others that Preston and de Waal (2002, p. 18) call *cognitive empathy*.[2] This requires a differentiation between subject and object where "the subject is thought to use perspective-taking processes to imagine or project into the place of the object." To be able to feel empathy, you have to be able to distinguish your own feelings from those of others. Cognitive empathy develops later in children, at about the same time they learn to represent the belief and knowledge of others.

Research concerning cognitive empathy in apes has focused on their *consolation* behaviour, which is well attested in apes (de Waal and Aureli, 1997), but has not been found in monkeys or any other mammal species. Consolation is cognitively more complex than simple empathy since the consoling individual not only feels that somebody else experiences a negative emotion, but also intends to help relieve this, implying an ability to imagine the more positive emotional state in the other individual.

Presumably empathy functions as follows: the sensory impressions that the other's bodily expressions arouse in me provoke a corresponding emotion in me as well. One can then, of course, envisage a more advanced stage at which one also creates an idea of the other's feelings. But the capacity for cognitive empathy is an independent stage that means a further step in the expansion of the inner world.

Not all people are good at empathy. One of the central criteria of a *psychopath* is that he (or, in rare cases, she) finds it difficult to feel empathy. Psychopaths do not have the normal blockage against hurting or tormenting other people. Nor can they understand the difference between the rules of behaviour that exist for conventional reasons and those which are intended to prevent us from harming other people.

6. Representing desires

Emotions concern the inner state of an individual, without reference to an external object. A desire is a positive attitude towards some external object or event. Understanding that somebody has a desire for something therefore involves more than understanding emotions.

Representing the desires of somebody else involves understanding that the other may not like the same things as you do. A typical test of this capacity (Wellman and Liu 2004) involves a toy character and pictures of a carrot and a cookie. The child is first asked whether he or she would like the carrot of the cookie best. If the child answers cookie, then he or she is told that the toy character really likes the carrot best (or vice versa if the child chooses the carrot). Then the child is told that the toy character can only choose one snack and is asked which snack the character would choose. Similar tasks have been performed by e.g. Flavell *et al* (1990). The results suggest that 3-year-olds can understand that others have different desires than they have themselves. As we shall see below, this occurs earlier in children's development than understanding the beliefs of others.

For another type of example, note that an act of *teasing* involves understanding that the human or animal that is being teased does not like the activity and desires that it stops. Children begin teasing already during their first year (Reddy, 1991) and in the animal world one finds teasing among mammals and birds, for example crows harassing birds of prey (Bekoff and Byers, 1998). This means that representing desires occurs at a fairly early stage of the ontogenetic or phylogenetic development of intersubjectivity.

It should be noted that understanding the desires of others does not require them to represent the belief of the others. Thus, thinking that "A wants X" does not require thinking that "A believes that X is most desirable". Thus, desires need not be "value beliefs" as they are called by Flavell *et al.* (1990). Similarly, ascribing a desire to somebody does not entail that one ascribes an intention. Thus, thinking "A wants X" does not require that one thinks "A intends to choose X."

7. Representing attention

Humans are very good at following the *direction* of other people's gazes (Emery 2000). Even very young children have the ability to understand where other people are directing their attention. In an experiment, mothers were asked to sit opposite their infants and look in different directions (sometimes the mother did this while turning her head, sometimes not) and then the observers followed where the children aimed their gaze. At the age of just six months a child can follow its mother's gaze if she turns her head at the same time, and at the age of twelve months it can manage this even if she just moves her pupils. Such young children, however, can only follow their mother's gaze if it falls within their own field of vision. It is not until the age of eighteen months that that they can turn to find a point behind them that the mother is looking at. This task is more difficult since it requires that the child can understand that space also continues beyond their field of vision. This ability calls for a detached representation of space in the inner world.

Chimpanzees are also good at understanding where others are looking. Povinelli and Eddy (1996) tested putting a human opposite a chimpanzee with a transparent screen between them. The apes had no trouble in looking at the spot to which the experimenter was directing his gaze, even if it was behind the chimpanzee. In a variant of the experiment, a screen with an opaque lower half was set up between the human and the chimpanzee. When the experimenter now looked at a point on the opaque part of the screen, the ape would lean forward to try to see what was on the other side. This shows that it is not just the direction of the gaze that the apes follow, but also that they understand that the gaze is directed towards a certain point in the surroundings. Apart from chimpanzees, the capacity to follow gazes is well attested for several species of primates, for dogs and for goats (Kaminski *et al*, 2005). In brief, apes and other animals *see that others see* the object. This is called second-order attention: I notice that you notice.

A more advanced form of behaviour is *making* someone else notice an object. Gómez (1994) tells of a young gorilla at the Madrid zoo that wanted to open a door in a cage that was closed by means of a catch at the top. The gorilla's first strategy was to use the person in the cage as an object: the gorilla nudged the person towards the door and then tried to climb up on him to reach the catch. At the age of eighteen months, however, the gorilla began to use a different strategy: she took the man by the hand, led him to the door while alternating her gaze between the man's eyes and the catch. The gorilla wants to succeed in getting the man's attention focused on the catch as well. The gorilla's behaviour was intentionally planned, but it does not presuppose that she has any theory of the human mind. It is sufficient that she has learned that there is a link between a person's attention being directed towards a locked door and the actions that humans usually perform when they encounter locked doors.

When two individuals become aware that the other is attending to the same object, what results is *shared attention*. Such second-order attention is well testified among great apes (Hare *et al*, 2000). This comes a good deal towards the construction of a "consensual reality" that can be communicated about, but does not quite reach it (cf. Brinck 2004). To make a given object X fully intersubjective between you and me, I would need not only to "see that you see X", (second-order attention, see Fig. 12.1), but also "to see that you see that I see X" (third-order attention, see Fig. 12.2) and vice versa – which is what it means to engage in *joint attention* (see e.g. Zlatev, Persson and Gärdenfors, 2005). Joint attention appears to be beyond the capacities of apes (Tomasello, 1999).

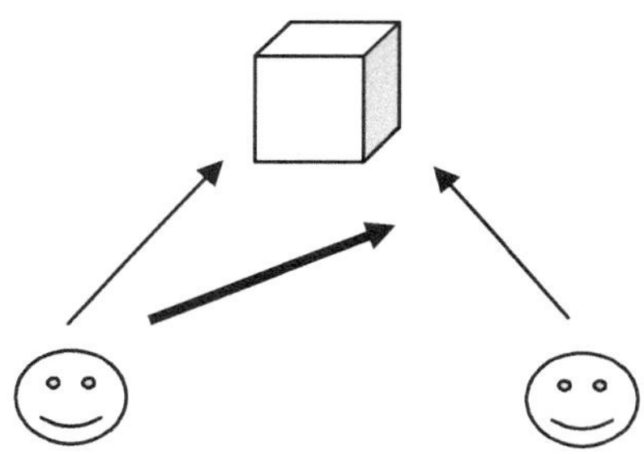

Fig. 12.1 *Shared attention*: Second-order attention: "I see that you see X" (and vice versa). (From Zlatev *et al*, 2005).

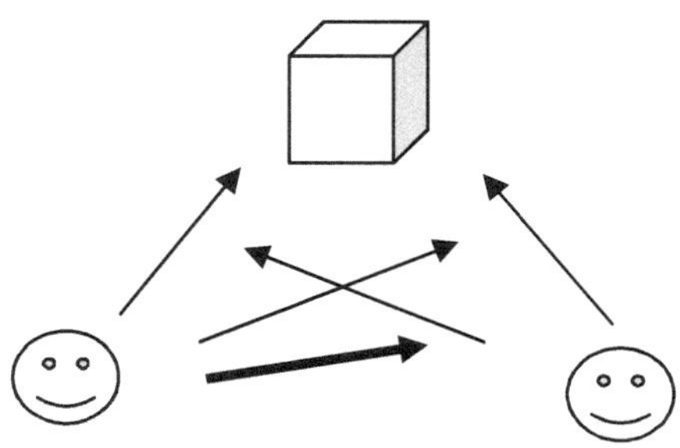

Fig. 12.2 *Joint attention*: Third-order attention: "I see that you see that I see X" (and vice versa). (From Zlatev *et al*, 2005).

To understand another individual's attention, one need only understand *what* he or she is looking at, not how he or she imagines the world. A further step towards having an idea of others' minds consists in being able to put oneself in their position to see *how they see the world.* This can be described as detaching oneself from one's own perspective. Young children are incapable of this, as many experiments have shown. A classical example is Piaget's "three mountain test" (Piaget and Inhelder, 1956). In this experiment one puts three "mountains", one of them much bigger than the other two, in a triangle on a table. The child who is to be tested sits in front of the small mountains and can see the big one behind them. A doll is placed on the other side of the table with its face towards the big mountain. The child is asked to draw what the doll can see from its side of the table. A child at the "pre-operational level" (Piaget's term) draws the scene from his own perspective, regardless of where the doll is sitting. However, a child at the "concrete operational level" (roughly from the age of seven), can imagine how the doll sees the mountains and draws the scene from the right perspective. When the child has reached this stage, then, it can *imagine* the world from different angles, regardless of the perceptions it has at a particular moment. Later experiments have shown that young children are better at imagining what others see in a certain situation if they themselves have previously been in that situation.

8. Representing intentions

Humans have a powerful inclination to look for *causes* in the world. Kant calls this disposition one of our basic categories, that is, a pattern of thought which we cannot abandon. There are good evolutionary reasons why we have to look for causes, since they help us understand why the world is the way it is. By reasoning about causes and effects, we become better at predicting the future.

The ability to distinguish phenomena caused by other agents (animals or humans) is fundamental for being able to understand intentionality in other individuals' behaviour. When we see that something is caused by an agent, our cognitive system *presumes* that there is some purpose for the act, in other words, that it is intentional. Humans thus find it very easy to create a representation of the objective of an action – we see other people's behaviour as being *goal-directed.* This is probably yet another of our fundamental categories. Mostly, of course, it is true that other people's actions are intentional, but it happens that people do things unintentionally.

Tomasello *et al* (2005) distinguish between three levels of representing intentions:

(a) Acting animately: The observer understands that an actor generates its movements autonomously. Premack (1996) has shown that even very young children are able to distinguish between effects caused by "natural" (physical) forces and those caused by "arbitrary" forces. The latter comprise movements and events caused by animals and humans. Premack presumes that different nervous systems in the brain are involved when we recognize the two types of causality.

(b) Pursuing goals: The observer understands that the actor behaves with persistence towards the goal. In the context of the present chapter, this meaning of representing intentions will be in focus. For example, in an experiment by Gergely *et al* (1995), infants were habituated to a large dot "jumping" over an obstacle and approaching a small dot. Then when the obstacle was taken away, nine months old children (but not six months old) looked more at the same jumping motion, even though its movement was the same, but they did not look more to the dot going directly to the small dot, even though this was a new motion. The argument is that they saw the large dot move in the same goal-directed way towards the small dot.

(c) Choosing plans: The observer understands that the actor considers and chooses between action plans to reach the goal. In an experiment, Gergely *et al.* (2002) showed 14-month-old children an adult touching the top of a box with its head to turn on a light. For half of the infants, the adult was shivering and holding a blanket around his shoulders so his hands were occupied. For the other half, the adult's hands were free during the action. Then when it was the children's turn, those who saw the demonstration with the hands free bent over and touched the box with their heads more often than the infants who saw the demonstration with the hands occupied. It seems that the children assumed that if the adult's hands were free and he still chose to use his head, he turned on the light with his head for some *reason*. In contrast, if the adult's hands were occupied, then the use of the head was interpreted as a convenient solution and in this case they could choose a simpler way of turning on the light.

It is important to notice that, even though one can interpret someone else's behaviour as goal-directed, this does not necessarily mean that one has any conception of the other's beliefs and knowledge. It is sufficient that one creates for oneself a *representation of the goal* of the action. The ability to understand intentions thus requires a less advanced inner world than the ability to represent the beliefs of others. When one understands that someone else has intentions, one sees that individual as an *agent*, but when one has an understanding of the other's beliefs and knowledge one sees that individual as a *subject*.

When it comes to the question of whether apes have any theory of intentionality, opinions are divided. Premack and Woodruff (1978) had noted that Sarah, one of their chimpanzees, was fond of watching commercials on television, which led to the thought that she understood something of the desires and objectives of the actors. In an experiment they let her watch video films in which a person had to solve various types of problem, such as getting out of a locked cage. The film was stopped in the problematic situation and Sarah had to choose between two pictures, one of them showing a solution to the problem, for example, a key, and the other showing something that provided no solution to the problem. She always chose the right picture. Premack and Woodruff interpreted her success at this task as indicating that she understood the *objectives* of the human in the filmed situation. Other scientists have criticized these experiments, however, and it is conceivable that Sarah merely made the right *associations* between the pictures and the problematic situations without having any grasp of the human's objectives.

Tomasello *et al* (2005) discuss the extent to which infants and apes can represent the intentions of others. One type of experiment involves an adult either holding out a toy in a teasing fashion or holding out the toy out but dropping it (seemingly) accidentally. The result was that nine months old infants, but not six months old, showed more impatience when the adult was teasing them than when he was simply dropping the toy, indicating that the infant could understand the difference in intention.

A similar experiment was tested on chimpanzees where a human began giving food to an ape through a hole in a transparent wall, sometimes refusing to give it to the ape and sometimes attempting to give it to the ape but failing. The chimpanzees gestured more and left the area earlier when the human was unwilling than when he was unable. In the latter case they tended to wait patiently throughout the unsuccessful attempts. The interpretation of the results is that the chimpanzees understood the behaviour of the human who "failed" as an intention to give them food.

This experiment has been interpreted as indicating that chimpanzees have some representation of the intentions of others. A limitation is that the experiment only concerns a basic test of whether an action is intentional or not and it does not show that the apes represent the *contents* of a specific intention. In my opinion, we have not yet found an adequate methodology for testing whether apes can understand the intentions of others, so we will have to wait for the development of new experimental paradigms before this can be decided.

Tomasello *et al* (2005) strongly emphasize the role of *sharing intentions*. They write that "the crucial difference between human cognition and that of other species is the ability to participate with others in collaborative activities with shared goals and intentions: shared intentionality" (2005, p. 675). At around nine to twelve months of age, infants begin to understand other individuals as intentional and then they can begin to interact with them in activities with shared goals. For example, an adult and an infant can create a shared goal to build a tower of blocks together.

An even stronger form of intersubjectivity is to achieve *joint intention*. This involves that a child can understand the plans of somebody else and coordinate its own intention with the goals of the other. This involves that "I intend that you intend" and that "you intend that I intend" and that both are aware of these second-order intentions. When one can coordinate roles in working towards a goal, then joint intention is achieved. For example, in building the tower of blocks the child may understand that the adult holds the tower steady while the child places new blocks. In human children, joint intentions seem to develop between twelve and fifteen months of age.

9. Representing beliefs and knowledge

The most advanced test of the intersubjectivity of animals is to find out whether they understand what others *believe* or *know*. Unfortunately, this is not easy to determine. A field observation of baboons in Ethiopia is interesting as an example (Fig. 12.3). An adult female spent twenty minutes slowly moving a few metres to a position behind a stone half a metre high. Behind this stone there was a younger male from the group whom she began to groom. This behaviour would not have been tolerated by the dominant male, if he had been able to see it. From his position, he could just see the head and back of the female, not the younger male and the fact that she was grooming him.

Fig. 12.3 The female baboon represents the attention of the alpha male, but does she represent what he believes? (From Gärdenfors, 2003).

The factor that makes it possible to interpret the female's behaviour as a desire to *deceive* the leader is her careful movement. It can be described as showing that she did not want the leader to *know* that there was another baboon behind the stone. This interpretation would ascribe a representation of the other's beliefs to the female baboon, which is a strong assumption. A weaker interpretation would be that she did not want the leader to *notice* that there was another baboon behind the stone. This interpretation presupposes only that she can represent someone else's attention. But she must also have understood that the stone was blocking the leader's vision. We will never know what went on in the female's head in this situation.

Tomasello and Call (2006) review the experimental evidence concerning whether chimpanzees know what others have seen. The most interesting results come from a series of experiments by Hare *et al.* (2000, 2001) where a subordinate and a dominant ape were placed into rooms on opposite sides of a third room. Each of them could see food placed in the open or behind barriers in the middle room. The problem for the subordinate is that the dominant will take all food she can see. Sometimes the subordinate could see one piece of food on the subordinate's side of a barrier that the dominant could not see. It was found that the subordinate went for the food that only they could see much more often than the food they could both see.

In another experiment the door to the dominant's room was sometimes open and sometimes closed so that she could not see when the food was hidden. In this case the subordinate more frequently went for the food the dominant had not seen hidden. Furthermore, in a condition where the dominant first saw the food hidden and the food was then transferred to the other barrier when the dominant's door was closed, the subordinate exploited this situation and went more often for the place where the food was now, but where the dominant did not know it was. Tomasello and Call (2006, p. 375) say that "this indicates that they know not only what others can and cannot see at the moments, but also what others have seen in the immediate past." This phenomenon can be phrased as "seeing is knowing" and it entails that the chimpanzees have some limited representation of the beliefs and knowledge of others.

It is easier to test whether young children can understand that "seeing is knowing", since one can communicate with them through language from a fairly early age. In one experiment, children aged between three and five were asked whether another person knew what was in a box. Some of the people had looked inside the box while the others had not. In the same way, some of the children were allowed to see what was in the box while the others were not. None of the children had any problem saying whether a particular person had looked in the box or not. But when the question concerned whether the person knew what was in the box, the three-year-olds consistently replied that the person knew if they themselves had seen what was in the box, and that the person did not know if they had not seen what was in the box. Whether the person had looked or not made no difference. The four- and five-year-olds, on the other hand, were able to connect "seeing" with "knowing". This is one of several experiments, which suggest that a representation of others' knowledge develops in humans at the age of about four.

Another type of tests of children's understanding of other people's beliefs concern whether they can understand that someone else has a *false* belief about what the world is like (see e.g. Perner *et al*, 1987, Gopnik and Astington, 1988, Mitchell, 1997). The best known of these experiments was about an oblong Smarties tube. The test was carried out on children aged between three and five. The children are first shown the tube and then asked what they think is in it. All the children reply "Smarties" (or "sweets"). When the tube is opened it is found to contain a pencil. Then the tube is closed. The children are now asked what a friend, who has not yet seen what is in the tube, will say that it contains. The three-year-olds generally answer "pencil" whereas most of the older children say "Smarties". The older children understand that the friend *does not have the same knowledge* as they do. They thus realize that the friend has a false belief about what is in the tube. This is a clear example that they represent the belief of others. The younger children, on the other hand, do not appear to be able to distinguish between their own beliefs and other people's (Perner *et al.* 1987). In a nonverbal version of the test (Call and Tomasello, 1999), children performed as well as in the verbal form, but all apes that were tested failed the test.

In an extension of the experiment, the children were also asked what they thought was in the tube when they had first seen it. Most young children replied "pen" while the oldest ones said "Smarties". The older ones could remember, that is, *create a representation* of having originally thought differently. Regardless of age, there was a strong correlation between the children who answered "pencil" when asked what the friend would say was in the tube and those who answered "pencil" about what they first thought was in the tube.

Wellman and Liu (2004) argue that children can understand other persons' "diverse beliefs" before they can judge false beliefs. They test the understanding of diverse beliefs by presenting a child with a toy girl and a drawing with bushes and a garage on it. Then the children are told the following: "Here's Linda. Linda wants to find her cat. Her cat might be hiding in the bushes or it might be hiding in the garage. Where do you think the cat is? In the bushes or in the garage?" This is called the own-belief question. If the child chooses the bushes it is told the following: "Well, that's a good idea but Linda thinks her cat is in the garage." Then the child is asked the target question: "So where will Linda look for her cat? In the bushes or in the garage?" (ibid., p. 538). Wellman and Liu found that many 3-year-olds who cannot manage the false belief test still answer the target question opposite from the own-belief question, which suggests that they understand that people have diverse beliefs that influence their actions.

Humans cannot only know that someone else know, that is have second-order knowledge, but they can also have higher orders of knowledge and belief, such as "I know that you know that I believe that there will be frost tonight." This capacity forms the basis for *joint beliefs*, which is often called *common knowledge.* Joint beliefs open up for many new forms of collaboration. For example, in human societies many forms of cooperation are based on *conventions*. The important point is that conventions presume joint beliefs. For example, if two cars meet on a gravel road in Uganda, then both drivers know that this co-ordination problem has been solved by driving on the left hand side numerous times before, both know that both know this, both know that both know that both know this, etc, and they then both drive on their left without any hesitation (Lewis, 1969).

In general, joint beliefs form the basis for much of human culture. They make many new forms of cooperation possible. For example, to promise something only means that you intend to do it. On the other hand, when you *commit* yourself to a second person to do an action, you intend to do it, the other person wants you to do it and intends to check that you do it, and there is joint belief concerning these intentions and desires (Dunin-Kepliz and Verbrugge, 2001). Unlike promises, commitments can thus not arise unless the agents achieve joint beliefs. For similar reasons, *contracts* cannot be established without joint beliefs.

10. Coevolution of intersubjectivity and cooperation

In the previous sections, five components of intersubjectivity have been presented. I consider that these capacities make up a series of layers in the cognitive development both from a phylogenetic and an ontogenetic point of view. Phylogenetically, one can detect a funnelling of species as one goes up the levels, even if the evidence in many cases is unclear and incomplete. Starting from the first level, empathy seems occur in all mammals (Preston and de Waal, 2002) and maybe in some further species. If teasing is a criterion for understanding the desires of others, we find this in mammals and birds. Understanding the attention of others in the form of gaze following has been shown in all primates that have been studied and in canines and goats. When we come to representing the intention of others, we seem to be confined to the apes. Finally, chimpanzees may have a limited understanding of "seeing is knowing" but they are far from the understanding of the belief and knowledge of others that an adult human exhibits.

From an ontogenetic perspective, children seem to represent the emotions of other before they are six months old. Again taking teasing a criterion for representing the desires of others, children seem to develop this capacity during their first year. As regards attention, they start following the gazes of others when they are about six months old, but the capacity develops over the first eighteen months. Then children learn to react to the intentions of others between when they are between nin and fourteen months old. Finally, they do not seem to understand the beliefs of others until they are about four years old, as evidenced by the false belief tasks. In brief, both the phylogenetic and the ontogenetic evidence support that the levels of intersubjectivity should be ordered in the way I have presented them here.

Giving a third perspective on intersubjectivity, I shall in this section focus on the *evolution* of the uniquely human forms of intersubjectivity. The key evolutionary question is what have been the selective forces that have given humans the capacity for joint attention and for representing the beliefs of others. In their "phylogenetic hypothesis," Tomasello *et al* (2005) propose that "humans evolved skills and motivations for collaborating with one another in various social activities involving shared goals and joint intentions/attention." I agree that increased collaboration is a key factor, but Tomasello *et al* do not present any substantial evolutionary arguments for their position (Bickerton, 2005). Saying that "some change in the ecology of *Homo* made it more likely that entire groups with many collaborators outcompeted other groups with fewer collaborators" (Tomasello *et al*, 2005, p. 687) is far from enough.

In the discussion on intersubjectivity, or "theory of mind," in ethology and animal psychology, the focus has been on animals' and humans' capacity for *deception*. This has also been dubbed the Machiavellian intelligence of animals (Byrne and Whiten, 1988, Whiten and Byrne, 1997). It has been suggested that one of the main reasons why hominids' brains grew so large is that deception leads to an "arms race" in reading one another's minds (Dennett, 1988): If the one who is deceived can read the deceiver's intentions, he can take counter-measures; but if the one who wants to deceive understands that the victim has taken counter-measures, he can choose a more sophisticated strategy, and so on. It is argued that this escalation of deceptive strategies and counter-moves creates a long-term evolutionary pressure that makes individuals better able to interpret each other's intentions and beliefs.

In contrast, I focus on the role of *cooperation* in hominid societies as a selective factor behind the evolution of human intersubjectivity. In particular, I shall discuss what were the aspects of the hominid ecology that promoted advanced forms of collaboration and consequently an understanding of the minds of others.

Apart from well-developed intersubjectivity, human cognition is also unique in its capacity to represent the future. Many animals can plan, but they plan for their current needs. In contrast, humans can imagine themselves in a different state than they are now: we can predict that we will be hungry tomorrow and save some food, and we can imagine that the winter will be cold, so we start building a shelter already in the summer.

More specifically, Bischof (1978) and Bischof-Köhler (1985) hypothesize that animals other than humans cannot anticipate future needs or drive states (see also Gulz 1991). Gulz (1991, p. 55) calls planning for present needs *immediate planning* while planning for the future is called *anticipatory planning*. This is a special case of what has been called "mental time travelling" (Suddendorf and Corballis, 1997; Suddendorf and Busby, 2003). The crucial distinction is that for an individual to be capable of anticipatory planning it must have a detached representation of its *future needs*. In contrast, immediate planning only requires a cued representation of the current need.

In general, the cognition of other animals concerns here and now, while humans are mentally both here and in the future. The squirrel or nuthatch that is gathering and storing food for the winter is not engaged in anticipatory planning because it is not planning at all. It has no detached *representation* of the winter, let alone its needs. The gathering behaviour is an innate complex behaviour pattern that is stereotypical without sensitivity to varying circumstances (cf. Gulz, 1991, p. 62).

Until recently, evidence as regards planning in primates and other animals suggested that they only plan for present needs and thus supported the Bischof-Köhler hypothesis. New results indicate that, under experimental conditions, great apes have a rudimentary capacity to act in order to fulfil future needs (Mulcahy and Call, 2006). They are however not known to use this capacity in the wild, which implies that anticipatory cognition plays a minor or non-existing role in the life of great apes. More studies are required to determine to what extent it can be said that the apes are on the cognitive brink of anticipatory planning.

When trying to understand the evolution of the uniquely human cognition, one must identify evolutionary forces that have only applied to the hominid line. One factor that has been surprisingly neglected is the difference in the *ecology* of the early hominids and the other apes (Bickerton, 2002). In brief, life on the savannah resulted in selective pressures on the hominids that lead them to change their diet from predominantly vegetarian to more protein and fat based. One can conclude that the day ranges of the early hominids must have been larger than those of extant apes. There is clear evidence that already more than two million years ago *transport* of the artefacts (at least the stone tools) was an important trait of the Oldowan culture (Toth, 1985) and it seems clear that there were selective advantages in *transporting* tools and carcasses. The hominid life on the savannah during the Oldowan era opened up for many new forms of cooperation for future goals (Plummer, 2004). Following Osvath and Gärdenfors (2005), I suggest that the Oldowan culture led to the co-evolution of transport and anticipatory planning.

For most forms of cooperation among animals, it seems that advanced forms of intersubjectivity are not needed. If the common goal is *present* in the actual environment, for example food to be eaten or an antagonist to be fought, the collaborators need not focus on a joint representation of it before acting – it is sufficient to share attention. If, on the other hand, the goal is distant in time or space, then a *mutual* detached representation of it must be produced before cooperative action can be taken. For example, building a common dwelling requires coordinated planning of how to obtain the building material and advanced collaboration in the construction. In general terms, cooperation about future goals requires that *the inner spaces of the individuals be coordinated.*

The upshot is that cooperative anticipatory planning generates a demand for a sharing of minds.

Boesch and Boesch (1989, 2000) make a distinction between four kinds of co-operation based in studies of the hunting behaviour of wild chimpanzees. The most advanced kind is the collaborative hunt in which the hunters perform different complementary roles directed towards the same prey. The collected evidence shows that collaborative hunting is frequently achieved by chimpanzees, but also by wolves and lions, although this occurs more rarely. Collaborative hunting, in this sense, is thus not unique to primates (Boesch and Boesch-Achermann, 2000, p. 186). Coordinated hunting requires shared attention among the hunters, but not joint intention. However, if joint intention can be achieved, even stronger forms of collaboration can develop. In all the kinds of hunting brought up by Boesch and Boesch, co-operation concerns the present, both in time and space. The long ranging life on the savannah makes cooperation involving joint intentions necessary for hunting and other cooperative tasks.

Another example of anticipatory thinking that had selective effects is division of labour. A division of labour within the group could solve a multitude of needs at once. Some individuals may carry throwing stones, some may carry sharp edges and others could carry water or wooden tools. It is a way of optimizing the carrying resources of the group, which is probably already burdened with carrying infants. Such cooperation requires joint intentions involving shared goals outside the scope of the immediate drive state.

To show the evolutionary importance of cooperation for future goals, Deacon (1997, pp. 385-401) suggests that the first form of symbolic communication is marriage agreements, that is, deliberate commitments to pair bonding. Commitments require joint beliefs and thus a well-developed form of intersubjectivity. A marriage is a special case of a *contract*. Creating contracts is a uniquely human activity. The reason for this that a contract presumes that both partners have joint beliefs: If we agree that I shall deliver a hen tomorrow in exchange for the axe you have given me now, I believe that you believe that I will deliver the hen and you believe that I believe that our agreement will then be fulfilled, etc. Furthermore, a contract depends on the possibility of *future* sanctions and thus on anticipatory cognition: If I don't deliver the hen, you or the society will punish me for breeching the agreement.

These examples suggest that there has been a coevolution of the uniquely human forms of intersubjectivity and the more future-oriented forms of cooperation that we first find evidence for along the hominid line. The evolutionary forces driving this coevolution emanated from the unique ecology of the hominids, perhaps beginning with the niche created by the Oldowan culture (Plummer, 2004, Osvath and Gärdenfors, 2005).

11. Conclusion

Before concluding, I will make some brief remarks on the development of self-consciousness, which I define as the ability to represent ones own consciousness. Do animals other than humans have self-consciousness? For one thing, Gallup's (1977) experiments show that chimpanzees and orangutans can recognize themselves in mirrors. But recognizing oneself in a mirror or on a photograph only requires awareness of one's own *body*, not of one's own mind. A final step in the evolution of higher-level inner representation is small but crucial for self-consciousness in its proper sense: I must realize that the inner world of my opponent does not only contain a representation of myself as a bodily agent, but as *an agent with inner representations as well.* I believe that it is only after this insight that the agent can become self-conscious in the sense that it can form representations of its own representations. Some support for this evolutionary point can also be obtained from results in developmental psychology (see e.g. Wimmer and Hartl, 1991 and Gopnik, 1993).

In other words, *self-consciousness* can develop as a shortcut in the representations involved in higher forms of intersubjectivity: I can in my inner world have a detached representation of my own inner world. The most important aspect of this form of detachment is that I can *attend* to different aspects of my inner world and use this attention in my planning. Other animals can only attend to the external world as it is represented to them. In particular, in anticipatory cognition I can attend to my own feelings and desires and form a representation of how these feelings may be different in the future.

However, I submit that this kind of self-consciousness could never develop without the previous establishment of a representation of the inner world of others. In other words, I claim that an "I"-experience must be preceded by a "you"-experience. In brief, self-consciousness can be seen as a spin-off from advanced forms of intersubjectivity.

It should be noted that just as there are different capacities involved in intersubjectivity, one finds different stages of self-consciousness. For example, Reddy (2003) argues that already during the first year, children develop an understanding of being objects of attention. She suggests that "before the infant has a conception of him or herself, he or she is aware of being an object to others" (ibid., p. 400).

In the literature, the question of whether an ape or a child has a theory of mind has often been presumed to have a yes-or-no answer. However, if one considers intersubjectivity more generally, it becomes clear that there is no simple answer In this chapter, I have divided intersubjectivity into five components: representing the emotions of others (empathy), representing the desires of others, representing the attention of others, representing the intentions of others, and representing the beliefs and knowledge of others. In this way, one obtains sharper tools for analyzing the mental capacities of animals and humans at various stages of cognitive development. I am not claiming that this is the only way to break down intersubjectivity. But I hope to have shown that this more fine-grained presentation can help us in asking the right questions when designing experiments with animals and children and when interpreting the outcomes of these experiments. A practical implication is that a division of intersubjectivity can help to more adequately capture individual children's intersubjectivity developments.

The five components can also be connected to various forms of collaboration, which may help us understand the coevolution of intersubjectivity and cooperation in the predecessors of Homo sapiens. Finally, the components can be connected to the development and evolution of *communication*. Each form enables a certain type of communicative behaviour. For example, the capacity for conveying communicative intent depends on sharing emotions and referential communication depends of joint attention (Brinck, 2006). By connecting intersubjectivity, cooperation and communication we can gain a much deeper understanding of the development and evolution of the human mind.

Acknowledgements

Research for this chapter has been supported by the Swedish Research Council, which is gratefully acknowledged. I want to thank Ingar Brinck and other members of the EU-project Stages in the Evolution and Development of Sign Use (SEDSU) for helpful discussions.

Notes

1. Baron-Cohen also argues that the components are "modules" in the sense of Fodor (1983). This is an assumption that I don't make concerning the five components of intersubjectivity. Furthermore, Tirassa *et al.* (2006) argue against partitioning a "theory of mind" into components. However their arguments mainly concern a modular approach, and in my opinion they do not contravene the theory presented here.

2. In Gärdenfors (2003), empathy was called compassion and cognitive empathy was called just empathy. Here I follow the terminology of Preston and de Waal (2002).

References

Baron-Cohen S (1994) How to build a baby that can read minds: Cognitive mechanisms in mind reading. *Cahiers de Psychologie Cognitive* 13:513-552.

Barsalou LW (1999) Perceptual symbol systems. *Behav Brain Sci* **22**: 577-609.

Bekoff M, Byers J (Eds) (1998) *Animal Play: Evolutionary, Comparative and Ecological Perspectives.* Cambridge: Cambridge University Press.

Bickerton D (2002) Foraging versus social intelligence in the evolution of protolanguage. In: A Wray (Ed) *The Transition to Language.* Oxford: Oxford University Press, pp 207-225.

Bickerton D (2005) Language first, then shared intentionality, then a beneficent spiral. *Behav Brain Sci* **28:**691-692.

Bischof N (1978) On the phylogeny of human morality. In: G Stent (Ed) *Morality, as a Biological Phenomenon.* Berlin: Abakon, pp 53-74.

Bischof-Köhler (1985) Zur Phylogenese menschlicher Motivation. In: LH Eckensberger and ED Lantermann (Eds) *Emotion und Reflexivität.* Vienna: Urban und Schwarzenberger, pp 3-47.

Boesch C, Boesch H (1989) Hunting behaviour of wild chimpanzees in the Tai National Park, Ivory Coast. *Am J Physical Anthropol* **78**:547-574.

Boesch C, Boesch-Achermann H (2000) *The Chimpanzees of the Tai Forest: Behavioural Ecology and Evolution.* Oxford: Oxford University Press.

Brinck I (2004) Joint attention, triangulation and radical interpretation: A problem and its solution. *Dialectica* **58**:179-205.

Brinck I (2006) The role of intersubjectivity for intentional communication, To appear in T Racine, C Sinha, J Zlatev and E Itkonen (Eds) *The Shared Mind.* Amsterdam: John Benjamins.

Brinck I, Gärdenfors P (2003) Co-operation and communication in apes and humans. *Mind and Language* **18**: 484-501.

Byrne R, Whiten A (Eds) (1988) *Machiavellian Intelligence: Social Expertise and the Evolution of Intellect in Monkeys, Apes, and Humans*. Oxford: Oxford University Press.

Call J, Tomasello M (1999) A nonverbal false belief task: The performance of children and great apes. *Child Development* **70:**381-395.

Cheney DL, Seyfarth RL (1990) *How Monkeys See the World: Inside the Mind of Another Species.* Chicago: University of Chicago Press.

Craik K (1943) *The Nature of Explanation*. Cambridge: Cambridge University Press.

Darwin C (1872) *The Expression of the Emotions in Man and Animals*. London: Murray.

Deacon TW (1997) *The Symbolic Species.* London: Penguin Books.

Dennett D (1978) *Brainstorms: Philosophical Essays on Mind and Psychology,*. Cambridge MA: The MIT Press.

Dennett D (1988) Why creative intelligence is hard to find. *Behav Brain Sci* **11**: 253.

de Waal FBM, Aureli F (1997) Conflict resolution and distress alleviation in monkeys and apes. *Ann N Y Acad Sci* **807**:317–28.

Dunin-Kepliz B, Verbrugge R (2001) A tuning machine for cooperative problem solving. *Fundamenta Informatica* **21**:1001-1025.

Emery NJ (2000) The eyes have it: the neuroethology, function and evolution of social gaze. *Neurosci Biobehav Rev* **24**:581-604.

Flavell JH, Flavell ER, Green FL, Moses LJ (1990) Young children's understanding of fact beliefs versus value beliefs. *Child Development* **61**:915-928.

Fodor J (1983) *The Modularity of Mind.* Cambridge MA: MIT Press.

Gallup GG (1977) Self-recognition in primates. *Am Psychol* **32:**329-338.

Gärdenfors P (1996) Cued and detached representations in animal cognition, *Behav Processes* **36:**263-273.

Gärdenfors P (2001) Slicing the theory of mind. *Danish Yearbook for Philosophy* **36**:7-34.

Gärdenfors P (2003) *How Homo Became Sapiens.* Oxford: Oxford University Press.

Gergely G, Nadasdy Z, Csibra G, Biro S (1995) Taking the intentional stance at 12 months of age. *Cognition* **56**:165–93.

Gergely G, Bekkering H, Kiraly I (2002) Rational imitation in preverbal infants. *Nature* **415**:755.

Glenberg A (1997) What memory is for. *Behav Brain Sci* **20**:1-19.

Gómez JC (1994) Mutual awareness in primate communication: a Gricean approach. In: ST Parker, RW Mitchell and ML Boccia (Eds) *Self-awareness in Animals and Humans.* Cambridge: Cambridge University Press, pp 61-80.

Gopnik A (1993) How we know our minds: the illusion of first-person knowledge of intentionality. *Behav Brain Sci* **16**:1-14.

Gopnik A (1998) Explanation as orgasm. *Minds and Machines* **8**:101-118.

Gopnik A, Astington JW (1988) Children's understanding of representational change, and its relation to the understanding of false belief and the appearance-reality distinction. *Child Development* **59**:26-37.

Grush R (1998) Wahrnehmung, Vorstellung und die sensomotorische Schleife. In: F Esken and H-D Heckmann (Eds) *Bewußtsein und Repräsentation*. Paderborn: Verlag Ferdinand Schoeningh.

Grush R (2004) The emulator theory of representation: motor control, imagery and perception. *Behav Brain Sci* **27**:377-442.

Gulz A (1991) The Planning of Action as a Cognitive and Biological Phenomenon. *Lund University Cognitive Studies* 2.

Hare B, Call J, Agnetta B, Tomasello M (2000) Chimpanzees know what conspecifics do and do not see. *Animal Behaviour* **59**:771–85.

Hare B, Call J, Tomasello M (2001) Do chimpanzees know what conspecifics know?, *Animal Behaviour* **61**:139–151.

Hesslow G (2002) Conscious thought as simulation of behaviour and perception", *Trends Cogn Sci* **6**:242-247.

Heyes C (1998) Theory of mind in nonhuman primates. *Behav Brain Sci* **21**:101-115.

Hughes C, Russel J, Robbins TW (1994) Evidence for executive dysfunction in autism. *Neuropsychologia* **32**:477-492.

Humphrey N (1993) *A History of the Mind.* London: Vintage Books.

Jeannerod M (1994) The representing brain, neural correlates of motor intention and imagery. *Behav Brain Sci* **17**:187-202.

Kaminski J, Riedel J, Call J, Tomasello M (2005) Domestic goats, *Capra hircus*, follow gaze direction and use social cues in an object task. *Animal Behaviour.* **69:**11-18.

Leslie AM (1987) Pretense and representation: the origins of 'theory of mind'. *Psychol Rev* **94**:412-426.

Lewis D (1969) *Convention.* Cambridge MA: Harvard University Press.

Masserman JH, Wechkin S, Terris W (1964) "Altruistic" behavior in rhesus monkeys. *Am J Psychiatry* **121**:584-585.

Mitchell P (1997) *Introduction to Theory of Mind: Children, Autism and Apes.* London: Arnold.

Mulcahy NJ, Call J (2005) Apes save tools for future use. *Science* **312**:1038-1040

Osvath M, Gärdenfors, P (2005) Oldowan culture and the evolution of anticipatory ognition. *Lund University Cognitive Studies* 122.

Perner J, Leekam S, Wimmer H (1987) Three-year-old's difficulty with false belief: the case for a conceptual deficit. *Br J Develop Psychol* **5**:125-137.

Piaget J, Inhelder B (1956) *The Child's Conception of Space.* London: Routledge and Kegan Paul.

Plummer T (2004) Flaked stones and old bones: Biological and cultural evolution at the dawn of the dawn of technology. *Yearbook of Physical Anthropology* **47**:118-164.

Povinelli D (2000) *Folk Physics for Apes: The Chimpanzee's Theory of How the World Works.* Oxford: Oxford University Press.

Povinelli DJ, Eddy TJ (1996) What young chimpanzees know about seeing. *Monographs of the Society for Research in Child Development*, Vol **61**.

Premack D (1996) Cause/induced motion: intention/spontaneous motion. In: J-P Changeux and J Chavaillon (Eds) *Origins of the Human Brain.* Oxford: Clarendon Press, pp 286-308.

Premack D, Woodruff G (1978) Does the chimpanzee have a theory of mind? *Behav Brain Sci* **4**:515-526.

Preston SD, de Waal F (2003) Empathy: Its ultimate and proximal bases. *Behav Brain Sci* **25**:1-72.

Proust J (1998) Can nonhuman primates read minds? *Rapport no. 9807.* Paris: CREA, Paris.

Reddy V (1991) Playing with other's expectations: teasing, joking and mucking about in the first year. In: A Whiten (Ed) *Natural Theories of Mind.* Oxford: Blackwell.

Reddy V (2003) On being the object of attention: implications for self–other consciousness. *Trends Cogn Sci* **7**:397-402.

Sjölander S (1984) *Nya Tankar om Gamla Hjärnor*. Stockholm: Brombergs.

Stern D (1985) *The Interpersonal World of the Infant*. Basic Books, New York, NY.

Suddendorf T, Busby J (2003) Mental time travel in animals? *Trends Cogn Sci* **7**:391-396.

Suddendorf T, Corballis MC (1997) Mental time travel and the evolution of human mind. *Genetic, Social and General Psychology Monographs.* **123**:133-167.

Tirassa M, Bosco FM, Colle L (2006) Rethinking the ontogeny of mind reading. *Consc Cogn* **15**:197-217.

Tolman EC (1948) Cognitive maps in rats and men. *Psychol Rev* **55**:189-208.

Tomasello M (1999) *The Cultural Origins of Human Cognition.* Cambridge MA: Harvard Unversity Press.

Tomasello M, Call J (2006) Do chimpanzees know what others see – or only what they are looking at? In: S Hurley and M Nudds (Eds) *Rational Animals.* Oxford: Oxford University Press, pp 371-384.

Tomasello M, Carpenter M, Call J, Behne T, Moll H (2005) Understanding and sharing intentions: The origins of cultural cognition. *Behav Brain Sci* **28**:675-691.

Toth N (1985) The Oldowan reassessed: a close look at early stone artefacts. *Journal of Archeological Science* **12**:101-120.

Wellman HM, Liu D (2004) Scaling of theory-of-mind tasks. *Child Develop* **75**:523-541.

Whiten A, Byrne RW (1988) Tactical deception in primates. *Behav Brain Sci* **11**:233-244.

Whiten A, Byrne RJ (Eds) (1997) *Machiavellian Intelligence II*. Cambridge: Cambridge University Press.

Wimmer H, Hartl M (1991) Against the Cartesian view on mind: young children's difficulty with own false beliefs. *Br J Develop Psychol* **9**:125-138.

Zlatev J, Persson T, Gärdenfors P (2005) Bodily mimesis as the 'missing link' in human cognitive evolution. *Lund University Cognitive Studies* **121**.

Chapter 13

The Phenomenon of Consciousness from a Popperian Perspective[1]

Peter Munz

I propose to make use of the philosophy of Karl Popper to arrive at least at a partial explanation of human consciousness. This is not supposed to be either an interpretation of what Popper thought or an attempt to find out what he really meant. Instead I am planning to make use of his thoughts to arrive at an explanation of consciousness which may well be other than what he had intended to offer. To be precise, a Popperian, though not necessarily Popper's, theory of consciousness is directly opposed to almost all other theories which, for the most part, start at the bottom and work upwards. In analogy with outmoded ideas of scientific method - according to which knowledge is obtained when sense observations are summarised inductively until a general law emerges - they start from the stimulus of the world upon the body's neurons and then seek to determine when, exactly, that neuronal behaviour generates or blossoms into consciousness, at which point a linguistic formulation may or may not arise as the end product. As against this, a Popperian theory of consciousness - in analogy with Popperian scientific method - goes from the top down. It suggests the conditions under which and how far an exosomatically conceived language reaches down into the body and via the body, talks about the world of nature.

1. Popper's view on consciousness

Popper's own thoughts on consciousness were casual and almost accidental and, as I will try to show, will have to be amended. He never addressed the problem of consciousness as such, but mentioned it en passant, as if no further explanation was required. He also talked about the idea of interaction between mind and body as if it could

[1] Regrettably, Peter Munz died before this chapter was finalised, with regard to the layout of the book. Thus, we have constructed subtitles to substitute his initial roman numerals, with kind permission of his wife.

Consciousness Transitions: Phylogenetic, Ontogenetic, and Physiological Aspects
Edited by Hans Liljenström and Peter Århem

be taken for granted that it took place, without giving a specific definition of what he meant by "mind". Here is an example of these shortcomings. In his *Objective Knowledge* of 1972 he wrote on p 251, that "consciousness...appears as just one of many interacting kinds of control; and if we remember the control systems incorporated for example in books – theories, systems of law, and all that constitutes the 'universe of meanings' - then consciousness can hardly be said to be the highest control system in the hierarchy. For it is to a considerable extent controlled by these exosomatic linguistic systems - even though they may be said to be *produced* by consciousness. Consciousness is, in turn, we may conjecture, produced by physical states; yet it controls them to a considerable extent. Just as a legal or social system is produced by us, yet controls us, and is in no reasonable sense 'identical' to or 'parallel' with us, but *interacts* with us, so states of consciousness (the 'mind') control the body, and *interact* with it". Popper's conception of such 'interaction' is misconceived. For interaction to be conceivable and take place, the subjective states of one's body would have to be definably distinct so that one could determine how they interact with the mind - but the subjectivities in one's body, being subjective and thus not capable of being referred to in a purely private language, are not distinctly definable. Last not least, he maintained that the kind of language required to make the all important hypotheses on which, he had shown so convincingly in his *Logic of Scientific Discovery* of 1934, all our knowledge depends, was a language capable of engaging in arguments, that is, capable of doing more than symbolise and describe. In fact, I will try to show that it follows from his own view of what the all important hypotheses are supposed to achieve, that we have a language which can not only go beyond symbolizing and describing, but which can also do more than just argue.

3. The Popperian three-world idea

First of all, one has to ascertain what one means by the term "consciousness" so that one knows what one is trying to explain. There is a philosophical as well as a semantic problem which must be clarified before any attempt at a scientific explanation is possible. In the case of consciousness, only a philosopher is able to tell us what it is we are wanting to explain. The scientist's job starts when the philosopher has finished. Nothing could be more misconceived than Jeffrey Gray's contention, in his recent, prize winning book *Consciousness: Creeping Up on the Hard Problem* (2004) that we need nothing but science to find out why and how we can be conscious and that philosophers need not apply themselves to this problem. I will, therefore, start with philosophy. Philosophically speaking, I understand by consciousness a state in which we can say *what* it is we are conscious of. If we cannot *say* what it is we are conscious of, we obviously are not *conscious* of it. This means that there is an intimate and essential connection between our language and our consciousness. I admit that this definition of consciousness is austere and demanding, because it excludes states which are felt vaguely and which, though noticeable, have no identifiable shape. A mere subjective feel of what one is doing or experiencing does not merit the designation of consciousness. Popper's contribution to the problem of consciousness is therefore of enormous importance because in his philosophy there is a telling explanation of how linguistically formulated theories and hypotheses relate to our bodies as well as to the world in which our bodies are living. It is these special kinds of linguistic formulations which are at the centre of the problem of consciousness. This conclusion implies that whatever it is many animals are aware of - and there is no denying that the higher animals are capable of some very telling awareness - this cannot be considered to be a form of consciousness. Both higher animals and humans are capable of subjective feels. They are aware of them without being able to say what it is they are feeling. But - and this is the difference between animals and humans - animals respond to them and act without further hesitation, the way humans, unconsciously, will blink an eyelid or duck when something is falling on them. But humans in many cases will and can do more than respond; they tend to attend to those subjective feels by seeking to tell themselves what it is they are noticing. This is where the difficulty begins, for those subjective, raw feels are fluid and have no shape. In many cases they can trigger a response but they can't be known. They are literally "raw", and when we have nothing but raw feels, we neither know what we are experiencing, nor whether the raw feel is an experience of. Popper himself distinguished "full" consciousness from "lower" consciousness (1994, p. 111) To merit the designation "consciousness", the raw feel has to be "cooked" by a verbal baptism, that is, it has to be named. The raw feel cannot be deciphered for, being vague, it does not carry any information one could decipher. But humans, with the help of their language, can invent a hypothesis about it and, in so naming it, can make us conscious of its special quality. To avoid misunderstanding, it has to be explained that "rawness", though it occurs, is not really sensed. For as soon as it is sensed, we are inflicting a name on it - for this is what we mean by "sensing": we are *saying* what we are sensing. When sensed, we become conscious of the feel and it ceases to be raw. To put it differently: when we introspect we discern nothing more than a flow of feels without shape. By he time introspection allows us to *see* something in our inside, we have imported a word which acts as a mould to parts of that flow. This is more than deciphering the flow. It amounts to bringing in a word from the outside. In short, when introspection is yielding consciousness of what we are feeling, it is no longer introspection. Such raising to consciousness does not occur because the rawness contains enough information to let us decipher what it is. Nor is such raising to consciousness achieved by introspection, for we cannot just concentrate on our inner self and look at what we find. When we are trying to look, i.e. when we are introspecting, we find nothing more than a waving sea of feels, something Baudelaire rightly compared to looking at he waves of the ocean. We cannot decipher its shape for the very simple reason that it has no shape. The state we become conscious of, is not identical with the feeling state we imagine to have accessed. This procedure, therefore, is very different from "introspection". We cannot introspect, because when we do not impose a name on the feel, we are not able to "see" it. We can only "see" it when it ceases to be something other than a feel. But by that time, we are no longer introspecting. Nor is that consciousness of the feel produced by a neuronal accompaniment of a special kind such as, fore example, quantum effects in microtubules or 40 Hertz oscillations. It is conceivable that such neuronal accompaniments of the neuronally generated subjective feels can intensify the feels. But no amount of intensification can give the liquidity of the feels a recognisable shape so that one can decipher it and say what the feel is a feel of. Without such a shape it cannot qualify as a *conscious* feel.

In his Objective Knowledge of 1972 Popper explained how certain linguistic formulations - theories or hypotheses - play a role which is independent of both the subjectivity of our bodily awareness and the natural world our bodies are part of. In that scheme there is a body; but the mind or the mental events or, more correctly, the verbal formulation of hypotheses about either body or the world the body is living in, are a world apart. They are outside the body and, what is more, they are accessible to bodies other than the body which had provided a material basis for them in the form of uttered words and the breath and tongue movements necessary for uttering them. They are not private and no longer internal, but "concerted" actions sometimes learnt from other people (Melser, 2004, passim.) and at other times speculatively or imaginatively invented. To anticipate, it was something Popper called world 3, while he called the natural world and the body in that world, world 1 and world 2 respectively. I will retain this terminology, even though I do not consider it felicitous. It would have been better if world 1 had been called the world of nature; world 2, the world of our bodies; and world 3, a world of autonomous and independent speculations. That world 3 was not part of the body, let alone a spiritual or soulful part of the body and, therefore, Dennett's contemptuous and jocular suggestion (1979, p. 93) that when Popper was speaking of interaction of body and mind, he must have been thinking of a "Poltergeist" jumping on the body's neurons, was completely beside the point. There is a very simple and compelling reason why world 3 which consisted of hypotheses about world 1 and world 2 was outside the worlds it was hypothesising about. For these hypotheses were not derived from or caused by either world 1 or world 2. They were neither reports nor protocols nor summaries of sense-data nor causal consequences of what had been observed to have gone on in world 1 or gone into world 2 from world 1. Being independent, they *appeared* to be something mental, because they were linguistic formulations which were neither pictures of reality nor reports of observations of reality, nor the sort of algorithmic calculations which a computer might accomplish about reality. They were, on the contrary, free and speculative inventions about the body and the world, waiting to be tested. Being literally outside the body, they appear as something mental - totally divorced from both world 1 and 2. Strictly, there was nothing "mental" (= spiritual?) about them, because they were constructed with the help of the brain and produced by a certain kind of way the breath goes through the larynx. But one must note that whatever precisely the brain's input is, it was not a transmission of signals from the world to the formulation of the hypotheses. It would be more correct to consider them verbal hypotheses which were invented and formulated freely, but not inspired, let alone causally determined, by events in either of the two other worlds. It is this lack of connection which made them free floating and which creates the *appearance* of mentality, whatever the concept of mentality might exactly mean. This mentality, as we will see, when we are speaking in so many words about events in worlds 1 and 2, helps to make us *conscious* of those events. We must take it that consciousness is directly linked to this special kind of verbal formulation because being divorced from world 1 and 2, we are conscious of these formulations because we do not experience them the way we experience a chair. In short, when we are hypothesising about the world or our feelings, we are *conscious* of doing so, even before we know what it is we are conscious *of*. We are so because those hypotheses are invented and made up. They are not induced from observations or inferred as sense-data from the world and from the body that is putting them forward. There is no need to postulate either the operation of a homunculus, much lampooned by Dennett or Crick's 40Hertz oscillations, or Edelman's neuronal re-entries or Penrose's quantum effects in microtubules, to bring this initial kind of consciousness about.

The hypotheses in world 3 are genuinely and literally exosomatic. One could define their apparent mentality in terms of their exosomaticity. Both Darwin and Einstein, to mention two striking examples, were aware of the exosomaticity of their hypotheses about evolution by natural selection and the Theories of Relativity. Darwin stated that "I am a firm believer that without speculation there is no good or original observation" and Einstein said that "I try to catch the rule of Law within a world of some objective reality in a wildly speculative way" (Munz, 2004, p.153). They described themselves independently as "speculators" rather than as observers. Incidentally, it should be noted, in the light of many recent beliefs, that such documented insistence on the role of speculation precludes the idea that we operate by algorithms and that *Artificial* Intelligence is their analogue. If Darwin and Einstein's confessions are correct - and I have no doubt that they are - it follows that the hypotheses which constitute world 3 can act on world 2 but, since they are *about* it without being derived from it, cannot *interact* with world 2. They can, as I will argue later, be imposed on world 2 and interpret events in world 2. But since they are not causally dependent on world 2, it makes no sense to speak of 'interaction'. When we speculate, we do not interact with whatever it is we are speculating about. If one must think of interaction, one can only mean not causal interaction, but unpredictable suggestions or allusions which could be taken to have "interacted" with the hypotheses or speculations that are being put forward. But let me begin at the beginning.

3. The three-world idea and the theory of knowledge

When the pursuit of scientific knowledge became institutionalised by the foundation of the Royal Society of London in 1660, it was laid down that for knowledge to be counted as scientific, it had to consist of or be derived from observations reported in "unadorned" language (Sprat, 1959, p. 113). At the time this was sound advice because for hundreds of years, scholars and thinkers had been used to derive their knowledge not by looking at he world, but from consulting famous books. The use of the word "unadorned" was also important because it was well known that it was possible to dress plain observations up by the employment of rhetorical, metaphorical, emotional and other suggestive adornments and thus obscure what had been observed. The directive was intended, while acknowledging the importance of language, to keep the role of language at a minimum. The directive given to aspiring scientists was well meant and a useful rule of thumb. What is striking, however, is that the really great scientists like Newton and Darwin though they paid lip service to it in order to appear to be politically correct, did not take the directive seriously. Newton proclaimed that he did not use hypotheses (*"hypotheses non fingo"*); but it is perfectly clear from his theories about gravity, that whatever observations he had found suggestive, there was no way in which he could have advanced his striking hypotheses as nothing but unadorned reports of what he had observed. Or take the case of Darwin. When he pondered how evolution might have taken place he eventually was inspired by Malthus' half baked theory about birth rates and diminishing food supply to formulate his theory of evolution by natural selection. But when, towards the end of his life, he came to write his Autobiography, he knew very well that to be scientifically respectable he had to pretend that he had followed the directive laid down by the Royal Society some two hundred years ago and done nothing but observe and added his observations up algorithmically.

While many scientists only paid lip service to the directive in order to appear politically correct, even though they were actually disregarding it (Munz, 1993, pp. 96-7.), it is a strange fact that time and again philosophers as well as the educated public, right down to the early 20th century, took the directive not only seriously but also literally and paraded it as the sound criterion for distinguishing science from superstition. They stressed that the directive was the sound and only method to arrive at knowledge which was entitled to claim to be called scientific. David Hume had had his doubts that one could get good science by following the directive, but being sceptical of the directive and the practice of induction it implied, he ended up being sceptical of science. He found fault with the directive, but took it that since there was no other method in sight one had to be sceptical of the claim that there could be such a thing as scientific knowledge. Unlike Hume, Jonathan Swift made a direct assault on the directive in his Ch. 5, Bk. 3 of his *Gulliver's Travels*. He satirised it and reduced it ad absurdum by introducing a scientist who had realised that there can be no such thing as an unadorned verbal report of an observation. That scientist, he wrote, had taken the bull by the horns. Realising that "unadorned" language was impossible, he avoided language altogether. Instead he carried all objects worth observing and talking about in a bag and simply reached for the one he was intending to "talk" about and displayed it silently. In this way he proved that language, even the most unadorned language, was superfluous and, in any case, never unadorned enough to be a reliable report of an observation of the object in question. In spite of Hume's scepticism and Swift's satire, at the beginning of the 20th century that old Royal Society directive was advanced in a new guise in Vienna under the name of Logical Positivism. The Logical Positivists proclaimed that only knowledge which was a "protocol" of an observation or at least reducible to a protocol of an observation, could count as knowledge. More or less at the same time, Ludwig Wittgenstein published his famous *Tractatus* in which he argued that only language which is a "picture" of reality (read: a report of observation) is genuine knowledge. The terminology had changed, but the original message of the Royal Society had remained in tact.

The moment one looks soberly at language one must concede Swift's point. And the moment one *looks* at observations, one must concede Hume's point that no matter how many identical observations one happens to be making, one is not entitled to conclude that one can derive a general law from them. The ice which had formed around Swift and Hume as well as the ice which ought to have been paralysing both Wittgenstein and the Vienna Logical Positivists, was finally broken by the philosophy of science of Karl Popper when he published his *Logic of Scientific Discovery* in 1934. Popper, like both Swift and Hume before him, realised that neither unadorned protocols of observations nor the nature of necessarily limited observations can yield scientific knowledge. But unlike Swift and Hume, he did not become a sceptic. He showed instead that the truth of the matter is not what philosophers had proclaimed and what scientists had been intimidated to pretend to have practiced, but what scientists had *actually* practiced. They had first formed daring hypotheses about the world or parts of the world and then tested them, because the hypotheses guided their observations in the right direction. As long as every such hypothesis indicated what would have to be the case for it to be false, it could be taken as good scientific knowledge until such time as somebody might succeed in falsifying it. In which case, one would have to come up with a different hypothesis. Above all, good knowledge did not *start* from observations but ended with them. To get knowledge, he explained, one had to start with a verbal hypothesis and then look downwards. It was wrong to, as official opinion had had it, to start at the bottom fro observations and work upwards towards a hypothesis.

All this is well known but I had to rehearse it in order to come to the point. The directive of the Royal Society, whether in its original or in its 20t century form took it that there was world 1 on one side and the human body (world 2), on the other. When the body observed and reported the observation in an unadorned way, the body got knowledge. It was a matter of taste whether one would call the verbal report the body's mind or whether one would simply be satisfied with the verbal reporting as part of the body's neuronal operations. In either case, the verbal report was a conscious operation and it was taken that consciousness accompanied that operation automatically. There were two entities: on one side the world and, on the other, the body or the body plus the mind. But the mind was not separate. It was the sum of the unadorned verbal reports about what had flown in. As I said, the process from nature to verbally formulated consciousness was a linear, wellnigh automatic, causal progression. The result was scientific in proportion to how unbroken this causal chain was. If there was a *gap* between observation and the formulation of the results, either by adorned language or by a speculation which went beyond what had been observed, the result was not entitled to be called knowledge. With Popper's revolutionary correction of how we get scientific knowledge, all this had to change. The mind, so called, was now really separate. It was the place or the activity where the hypotheses, which could not be derived or inferred from observations, were invented. They had to be tried out in the world. The body, which was in between, was nothing more than the technique by which the predictions deduced from the hypotheses were tried out against what the hypotheses instructed the body to take in from the world. The body, far from amassing observations, took in and observed only what the hypotheses dictated. In 1934 Popper left it at that. There was one shortcoming, though. If the "mind" produced the hypothesis, where did the language in which the hypotheses were formulated, come from when it was so much more than an unadorned summary of observations? Better, how come that that language was intelligible when it was not possible to define its meanings in terms of what had first been observed or in terms of events which has been noticed *before* the language was formed? Obviously it did not come from the body, because it was anything but the unadorned report of what the body was sensing or observing. But Popper had nothing to say on this matter and, as I tried to show in my *Beyond Wittgenstein's Poker* of 2004 on page 33, it would have made it necessary for Popper to fall back on the thought of the later Wittgenstein whose theory of language production in and language control by speech communities would have explained how it is possible to invent intelligible hypotheses the words of which were not derived from prior bodily sensations. Popper never availed himself of Wittgenstein.

Eventually Popper worked out the implications of his rejection of the Royal Society's directive and its modern versions. If the simplistic idea that there is a body which picks up signals from the world and formulates knowledge by summarising them, is rejected, the body/world scheme falls to the ground. In 1972 in His *Objective Knowledge* Popper enlarged his philosophy of science first presented in 1934 by a new scheme. In the new scheme, in harmony with the idea that hypotheses *preceded* observations, there had to be three, not two, worlds - world 1, the world of nature; world 2, our bodies' subjective awarenesses, possibly, their minds; world 3, the exosomatic world of the hypotheses we make up about worlds 1 and 2. This new threefold division followed strictly from his revolutionary idea that hypotheses *precede* observations and, far from resulting from or being summaries of observations, tell us what it is we have to observe, with a view of attempted falsifications. In short, there had to be a third world, for the formulation of verbal hypotheses could no longer be seen as the result of a chain reaction which started in world 1 and led to world 2 in which the observations were being summarised, almost mechanically so or as modern cognitive scientists would say, algorithmically so. In the new scheme, world 3 was detached and not the result of a process which started when world 2 was watching world 1. In the old scheme, if anything, knowledge in the shape of linguistic formulations had been very much part of world 2. Popper's main revolutionary argument was that world 3 is *not* causally connected either with the subjective feelings in world 2 or with the physicality of the relations between nature and our bodies. Whatever observations by world 2 of world 1, they neither yield nor lead to the hypotheses which make up world 3. This was a crucial innovation. In the old scheme there were not only merely two worlds, but the first world (the outside world) was acting causally upon the world of the body inside which the verbal formulation of the experience of the first world was taking place. In this old scheme the unadorned verbal formulation of those subjective sensations or feelings was nothing new. It was merely the way the subjective part of the body which had been causally influenced by the outside world, presented itself. It could be argued - as indeed it was from time to time - that the way the body and even the body plus the mind summarised the impacts made by the world of nature on the world of the body, was a purely mechanical and not a conscious process. If anything, consciousness could be held to interfere with the mechanical or algorithmic summarising that was taking place. The less consciousness, the better the knowledge. As against this, in Popper's new scheme, world 3 was totally detached from the subjective feelings of world 2 and did not depend for its formulation of hypotheses about world 1 on world 1's impact on world 2. In other words, the "mind" was now seen as something sui generis and something separate. It was not produced or generated by events in world 2, but, on the contrary, set in motion events in world 2 which then went to see whether deductions from the hypothesis in world 3, could or could not be falsified in world 1 and, possibly in world 2. There was a one-way connection between world 3 and world 2 and it went down from world 3 to world 1 via world 2, not the other way round. It is strange that in spite of this fact, Popper always spoke of an "interaction" between body and mind, that is, between world 2 and world 3.

We might get a glimpse of what was in his mind when we look at a short paragraph on page 156 of his *Objective Knowledge*. Since the world 2 of subjective experiences and sensations was a reality, though not a reality that led to, let alone caused world 3, Popper suggested that "one day we will have to revolutionise psychology by looking at the human mind as an organ for interacting with the objects of world 3; for understanding them, for contributing to them, participating in them; and for bringing them to bear on the first world". In the first instance he meant that our subjective awareness (world 2) enables us to formulate and grasp the exosomatic hypotheses in world 3, in spite of the fact that the latter are not caused by the former. In he second instance we can take him to have meant that a future psychology would start from world 3 by making hypotheses about the subjective feels in world 2: "An objectivist epistemology which studies the third world can help to throw an immense amount of light upon the second world of subjective consciousness (my comment: he ought to have written "awareness", not "consciousness"), especially upon the subjective thought processes of scientists; but *the converse is not true.*" (Popper, _ 1972, p. 112; italics in the original) It is important to stress that Popper stressed that there is no logical way which can lead from world 2 to world 3, even though the subjective feels of world 2 can throw a psychological light on world 3. He saw this correctly as a revolution in psychology, because so far, psychology had started from the experiences which were induced by world 1 in world 2. Now, with the revised scheme of a three-fold division, the suggestion was that psychology, like natural science about world 1, starts in world 3 with a hypothesis and then tests whether such a hypothesis could be falsified by the behaviour or feelings (all subjective) in world 2. Popper thought of this relationship between world 2 and world 3 as an interaction. However, he should not have used that term, because there is a catch. Popper was far too sanguine n his belief that the hypotheses in world 3 could be *tested* in world 2. In reality, the subjective states in world 2 are far too vague, too undefined, to allow such testing. A hypothesis from word 3 can be used to suggest what such raw subjective feels in world 2 might amount to, but no raw feel can either falsify or verify such a hypothesis. For such testing to be possible, the subjective states in world 2 would have to have a shape ascertainable without the intervention of the hypothesis so that one could look upon the subjective state and then compare it to the hypothesis to see whether the hypothesis is true or false. The presence of such a shape is ultimately and explicitly denied by Popper himself when he wrote correctly (Popper, 1972, p. 74) that the ascertainable shape of the subjective states "depends" on the hypotheses.

4. On the difference between psychotherapy and drug therapy

Incidentally, this revision of the sequences throws a telling light on the difference between psychotherapy, which goes by words into world 3, and drug therapy, which goes by chemicals into world 2. The former can be therapeutic not by changing the subjective feels in world 2, but by suggesting different, more salutary interpretations of those feels - interpretations which come in so many words from world 3 and which have, if their intention is therapeutic, been called, felicitously, the "talking cure" (Munz, 1999, pp. 173, 181). By contrast, drugs act on the neurons which can change the subjective feels, without going through or affecting consciousness. This change may, at times, encourage a new interpretation which comes from world 3, though there is no guarantee that it will do so. But if there is a stubborn or compulsive adherence to the hypotheses which come from world 3, the new neuronal churnings and their subjective feels in world 2 affected by drugs, can, often enough, continue to be interpreted by the old hypotheses which come from world 3.When there is such a stubborn adherence to old interpretations, we speak of neurosis - and, possibly at times, of psychosis.

5. Popperian psychology

On the face of it, this Popperian psychology was a very promising revolutionary idea. It aligned psychological knowledge of world 2 with our scientific knowledge of world 1 and put both on the same footing as being derived from the exosomatic hypotheses in world 3. Popper thought of these connections as an interaction of mind and body; or, in the new terminology, of an interaction between world 2 and world 3. But "interaction" suggests a two way process. Both worlds contribute something and influence one another. It seems that Popper could not have been thinking of this kind of literal interaction. For he had made it quite clear and, in any case, it followed directly from, his philosophy of science, that the hypotheses made in world 3 (the mental events) were not dependent on, let alone caused, by events in world 1 or world 2. In the new scheme, whatever the subjective feelings in world 2, they do *not* lead to the formulation of theories or hypotheses about them in world 3. There is a downward process; but not an upward process (Popper, 1972, p. 74). Interestingly, Wittgenstein had come to a similar conclusion when he came to write in his book *On Certainty*, that he had reached the bottom of his convictions and realised that they all were hanging down from the roof (Munz, 2004, p. 106).

In short, there is no interaction, but only a downward path from world 3 to world 2. In this downward path world 3 hypotheses are tested in world 2. For real interaction to be possible, the subjective events in world 2 would have to stand out with sufficient clarity and precision to be pinpointed and named, so that they could be used to test, possibly to falsify, the hypotheses coming down from World 3. And, if not falsify, at least to confirm that events predicted by a hypothesis contained in World 3 could be instantiated in world 2. But in the absence of such precision and clarity, they cannot be used to test the hypotheses coming from world 3. The subjectivities can be made to stand out with precision only with the help of hypotheses from world 3. Popper's own statement eliminates the possibility that the hypotheses in world 3 can be tested in world 2. For he writes on p 74 of Objective Knowledge that the subjective knowledge of events in world 2 *depends* on world 3, i.e. our awareness of our purely subjective knowledge of our inner feelings cannot come in its own right or by itself, but is generated by our knowledge of some hypothesis in world 3. It is the hypotheses in world 3 which give our subjective feelings a local habitation and a name. Since there is no clear subjective knowledge, there can be no test of a hypothesis made in World 3. There can only be a suggestion that what is hypothesised in world 3 is a definition of the subjectivity in World 2. For example, if one hypothesises in world 3 that the sight of a baby produces a feeling of joy in world 2 and if there is a sight of a baby in world 1 taken in by world 2, then whatever silent and ineffable feels there are in world 2, are identified and named "joy" by the hypothesis in world 3. There is no *inter*action, but merely an action of world 3 on world 2. There is nothing sufficiently clear in world 2 to test whether the action is correct or not, for we cannot observe, refer to determine and name the feel. Therefore we can never tell whether the hypothesis that was used to "act" on the feel is true or false. Hence, we can only speak of an *interpretation* of the feel. The downward action is nothing more than an *interpretation* of the subjective feel in world 2 by a hypothesis in world 3. This means that the suggestion which comes from world 3 that what the sight of a baby causes in world 2 is a feeling of joy, is not a certainty, but a likely possibility we are left with because, in the absence of an alternative hypothesis from world 3, this is *all* we are left with. This indicates that our "knowledge" of psychology, so called, is not genuine knowledge in the sense of Popper's philosophy of knowledge, but a surmise beyond which we cannot go. Surmise or not, in imposing the word 'joy' on the subjective feeling which may well have originated by the sight of a baby, one becomes *conscious* of the silent, subjective, feeling, i.e. of joy. In this way, the verbal interpretation of a subjective feel, makes us conscious of the subjective feel, because we can now *say* what it is, i.e. what we are feeling. In this way, knowledge of the feel and consciousness of the feel come together and it is this knowledge which amounts to consciousness. Both knowledge and consciousness result from the verbal interpretation of the subjective feel suggested by the verbally formulated hypothesis in world 3. This highlights the crucial and decisive role of language for the production of consciousness. Recourse to special neuronal events like 40 Hertz oscillations or quantum effects in microtubules, to mention two proposed explanations of the genesis of consciousness, becomes superfluous. Moreover, this is in striking contrast to the appearance of consciousness in the old, Royal Society inspired scheme, where the appearance of consciousness is a disturbing accompaniment, likely to detract from knowledge; for when we are conscious of a tree in addition to knowing that there is a tree on our left, we are adorning the observation of the tree with a word and thus, in adorning it, detract from our naked knowledge of it. But let us try a thought experiment. Let us assume that we can identify the silent feeling well enough to be able to give a *literal* description of it. Such a literal description would be nothing but a verbal picture which would duplicate what is already there because the feeling by itself is a feeling and would not supply the word "joy" as a suitable name. In other words, it would not let us *know* that there is a feeling of joy (as against other possibilities) and in not letting us know the *nature* of the feel, we would not become conscious of it

For true interaction to take place, the events in world 2 would have to be able to stand on their own feet and in their own right, as they occur to one's subjective feel. They would have to supply information to world 3, information which would then help us to formulate a hypothesis. But this possibility is denied by Popper's own statement on p. 74 of his Objective Knowledge, cited above where he says that our knowledge of those subjective feels *depends* on our hypotheses about them in word 3. We are therefore left with the conclusion that there is no more than a downward action. The conclusion is inevitable: the events in world 2, events of a purely subjective nature and accessible exclusively to the body (world 2) cannot be named. They are too vague, too obscure, too wobbly for the person who is having them to choose a suitable word which, in any case, would have to be first fished out from World 3, the so called mind. These subjective feels, noticeable though they are, do not carry enough information which could indicate what should be selected from World 3 as a suitable "description" of the feel. In all cases, the naming comes from above, from the hypothesis in world 3, and in being named, we become conscious of them.

6. The tone of subjective states

The view that there is an absence of information in these subjective states of world 2 has to be qualified a little. They often contain a minimum of uncertain information. Often subjective states have a certain colour or, perhaps better, a certain tone. Such tone, not clearly identifiable as it is being felt, can act as a tentative guide to a seemingly compatible hypothesis from world 3 about the subjective state. It can do no more than that. If such a state results, for example, from seeing five children killed in an accident, the tone of the subjective state is unlikely to solicit an interpretation of, say, joy or contentment. The tone, however, is vague. It can point, in world 3, to horror, or to pity, or to indignation at the lack of care - all of which are compatible with the subjective feel. Given the uncertain vagueness of the feel, the class of compatible hypotheses is very large. These words do all come down from world 3; but the tone of the state in world 2 is likely to exclude a number of words, such as, in the case of the accident of the children, joy or contentment. There is just enough information in the subjective state to exclude a range of words and favour a class of other words. But that is all, but even so, one has to be wary. People we designate as sadists, watching that same accident which is happening in world 1, although prone to a very similar subjective state of, say, discomfort or even horror in world 2, will incline to use it as a guide to a very different hypothesis in world 3 - i.e. a hypothesis which interprets that subjective state as, for example, pleasure. The state by itself is sufficiently vague to tolerate all these and many other interpretations. In short, while there is a detectable tone, it cannot be taken as a reliable guide to a word from world 3 because it is no more than a tone. Given this circumstance, it would be unsafe to think that the events in world 2 have much constructive input into world 3. At most, they may provide a hint; but even such hints, as in the case of a sadist, are possibly misleading. All in all, the view that there is no constructive interaction must stand.

7. The gap between neural events and conscious states

It is unfortunate that this conclusion is not only not accepted, but explicitly contradicted by many leading neuroscientists. Admittedly, the subjective feel has to be seen as the direct, causal result of neuronal activity. But it is wrong to assume that, as so many neuroscientists are wont to do, that these neuronal activities can not only be described as so many chemico-physical events, but also translated without further ado into so many words which describe not just the behaviour of neurons, but the subjective feel these behaviours are causing. In making this assumption, they are making a totally unwarranted leap. For example Changeux links the experience of pleasure to dopamine synapses and Patricia Smith Churchland does not hesitate to state that verbal expressions of, say, colour "have as their physical substrate a neuronal phase space whose axes are...three streams of neurons carrying information about reflectance efficiency value at three wavelengths". Stephen Stich maintains that belief-state tokens are brain-state tokens (Munz, 1999, p. 9). It is a mistake to look upon these alleged links between neuronal behaviour and conscious states as a simple attempt at a shortcut, because, admittedly, something comes in between the neurons and the consciousness of colour or pleasure. The point is that the event which comes in between, though causally linked to the neuronal happening, is not causally linked to the outcome, i.e., the consciousness of colour or pleasure. The raw feel produced by the neurons is lifted to the level of consciousness by the name, say "red" or "pleasure", we chose to impose upon the raw feel. Shortcut or not, the mistake made by all these neuroscientists is to present the process from neuronal event to conscious state as if it were a series of causal links. In reality there is a gap. Many neuroscientists admit the gap, but consider it to be a gap in our present knowledge, assuming that sooner or later the causal link will be discovered and the gap closed (E.g., Horgan, 1999, p. 43.). As against this, I would argue that the discovery of the gap is an enormous addition to our knowledge as to how consciousness is actually generated. It is worth mentioning that this gap is very different from the gap which exists between realities and language. That gap results from the fact that no linguistic description of a reality is a picture of that reality. The gap here under discussion results from a different situation. This gap is due to the fact that the neuronally generated subjective feel, - "formlessly private" as T.S. Eliot put it in his Tradition and the Individual Talent - like liquid water - has no shape and, therefore, cannot be pictured, even if it were possible to use language pictorially. For this reason, the verbal label which raises it to the level of consciousness, though, in the end, is taken to refer to it, does not stand in a causal or pictorial or descriptive relation to it. Hence the gap between feel and consciousness of that feel. The first gap is due to the non-pictorial character of language; the second gap comes from the formlessness of subjective feels.

8. Somatic markers

We can get a better understanding of what is involved in this kind of interpretation and what the nature and status of psychological, a distinct from natural knowledge is, when we go forward to the work of Antonio Damasio. Popper was talking in very general terms when he wrote on p. 74 of his Objective Knowledge that our "subjective knowledge (world 2 knowledge) depends on world 3, that is to say on (at least virtually), *linguistically* formulated theories". But he did not indicate the nature of this dependence, nor did he so much as intimate, that we become conscious of the subjective knowledge only when it is interpreted by linguistically formulated hypotheses in world 3.

Damasio's analysis, though prone, as will be explained presently to an error, helps us to understand the exact nature of the interpretation and the kind of consciousness which results from this interpretation. Damasio discovered that our neuronal system which we cannot be aware of, produces "somatic markers". He described their relation to the body's neuronal system. The presence of such markers, rather than the neuronal churnings alone, help us to understand better what happens when one is thinking of the *interpretation* of subjective bodily events in world 2 by the mental hypotheses in world 3 than Popper's fairly bland statement quoted above. In several books and papers published towards the end of last century, Damasio produced a careful analysis of the way neuronal events and the body's awareness of these events hang together. He called the feels which the neuronal events produce "somatic markers" in order to distinguish them clearly and substantially from neuronal events. The neuronal events by themselves make no noise and are not experienced, even though they can be analysed by physics and chemistry and observed in PET scans and other forms of magnetic imagery. But they generate in a clear causal way distinctly noticeable feels such as changes in blood pressure, heart beats, tensions in the solar plexus and many others. As Antonio Damasio has shown these "somatic markers" are generated, in the first instance, by the prefrontal cortex. They are raw feels which are noticed, but which, as they stand, remain undefined because they cannot be articulated in so many words. Unarticulated, they cannot be talked about and not being able to be talked about, we cannot be conscious of them, except that we may be dimly aware of them and perhaps feel their tone without being able to say what they are. It has to be stressed that a dim awareness of a tone which cannot be identified, is *not* a state of consciousness. As they appear, they do not carry sufficient information to allow the person who is experiencing or feeling them, to find a suitable word which can bring them into full consciousness. (Damasio, 1994; see also Damasio, 1999).

Damasio, it has to be admitted, writing as a neuroscientist, is not given to philosophical reflection. He takes it that somatic markers can readily be identified. All one has to do is to look at them and name them. If one objects that those markers cannot be "looked" at, he would say that one has to "see" what they feel like in order to identify them verbally. He states without hesitation that "humans...can engender verbal narratives out of nonverbal ones" (Damasio et al., 1996, pp. 169, 173). In his major work, *The Feeling of What Happens* (Damasio, 1999, p. 108) he states explicitly, but erroneously, that "language gives us names for things" and that those "things" are known identifiably before we use language to designate them. This picture theory of language seems to be taken straight from Wittgenstein's *Tractatus*, and rejected in Wittgenstein's later work, the *Philosophical Investigations*. For instance, surprisingly, Damasio states, nearly fifty years after the appearance of Wittgenstein's *Philosophical Investigations*, that the fact that we can say "I", proves that there exists a non-verbal self of which the word "I' is a picture (loc. cit.). But only a little philosophy is required to understand that the use of the word "see" when one is vaguely aware of a feeling or somatic marker, is a metaphor and that the feeling stands as it is being *felt* - no more and no less. Most certainly, it cannot be "seen" anymore than the mere use and intelligibility of the word "I" proves that there is a pre-lingual "self". The ready assumption that somatic markers do not pose an identity problem because they can be named if one looks at them, is, I would suggest, Damasio's error, which replaces what Damasio has called Descartes' error (Munz, 1997, p. xvi). In fairness to Damasio one must note that most other neuroscientists tend to commit the same error. As mentioned above, they all tend to assume that once the purely physical and chemical behaviour of our neurons is identified, that behaviour can be designated as joy, fear, hatred, enthusiasm, etc.

As soon as one understands that the somatic markers cannot readily be identified verbally, one must look for the role of world 3 hypotheses which are able to interpret them. As these somatic markers are being been aware of, they cannot be named and we are not conscious of them. The closest reference to some such nameless and unnameable markers I have ever come across is by Baudelaire who referred to a somatic marker as an "a violent disturbance at the base of the brain" (Snell, 1997). While this sensation was hard enough, he also noticed in his sonnet *L'Homme et la Mer* that it was also as soft as the depths of the sea: "nul n'a sondé le fond de tes abîmes." But once we know that these somatic markers exist and are distinct from the physics and chemistry of our churning neurons, we know what it is that has to be brought into consciousness, that is, what role world 3 has to perform on world 2. This way of putting it has a great advantage over Popper's more simple statement that the subjectivities of world 2 "depend" on the hypotheses in world 3. While Damasio's own view that his somatic markers are available for verbal description, is erroneous, he has brought our understanding of how consciousness emerges a great step forward by showing what the hypotheses in world 3 are acting on (i.e., interpreting) and what it is they achieve by interpreting the somatic markers they are acting on. In naming them, albeit interpretatively so, they are bringing the somatic markers into full consciousness. They are turning blind feelings into the articulated emotions we are conscious of.

9. Supervenience of linguistic hypotheses: Between dualism and physicalism

While this explanation of consciousness from a Popperian perspective may not be the last word, it has an enormous advantage of all other theories about the genesis of consciousness which have been offered in recent times. When we are not just conscious but conscious *of* something, there is more than just a physical event in our bodies. Consciousness of something, like consciousness by itself, is something mental, something exosomatic. This has prompted many people, for thousands of years, to believe that there is a soul, that is, something spiritual in our body, which is active or being felt when we are conscious. This kind of dualism, for a number of good reasons, is now totally rejected. I agree with Derek Melser that there is now a prevailing opinion about it. (21 p. 134)) But the alternative view, that whatever it is that happens when we are conscious is nothing other than a special physical activity of our body is equally implausible. For a physical event to cause a seemingly mental state of consciousness, one has to wonder with Edelman (Edelman et. al., 2001. p. 6.) "how matter becomes imagination?" Or, to put it more poetically, Edelman asked himself "how the water of the physical brain is turned into the wine of consciousness". Nevertheless, the suggestion has been made that there are some special physical events in the body which cause this kind of transubstantiation by which neuronal events are transformed into conscious states. For Crick (Crick, 1995, pp. 244f.) these bodily events are 40 Hertz oscillations of special parts of the brain and for Penrose (Penrose, 1995, pp. 369-71) they consist of quantum effects in certain microtubules. For Edelman (Edelman, 1989.) and Humphrey (Humphrey, 1992) the magic is achieved by neuronal repetitiveness. Most recently, R. Llinás has agued hat consciousness is based on the simultaneity of activities in the thalamocortical system - that is a superposition of spatial and temporal mapping via thalamocortical resonant interactions (Llinás, 2001). Daniel Dennett, who, unlike Crick, Llinás and Penrose, does not believe in this kind of transubstantiation, is the only thinker who has taken this bull by the horns and argued that since there are only bodily events, there is no consciousness, or, if we think we are experiencing it, we are deluding ourselves because it is inconceivable that purely bodily events could usher in such a mental event as consciousness. As he puts it: "There is no single definitive 'stream of consciousness', because there is no central Headquarters...where 'it all comes together'..." (Dennett, 1991, p.253). But Dennett must be wrong, for if it is true that we are deluding ourselves when we think we are conscious of something, we cannot, at the same time, be deluding ourselves when we are conscious of being deluded. David Chalmers, like Dennett, does not believe in such transubstantiation either. But unlike Dennett, he accepts that we are, often enough, conscious. In the absence of a real explanation, he believes that consciousness, mysteriously and luxuriously, simply "supervenes" - but unnecessarily so, because we could function perfectly well without it (Chalmers, 1996).

From a Popperian perspective there is supervenience. But it is not consciousness which supervenes. The things that supervene are the verbal hypotheses in world 3. Our subjective feels are made conscious when those hypotheses supervene to give them a name. This kind of supervenience has a useful function. As against this, Chalmers' *thought* of "supervenience" as a wasteful accompaniment, is a confession of ignorance and comes close to throwing the towel in even though it takes him 400 odd pages to do so. No wonder that there are thoughtful people like McGinn (1989) who state more tersely that an explanation of consciousness is logically impossible and Fodor (1992, p. 5) who summed it up by saying that "nobody has the slightest idea how anything material can be conscious". The reason why nobody has the slightest idea is that consciousness is not the result of any chemical and physical occurrences in our neurons. It results from the supervenience of linguistic hypotheses which are reaching down from world 3 into world 2 - but stop short of reaching further down to the neurons which generate the somatic markers which are being reached by the hypotheses. I would further comment that the reason why nobody has the slightest idea how the physical and chemical events in neurons produce consciousness, is that the physical and chemical events *do not produce* consciousness. Seeing that they don't, it is hardly surprising that we have not the slightest idea how events that are not taking place, are taking place. As against all this, one can offer a constructive suggestion of how it is we are conscious of this or that in terms of the Popperian view of the role of language, though, perhaps not in literal terms of Popper's own thought. In those terms, consciousness does not result magically from physical and chemical events, but results from a seemingly baptismal occurrence.

10. The key role of language and world 3

The great advantage of Popperian thinking is that it focuses on the role of language for the way neuronally generated events like the somatic markers are turned into states of articulated consciousness. When one takes in the role of language, there is no transubstantiation, let alone magic. The role of language in world 3 is exosomatic and by using language to make us conscious of somatic markers by imposing a name on them or by interpreting them verbally, the physicality of those somatic markers is given a local habitation and a name. And in being given a name, they become conscious - or, better, we become conscious of them, because we now can *say* what those somatic markers (Popper's subjectivities) are. By saying so, we transform the vague sensation into fully blown consciousness. Many years ago, Nagel, in a famous paper kept wondering what it might *feel* like to be a bat (Nagel, 1974). But, in addition, he ought to have asked what it feels *like* to be a bat. No matter how virulent a feeling, if we cannot *say* what it is to be like a bat, we are not conscious of it. We are, at most, dimly aware of it. By the same token, the moment we can say, in so many words, what we are feeling, we are conscious of it. This way the obscure and vague somatic marker we are aware of even though we are not aware of its precise nature, is transformed, without transubstantiation and without any other form of magic, into full consciousness. But, to repeat, the action which results in consciousness comes from *above* and goes down, not all the way, but a certain way until it reaches the somatic markers. Consciousness is not the result of a process which starts at the bottom when we see or smell something that affects our neurons and the climbs up further until it reaches something we chose t call our mind

The emergence of consciousness - or perhaps one ought to say, the emergence into consciousness - is brought about by the employment of the kind of language of world 3. But here there must be a proviso. Popper himself amended Bühler's classification of language as symbolic and descriptive. (Popper, 1972, p. 120; cp. Munz, 2004, p.32) He showed that language can also be argumentative. This argumentative language was very much in world 3. Popper, and I am speaking here of personal experience, was obsessed by the need for argument and criticism. For example, when I gave him a copy of my little book on the Coronation of Charlemagne, his only reaction was : "here is my criticism" – even though he knew next to nothing abut Charlemagne. He was emotionally compelled to argue and criticise - which, in his case, meant that he wanted to show that he was right and his companion wrong. Therefore he ended up by seeing the language in world 3 as a never ending argument. True, it was the language used in arguments in that it was more than a mere description or copy or picture or protocol; but it was also *more* than merely the language used in arguments. It was a language which could be understood even though there was no prior experience of what it was referring to, so that one could not explain its intelligibility by taking it to be a picture or a protocol sentence of one's experience of what one is talking about - an experience which *precedes* the talking. It was this kind of language - non-pictorial and unargumentative - which enabled Newton to formulate his Laws of Gravity; Darwin, his theory of evolution by natural selection; and Einstein, his Theories of Relativity. "Without the development of an exosomatic descriptive language - a language which, like a tool, develops outside the body - there can be *no object* for our critical discussion....and it is only in this way, and only in this third world, that the problems and standards of rational criticism can develop" (Popper, 1972, p. 120). Precisely, though Popper did not spell this out, in this world 3, such language can also be understood, even though it does not always report, not even in an *adorned* way, what people can observe. Incidentally, it was this same kind of language which had enabled people to invent intelligible, though false, mythologies about God, gods and other spooks. I suggest that we add to Popper's concept of argumentative language the term "3-dimensional language" to refer to the kind of language which is able to invent and speculate about matters which cannot or have not yet been experienced or observed but which is, nevertheless, intelligible. It is this kind of language which is able to impose names and/or interpret somatic markers which, as they stand, do not have enough information built in to enable us to look at them and, by looking, identify them, and, in identifying them, instruct us to choose a suitable word for them. Again, in addition to argumentative language, it is this kind of 3-dimensional language of world 3 which changes dimly felt somatic markers into conscious states of mind. All this is an application and extension of Popper's terminology by Damasio's concepts. Popper himself phrased it in a more subdued form. Where Popper spoke of the "inborn dispositions" which must be the result of neuronal churnings and called them "subjective knowledge", Damasio spoke of somatic markers produced by our neurons. But where Damasio believed that somatic markers can be known and described by so many words, Popper, more correctly, realised that we need a special kind of world 3 which can form hypotheses about what he called subjective knowledge and what Damasio called somatic markers. This Popperian perspective of the role of world 3 language, augmented by Damasio's concepts, leads to a more fruitful understanding of consciousness than all those other theories which, disregarding the role of language, confine themselves to the physicality of the body and then rely on powerful transubstantiation to explain how consciousness emerges from such physicality. As against this, in Popperian perspective, we can understand, without recourse to transubstantiation, that we become conscious when we *name* the physicality of somatic markers with the help of invented verbal hypotheses. The hypotheses provide the name and the body provides the physical event in the shape of somatic markers. One must note that the hypotheses from world 3 do not act directly on the physical and chemical churnings of our neurons in our bodies, but on the dimly noticeable physical somatic markers generated by those churnings. Consciousness comes when name and somatic markers are brought together even though they do not hang together as cause and effect.

References

Chalmers DJ (2004) *The Conscious Mind.* Oxford: Oxford University Press.

Crick F (1995) *The Astonishing Hypothesis*. London: Touchstone Books.

Damasio A (1994) *Descartes' Error. Emotion, Reason and the Human Brain*. New York: Grosset/Putnam.

Damasio A (1999) *The Feeling of what Happens*. New York: Harcourt, Brace and Co.

Damasio AR, Damasio H (1996) Images and Subjectivity: Neurological Trials and Tribulations. In: RN McCauley (Ed) *The Churchlands and their Critics*. Oxford: Blackwell.

Dennett D (1979) *Review of Karl Popper and John Eccles, 'The Self and its Brain'* Journal of Philosophy **76**:91-97.

Dennett D (1996) *Consciousness Explained.* Boston: Little, Brown and Company.

Edelmann GM (1989) *Neural Darwinism.* Oxford: Oxford University Press.

Edelman et al (2001) *Consciousness. How Matter Becomes Imagination.* London: Penguin Books.

Fodor J (1992) Can there be a science of mind? *Times Literary Supplement* July 3 p. 5.

Gray J (2004) *Consciousness: Creeping Up on the Hard Problem*. Oxford: Oxford University Press.

Horgan J (1999) *The Undiscovered Mind: How the Brain Defies Explanation.* London: Weidenfeld & Nicolson.

Humphrey N (1992) *The History of the Mind.* New York: Simon and Schuster.

Llinás R (2001) *I of the Vortex. From Neurons to Self.* Cambridge MA: MIT Press.

McGinn C (1989) Can we solve the Mind-Body Problem? *Mind* **98**:349-366.

Melser D (2004) *The Act of Thinking.* Cambridge MA: MIT Press.

Munz P (1993) *Philosophical Darwinism.* London: Routledge.

Munz P (1999) *Critique of Impure Reason.* Westport CT: Praeger.

Munz P (1997) The Evolution of Consciousness – Silent Neurons and Eloquent Mind. *J Social and Evolutionary Systems* **20**:4.

Munz P (2004) *Beyond Wittgenstein's Poker. New Light on Popper and Wittgenstein.* Aldershot: Ashgate.

Nagel T (1974) What is it like to be a Bat? *The Philosophical Review* **83**:435-450.

Penrose R (1995) *Shadows of the Mind.* London: Vintage.

Popper KR (1972) *Objective Knowledge.* Oxford: Clarendon Press.

Popper KR (1994) *Knowledge and the Mind-Body Problem.* London: Routledge.

Snell R (1997) The Folly of Allegory and Interpretation. *Times Literary Supplement,* April 18 p 17.

Sprat T (1959) *History of the Royal Society.* London: Routledge.

Index